Liquid Sample Introduction in ICP Spectrometry

Liquid Sample Introduction in ICP Spectrometry

A Practical Guide

José-Luis Todoli
University of Alicante
Alicante, Spain

Jean-Michel Mermet
Spectroscopy Forever
Tramoyes, France

ELSEVIER

Amsterdam • Boston • Heidelberg • London • New York • Oxford
Paris • San Diego • San Francisco • Singapore • Sydney • Tokyo

Elsevier
Radarweg 29, PO Box 211, 1000 AE Amsterdam, The Netherlands
The Boulevard, Langford Lane, Kidlington, Oxford OX5 1GB, UK

First edition 2008

British Library Cataloguing in Publication Data
A catalogue record for this book is available from the British Library

Library of Congress Cataloging-in-Publication Data
A catalog record for this book is available from the Library of Congress

ISBN: 978-0-444-53142-1

For information on all Elsevier publications
visit our web site at books.elsevier.com

Transferred to Digital Printing in 2011

Working together to grow
libraries in developing countries
www.elsevier.com | www.bookaid.org | www.sabre.org

ELSEVIER   BOOK AID
International   Sabre Foundation

Preface

Inductively coupled plasma emission and mass spectrometry are routinely used for elemental analysis and are considered as routine and mature techniques. The most common way of introducing samples to be analyzed into the plasma is the production of an aerosol of fine droplets from a solution, through the use of a pneumatic nebulizer generally associated with a filtering chamber. Although the principle of pneumatic nebulization was established in the eighteenth century, sample introduction systems, although crucial, remain certainly the weakest point of the instrument. Pneumatic nebulization efficiency is still related to the physical and chemical properties of the solution. As a result, analytical results may be degraded because of possible bias, significant signal noise and drift. It could be thought that a plasma is highly tolerant to sample introduction because of its high temperature. However, an inductively coupled plasma (ICP) exhibits severe constraints in terms of droplet diameter and solvent loading, which makes the design of sample introduction systems rather complex and challenging. Over the total number of published work devoted to ICP spectrochemistry, a significant number of papers dealt with design, modeling and applications of sample introduction systems. However, some commercially available ICP spectrometers are still making use of nebulizer and spray chamber described in the 1970s, such as the concentric nebulizer and the double pass spray chamber. This can be probably explained because most of the published work described minor improvements but no significant gaps in terms of sensitivity, precision and non-spectral interferences. Actually, the principle of pneumatic nebulization remained relatively unchanged, but some trends were nevertheless observed towards micronebulization and total sample consumption systems. Considering work about nebulization, it may be seen (Figure 1) that the number of publications used in this book has been growing up to the 2000s, with then a plateau, or perhaps a decrease. It will be interesting to verify whether this recent trend will be confirmed in the forthcoming years.

Although some alternatives to pneumatic nebulization were described and made commercially available, such as ultrasonic nebulization, jet-impact nebulization and electrothermal vaporization, this book is devoted to pneumatic nebulizers as they remain the standard device for routine analysis. It has been tried to cover the fields of aerosol specifications and their consequences on nebulizer, spray chamber and desolvation design. Possible non-spectral matrix effects were also discussed in detail along with selection, diagnosis and maintenance of the introduction systems. Some key applications were quoted to illustrate the potential of some nebulizer-spray chamber combinations. Contents of this book were drawn from the international literature and from the experience of the authors.

Nebulization is a complex phenomenon, and we hope that this book could be of some help to the analysts, although there are not always simple solutions to the numerous problems raised by the use of pneumatic nebulization. It is obvious that work on

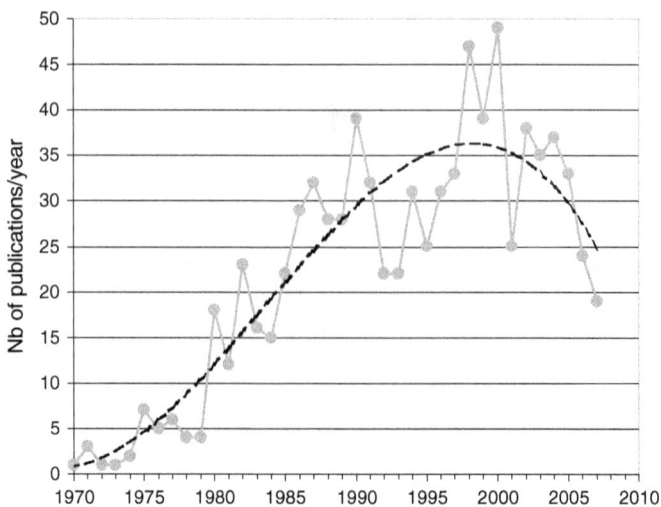

Figure 1 Distribution of the publications quoted in this book as year function.

nebulizer/spray chamber design and understanding is still required to take full benefit of the photon or ion plasma source.

We are indebted to authors and companies who have granted permission to use materials from their publications, brochures or presentations.

Contents

– 1 –

Introduction

Since the first publications on inductively coupled plasma (ICP), first for plasma processing [3, 4] and then for analytical applications [5–9], significant improvements have been observed for most parts of the instrument: more stable and compact hf generators, optimized torch design for sample introduction and plasma gas consumption, axial viewing mode in atomic emission spectrometry (AES), interface efficiency and ion optics in mass spectrometry (MS), reproducibility of wavelength positioning of AES sequential dispersive optical systems, use of multichannel detection to increase the amount of information in AES, high resolution of optical systems, and a large variety of mass spectrometers such as quadrupole, sector-field, time-of-flight and ion trap mass spectrometers. Besides, sophisticated softwares have been developed to drive the instrument and to optimize data acquisition and processing.

In marked contrast, although a significant number of publications have been devoted to sample introduction systems, this part of the instrument remains the weakest point of the analytical chain. Basically, the pneumatic concentric nebulizer does not differ from the nebulizer described by Gouy at the end of the nineteenth century [1], and it is still the most widely used device [44, 47]. However, to overcome the limitations of a concentric nebulizer, a significant number of alternative designs has been published, very often extrapolated from flame-AES or atomic absorption spectrometry. An example is the cross-flow nebulizer [22, 50, 130, 287, 642], with the possibility of operating at high pressure to improve precision [78].

So-called micronebulizers were basically of the concentric type but were slightly modified to operate at low delivery rates, i.e. down to tens of microlitres [410, 419, 454, 455, 457, 489]. They could cope with low liquid flow rates, for instance when an ICP was coupled with separation methods, or with limited amount of samples, for instance in biology or forensics.

One of the limitations of the concentric nebulizer is the risk of blocking with solutions having a high salt concentration, or solutions with suspensions. From the original design of Babington [13], several nebulizers were described that made use of a V-groove to allow the solution to flow down and to pass in a front of a gas orifice for nebulization [58, 66, 88, 92, 221, 243, 712]. An alternative design was described, replacing the V-groove by a cone [339]. Basically, these types of nebulizers remain of the cross-flow type.

Different designs were also published and evaluated such as the glass-frit nebulizer [82, 143, 245, 465] and the grid nebulizer [196, 200, 252, 319].

Totally different designs were proposed where the production of the aerosol was no longer produced by a gas–liquid interaction, but by an alternative process such as the oscillating nebulizer [463], the thermospray nebulizer [157, 181, 182, 193, 195] and the jet-impact nebulizer [126], with a further high-pressure version using a high-performance liquid chromatography (HPLC)-type pump [250, 412, 497].

All these types of nebulizers make use of a spray chamber to remove too coarse droplets. An alternative is the use of direct injection nebulization, that is to say the tip of the nebulizer is located very close to the base of the plasma, without the need of an injector [154, 185, 281, 312, 505, 507, 565].

Some nebulizers may be used for very specific applications, but the vast majority of ICP users will employ either a concentric nebulizer or a cross-flow nebulizer, with sometimes the need for a high-solids nebulizer.

Alternative designs are evidence that the most commonly used nebulizers, although satisfactory for routine analysis, do not fulfil the requirement for an ideal analytical system. A sample introduction system may be a significant, even limiting source of noise, drift and non-spectral interference effects leading to a bias and an inappropriate repeatability. Nebulizer blocking can obviously only occur during an experiment. The delay time between two samples may be significant compared to the analysis time, because of compulsory wash-in and wash-out times. Problems of maintenance are mainly related to cleaning or to a total or partial blocking of the nebulizer. The aim of this book was, therefore, to have a detailed overview on sample introduction systems based on pneumatic nebulization, so as to help the analyst in improving the analytical performance of the instrument.

Chapter 2 will describe the specifications of an ideal introduction system based on the use of pneumatic nebulization and taking into account current characteristics of an ICP.

Chapter 3 will describe the various pneumatic nebulizer designs.

Chapter 4 will deal with spray chamber design.

Chapter 5 will describe various means of desolvatation.

Chapter 6 will be devoted to matrix effects.

Chapter 7 will deal with selection and maintenance.

Chapter 8 will describe some typical applications.

– 2 –

Specifications of a Sample Introduction System to be Used with an ICP

2.1 INTRODUCTION

Like for a flame, liquid sample is usually introduced into the plasma in the form of an aerosol consisting of fine polydisperse droplets, along with some vapour produced during the droplet transport in the spray chamber. A vapour is a gas phase in a state of equilibrium with a liquid of identical matter below its boiling point. When the droplets enter the plasma at a rate of several millions per second, they undergo three major processes – desolvation, volatilization and atomization – before reaching the state of free atoms. Desolvation is the evaporation of the solvent from the droplets that leads to a suspension of a dry aerosol, that is desolvated particles. During the evaporation process, droplets and dry particles are surrounded by a solvent vapour cloud. When solvents evaporate, they remain near their boiling point, that is far from the plasma temperature. Volatilization produces the conversion of the dry aerosol into a gas or a vapour. Solute particles have a broad range of boiling points with values comparable to what can be observed within the plasma. Two transfer mechanisms are involved: heat transfer from the plasma that leads to surface boiling and followed by mass transfer from the surface of the droplet or particle. Atomization is the conversion of volatilized analytes into free atoms. It should be noted that the word atomization is also used to describe the conversion of a liquid into a spray or a mist. The sample introduction system is, then, called an atomizer, which is obviously wrong because no atoms are produced. Nebulizer is most appropriate as derived from the Latin *nebula* (mist).

Once free atoms are obtained, they undergo partial ionization, and both atoms and ions are excited. Timescale for the first three processes is of the order of *ms*, whereas excitation and ionization are significantly faster. Because an inductively coupled plasma (ICP) uses hf power of the order of kilowatts and reaches kinetic temperature in the range 4000–7000 K, it could be thought that this source can absorb a large amount of sample without any problem of evaporation and atomization. However, an argon-based ICP suffers from some severe limitations. A high-frequency field is mostly coupled with the periphery of the plasma, because of the so-called skin effect. Energy diffuses to the plasma centre through collisions with losses, and only a small fraction is available along the axis.

The energy consumed for desolvation and vaporization is actually a small fraction of the forward power. Besides, argon has a poor thermal conductivity, and residence time of the sample is rather short. Moreover, there has been a constant trend to decrease the available power of the hf generator for cost and size reasons.

An ideal sample introduction system should lead to complete desolvation and volatilization in the plasma. For that, its design must consider the efficiency of the heat transfer from the plasma to the droplets, which is related to the power reaching the sample and the residence time of the droplets. The residence time is, in turn, linked to the carrier gas flow rate, the injector's i.d., and the temperature along the axis of the plasma.

Practically, the design must take into account the answers to some key questions such as:

(i) What is the maximum droplet diameter ensuring a complete desolvation and volatilization?
(ii) What is the maximum amount of aerosol, in other words, the plasma loading, acceptable by the plasma?
(iii) Is it possible to avoid any change in the sample stoichiometry?
(iv) Is the system sensitive to any change in the physical properties of the solution, for example viscosity, surface tension, density and volatility in the case of organic solvents?
(v) Is desolvation and vaporization sensitive to any change of the element concentrations in a droplet, particularly those of the major elements?
(vi) Is there any benefit to keep water, or to remove it before introduction into the plasma?
(vii) How to minimize noise because of liquid delivery, aerosol production, filtration and transport in the introduction system, drain and processes in the plasma?

Obviously, the answers to these questions depend upon plasma characteristics, torch design and operating parameters such as:

(i) the physical properties of the plasma gas such as ionization energy and thermal conductivity, which in turn are related to the selected plasma gas,
(ii) the energy coupled to plasma, which depends on the generator design and frequency, and on the available hf power and the coupling efficiency,
(iii) the plasma gas flow rate, that is the amount of gas per time unit, usually expressed in $L\,min^{-1}$,
(iv) the residence time in the plasma that depends upon the carrier gas velocity resulting from the gas flow rate and the injector i.d., and upon the temperature along the central channel,
(v) the desolvation, evaporation and atomization timescale,
(vi) the solvent loading accepted by the plasma.

2.2 PHYSICAL PROPERTIES OF A PLASMA

The physical properties of a plasma, such as ionization energy and thermal conductivity, depend obviously upon the gas. Rare gases are usually used to generate plasma, because they emit only atomic spectra in emission spectrometry, and relatively simple spectra in mass spectrometry, although some rare gas–based molecular ions can survive during their transport to the ion detector. The highest ionization energy is for He (24.6 eV), followed by Ne (21.56 eV). The He gas has been used to sustain ICPs, but the cost can be a limitation, kinetic temperature is lower than that of an Ar plasma, and introduction of a sample is more challenging [153, 190, 227, 337, 367, 416, 449, 513]. Although providing interesting results [674], the Ne gas cannot be used on a routine basis because of its unacceptable cost. The argon gas, which has lower ionization energy (15.76 eV), exhibits some advantages. It is the most common of the rare gases, with 1% in air, which means that its cost is acceptable as a subproduct of air distillation. However, it is only easily available in industrialized countries. Some properties are important for sample injection and atomization. Because of the increase in the number of free electrons when the temperature goes up, the viscosity will move from $2 \times 10^{-5}\,\mathrm{kg\,m^{-1}\,s^{-1}}$ at room temperature to $20 \times 10^{-5}\,\mathrm{kg\,m^{-1}\,s^{-1}}$ at 6000 K, that is a factor of 10. This means that it would be difficult to inject a sample by using a cold Ar gas because of the penetration in a medium that is significantly more viscous. However, thanks to the use of a hf field, a skin effect is observed; that is the energy is mostly deposited at the periphery of the plasma, and then translated by particle collisions to the axis of the plasma. In other words, the energy, and therefore the temperature, is lower along the axis of the plasma, and so is the viscosity. The axis of the plasma is a zone where it is easier to introduce a sample by forcing the penetration with a gas having a sufficient speed. There is creation of a virtual central channel, which explains the success of an ICP. Sample introduction in other types of plasma such as microwave-induced plasmas or direct-current plasmas is far more challenging.

Another crucial parameter is the thermal conductivity, that is the capability to transfer heat. Molecular gases, air, N_2, O_2 and H_2 exhibit thermal conductivities at 5000 K that are 5.8, 5.4, 4.5 and 31.6 times higher than that of Ar, respectively. Even He has a thermal conductivity that is 10.1 times higher. This means that when using Ar, residence time of the sample must be longer to compensate for a lower thermal conductivity.

2.3 ENERGY DELIVERED BY THE PLASMA

There has been a constant trend to decrease the size and the cost of ICP spectrometers. One means of obtaining this reduction is through the use of low-power hf generators. Currently, the maximum power that can be used on commercially available ICP systems is in the 1.5–2 kW range, and the operating power recommended by ICP manufacturers is typically below 1.5 kW (1000 W in ICP–AES and 1200 W in ICP–MS). This has to be compared with early work in ICP–AES, where 7 MHz, 15 kW (Radyne) and 5.4 MHz, 6.6 kW (STEL) hf generators were used. Increase in the generator frequency was also observed along with the decrease in the power: 36 MHz, 5 kW (Radyne), 50 MHz, 2 kW (Philips), 64 MHz, 4.3 kW (Durr) and 27 MHz, 2.5 kW (PlasmaTherm). Currently, commercially available ICP instruments make use of 27 or 40 MHz (Tables 2.1 and 2.2). The operating frequency determines the skin depth, that is the penetration of the energy at the periphery of the plasma, and the coupling efficiency [265]. Coupling efficiency is also related to the ratio of the torch radius to the skin depth. It has been claimed that a 40 MHz

Table 2.1

Maximum power, generator frequency and inner diameter (i.d.) of the injector for radial and axial viewing for some commercially available ICP–AES systems

System	Maximum power (kW)	Frequency (mHz)	Injector i.d., radial viewing (mm)	Injector i.d., axial viewing (mm)
Horiba JY Activa	1.5	40	3	
Spectro Arcos	1.7	27	1.8	2.5
Thermo series 6000	1.5 (1.35 for Duo)	27	1.5	2.0
Leeman Prodigy	2	40		
PE series 5000	1.5	40	2	2
Varian series 700	2.0	40	1.4	2.3
Shimadzu ICPE 9000	1.6	27		

Table 2.2

Maximum power, generator frequency and inner diameter (i.d.) of the injector for some commercially available ICP–MS systems

System	Maximum power (kW)	Frequency (mHz)	Injector i.d. (mm)
PE 5000	1.4	32	
PE 6000	1.4	40	2
Agilent 7500	1.6	27	2.5
VG	2.0	27	
Varian	1.6	40	2.3

frequency facilitates sample injection because of a slightly lower temperature at the axis level [90, 609]. Thomson scattering has been performed to compare the characteristics of 27 and 40 MHz argon ICP [501]. The higher frequency led to lower values of the electron number density and the gas temperature in the central channel. As a result, a larger central channel was obtained, which is advantageous for sample introduction.

Theoretically, frequency may vary slightly but must remain within limits: 27.12 MHz ± 160 kHz and 40.68 MHz ± 20 kHz. However, modern instruments are carefully shielded to avoid any electromagnetic leaks, and some intermediate frequency (32 MHz) was used in ICP–MS.

Note that the power generally quoted in publications is the power leaving the hf generator, not that reaching the plasma. A crystal-controlled generator has a coupling efficiency in the range 60–70%. It is then difficult to compare similar forward powers from generators with different technologies.

2.4   CARRIER GAS FLOW RATE AND DROPLET VELOCITY

The gas has a double role with pneumatic nebulizer – production of the aerosol and transport. Traditionally, the gas is called carrier gas, but it would be more appropriate to call it aerosol production and transport gas. Keeping the same injector i.d. would lead to a higher carrier gas velocity. In early ICP–AES work, the conventional standard injector i.d. was 1.5 mm, with carrier gas flow rates about $1 \, L \, min^{-1}$. Because of the decrease in the forward power, it was necessary to also decrease the amount of aerosol, and so was the gas flow rate. Currently in ICP–AES, the carrier gas flow rate is in the range $0.5–0.9 \, L \, min^{-1}$. It should be noted that the injector i.d. is usually higher when the axial viewing mode is used (Table 2.1) to compensate for an apparent higher sensitivity to matrix effects resulting from the probing of the atomization zone within the coil. In marked contrast, because it is necessary to probe ions out of the coil, carrier gas flow rates used in ICP–MS are rather in the range $0.9–1.1 \, L \, min^{-1}$. Combination of various injector inner diameters and carrier gas flow rates can lead to a significant change in the gas velocity at the tip of the injector (Table 2.3).

Based on a classical double-pass spray chamber and a concentric nebulizer and using a power of 1000 W and a relatively low carrier gas flow rate of $0.66 \, L \, min^{-1}$, a velocity in the range $20–40 \, m \, s^{-1}$ was measured 3–35 mm above the load coil (ALC). Note that the droplet velocity is lower than that of the gas because of the drag effect. Assuming an injector i.d. of 1.5 mm, the velocity of the cold carrier gas at the tip of the injector was $6.2 \, m \, s^{-1}$, which means there is a significant acceleration of the gas within the coil. Although the gas speed at the injector exit varies with the carrier gas flow rate, the velocity of the gas ALC was found to be relatively independent of the carrier gas between 0.6 and $1.2 \, L \, min^{-1}$, whereas it was dependent of the power between 800 and 1200 W [266].

Using a high-speed camera at $4000 \, frames \, s^{-1}$, it has been reported [492] that a droplet has velocity of $30 \, m \, s^{-1}$, for a power of 1300 W and a carrier gas flow rate of $1.0 \, L \, min^{-1}$, which is in good agreement with the work of Cicerone and Farnsworth [266].

The concept of residence time is, therefore, related to the acceleration of the carrier gas within the coil region, from a few metres per second up to $40 \, m \, s^{-1}$. Moreover, a low carrier gas flow rate results in a longer path in the discharge since the atomization begins lower in the plasma.

A different case is the use of so-called direct-injection nebulizer (see Section 3.8.11). The aerosol is directly introduced into the plasma without passing through a spray chamber and an injector. Such a nebulizer produces a larger aerosol than that obtained with a conventional nebulizer associated to a spray chamber. Droplet velocity measured by phase–Doppler light scattering at the tip of the nebulizer was in the range $12–22 \, m \, s^{-1}$, significantly faster than at the tip of an injector [371].

Table 2.3

Effect of the injector i.d. and the carrier gas flow rate on the velocity $(m \, s^{-1})$ of the carrier gas at the exit of the injector

Gas flow rate	1.5 mm	2.5 mm	3 mm
$1.0 \, L \, min^{-1}$	9.43	3.40	2.36
$0.5 \, L \, min^{-1}$	4.72	1.70	1.18

2.5 DESOLVATION AND VAPORIZATION

It is important to try to ensure a complete droplet desolvation and particle vaporization. Incompletely desolvated droplets and vaporized particles cause time-dependent signal fluctuations, that is noise [244, 335, 336, 343]. Moreover, residual particles can block the sampling orifice in an ICP–MS [329].

Desolvation is a time-consuming process and is highly dependent upon the plasma-operating parameters, as it can be demonstrated when compared to a system introducing a dry sample, for instance, electrothermally generated [174]. A change in the carrier gas flow rate has a double consequence – increase in the amount of aerosol and decrease in the residence time. Note that desolvation seems to result in longer processes to obtain free atoms but does not necessarily correspond to a decrease in the local temperature, as seen later. For near-intact droplets, ion intensity falls, whereas the atom intensity increases, because of local cooling [244, 385]. Both theoretical and experimental studies have been published to describe the maximum diameter for which a droplet would be totally desolvated in an ICP [174, 385, 115, 134, 399, 452, 479, 480, 481, 512, 548, 560, 561, 606, 664, 698, 734, 737, 779] as a function of the operating parameters, power and carrier gas flow rate.

Early work [115] indicated that the droplets should be less than 8 μm to be totally desolvated. Using a model based on a Monte–Carlo technique, Yanping et al. [399] have shown that, for several nebulizer designs, the maximum evaporated droplet diameter decreased when the carrier gas flow rate was increased. Depending on the nebulizer design, the maximum droplet diameter was in the range 11.3–18.2 μm at a carrier gas flow rate of 0.65 L min^{-1}, while it was in the range 7–12.7 μm at a carrier gas flow rate of 1.10 L min^{-1}.

A computerized simulation of droplet desolvation was performed [698] to compare with the experimental results [335]. Actually, the gas temperature along the plasma axis is relatively insensitive to change in power, but significantly sensitive to changes in the carrier gas flow rate [698]. The crucial role of the carrier gas can be summarized (Table 2.4.) for the results obtained at a power of 1000 W and an observation height of 17.5 mm ALC.

A stochastic spray model taking into account plasma–droplet interactions and droplet collisions was used for both monodisperse and polydisperse droplets. The model confirmed that the height of complete desolvation increased linearly with the carrier gas flow

Table 2.4

Maximum diameter (μm) to obtain a complete desolvation at a power of 1000 W and an observation height of 17.5 mm above the load coil (ALC) as a function of the carrier gas flow rate

Carrier gas flow rate (L min^{-1})	Experimental [335]	Theoretical [698]
0.8	22.9	24.2
0.9	21.4	22.0
1.0	15.3	16.8

rate; whereas it decreased by an inverse relationship with the increase in the power [737]. Predictions were in good agreement with experimental results [335]. Note that a special device, called monodisperse-dried microparticulate injector (MDMI), was developed to generate a train of monodisperse droplets that were dried in a furnace, so that the original droplet diameter of 60 μm could be reduced on demand [387]. Using such a device, it was confirmed [390] that 23-μm droplets survived to 18 mm ALC, at a carrier gas flow rate of $0.8 \, \text{L min}^{-1}$. More recently, Mie scattering was used for monodisperse droplet desolvation rate measurements [831]. The gas velocity at the centre of the plasma was $14 \, \text{m s}^{-1}$ at a power of 1000 W. A rate of $0.65 \, \text{mm}^2 \, \text{s}^{-1}$ at 1000 W was measured and compared with a computerized rate of $0.41 \, \text{mm}^2 \, \text{s}^{-1}$ [698]. An important conclusion, when the concentration of solids was higher, was that the earlier appearance of analyte emission was due to the significant increase in the size of the desolvated particle so that droplet desolvation is completed earlier [831].

In the case of a direct-injection nebulizer, theoretically, it was found that a small fraction of the aerosol was still present after a distance of 20 mm in the plasma. This fraction contained droplets larger than 15 μm, that is a large amount of analyte [664].

2.6 PLASMA LOADING

It has been already mentioned that the power available in the centre of the plasma is rather weak, and is significantly less than 100 W for a forward power of 1000 W [114, 283]. The amount of aerosol will be limited for complete desolvation and vaporization. In any case, water has been reported as having a drastic effect on ICP excitation conditions [63, 146, 168, 187, 193, 198, 199, 208, 233, 239, 242, 264, 283, 292, 304, 470, 600]. The influence of water is difficult to assess as different operating parameters, such as power, carrier gas flow rate, i.d. of the injector and generator frequency, have been used in previously published work. Nevertheless, the role of water can be summarized through two different steps [146, 168, 208, 242]: (i) the absorption of plasma energy during the solvent decomposition process, as vaporization and dissociation will absorb the most energy and (ii) if the residence time is long enough to permit the conduction from the outer part of the plasma to replace loss in the central channel, dissociation products, that is molecular hydrogen and oxygen, will contribute to a local increase in thermal conductivity and heat transfer efficiency. Water, once dissociated, can play a beneficial role in vaporization. Operating conditions leading to an efficient energy transfer are called robust conditions. They are obtained by increasing the power and the injector's i.d., and reducing the carrier gas flow rate [268, 318, 383, 422, 660]. A rough estimation of the robustness of a plasma can be obtained by measuring the Mg II 280.2 nm/Mg I 285.2 nm line intensity ratio [318].

Concerning solvent loading, some general conclusions may be drawn: (i) the role of solvent loading is related not only to the amount of aerosol but also to the efficiency of energy transfer, which is in turn related to the residence time within the coil through the carrier gas flow rate [304]; (ii) although the energy needed for volatilization is low compared with the power [114], the role of the forward power is crucial; (iii) the influence of water is not the same if the sample is introduced in the form of liquid aerosol or vapour, the tolerance to water loading being higher with vapour than with aerosol [242]; (iv) the nature of the plasma–water interaction is complex [212] and (v) the influence of desolvation seems to depend on the plasma-operating conditions [208, 233, 292, 304], and has relatively little effect on atomic lines with excitation energies less than 5 eV [212, 292]. It should be noted that the tolerance to solvent loading has often been studied under fixed operating conditions.

With regard to the beneficial role of water, under robust conditions, the use of a desolvation system at the exit of the sample introduction system has no effect or results in a slight degradation of the excitation temperature and the electron number density, which is an evidence of the benefit of having water dissociation in the central channel. In contrast, under nonrobust conditions, external desolvation leads to an improvement in plasma characteristics, which means that water loading was too high.

At the exit of the spray chamber, a mixture of liquid aerosol and vapour is obtained, including residual water in argon. The amount of water can be expressed either in mg min^{-1} or mg L^{-1} of argon. Values between 0.3 and 40 mg min^{-1} have been obtained for carrier gas flow rates in the range 0.4–1.0 L min^{-1} [146, 187, 199, 242, 283, 304], and between 20 and 40 mg min^{-1} in the range 0.6–1.0 L min^{-1}. For a given carrier gas flow rate, it is possible to modify the solvent loading by changing the uptake rate [187, 283] or the temperature of the spray chamber [199]. As mentioned previously, solvent loading can be drastically reduced by using a desolvation system in conjunction either with an ultrasonic nebulizer [168, 233] or a pneumatic nebulizer [292]. Complete desolvation is also obtained using an electrothermal vaporization system [146, 242]. An alternative method to obtain dry aerosol is the use of spark or laser ablation systems.

2.7 ORGANIC SOLVENTS

Most of the studies concerning desolvation have been performed with aqueous solutions. However, organic solvents may be used for the direct analysis of organic matrices, for example polymers, organic fluids, petroleum products or when sample preparation makes use of solvent extraction. An ICP can also be used as a detector in HPLC. In contrast to water, the properties of which are only slightly modified with the addition of acids, organic solvents exhibit a large range of volatility, density and surface tension [83]. The most commonly used organic solvents are xylene, kerosene, toluene and hexane. Nevertheless, some general conclusions can be given. Because of the higher volatility leading to high vapour pressure, an increase in solvent loading is observed, consisting of a mixture of aerosol and a significant amount of vapour [256]. Besides, lower droplet diameters are produced [65, 796] with a higher desolvation rate. Consequently, the ICP must be operated at a higher power when compared to water. Note that in the 1980s, 1600–1800 W was considered as a low power. Even using a power of 1750 W, that is higher than what is currently used, Boorn and Browner [83] found that only chlorinated solvents and low evaporation rate solvents did not extinguish the ICP with a 1 L min^{-1} carrier gas flow rate. At 1600 W and 0.7 L min^{-1}, Miyazaki et al. [79] verified the ease of introduction of organic solvents. Introduction was very easy for hexanol, diisopropylketone, diisobutylketone, dimethylsulfoxide, xylene and relatively easy for methylisobutylketone, but almost impossible for toluene, chloroform and methanol. Boumans and Lux-Steiner had to move from 1100 W and a carrier gas flow rate of 1 L min^{-1} to 1700 W and 0.8 L min^{-1} for the introduction of methylisobutylketone [95].

It was found that to obtain similar ICP temperatures, it was necessary to increase the power by about 500 W with xylene at a carrier gas flow rate of 0.8 L min^{-1} [147], for example a jump from 1250 to 1750 W. With the current maximum power that is achievable, direct injection of highly volatile solvents such as methanol can even lead to plasma extinguishing.

Solvent loading may be reduced by cooling the spray chamber [67], reducing the delivery rate and decreasing the carrier gas flow rate [256], or by mixing the organic solvent with water, as for ethanol. In the case of ethanol–water mixture, an enhancement in sensitivity can be observed [399, 447]. Methanol–water mixtures were also studied with a glass frit nebulizer [143]. A desolvation system could be added to a pneumatic nebulizer consisting in a furnace followed by a condenser [331]. It was then possible to run a 1500 W ICP up to 95% ethanol with a 0.6 L min^{-1} carrier gas flow rate. A stronger removal of organic solvents could be obtained by using cryogenic desolvation [401].

Solvent loading (xylene) was also studied using the MDMI [477]. Control of the loading was obtained by modifying the frequency of droplet introduction into the plasma.

Note that even concentrations of organic solvents less than 2% in water may affect plasma characteristics and analytical results [273].

It appears that for most applications, xylene and kerosene are the most commonly used solvents because of their relatively low volatility, corresponding to power usually in the range 1300–1500 W. In any case, introduction of organic solvents remains a challenge with the current hf generators.

For more details about organic solvents, see Section 6.4.

2.8 IDEAL AEROSOL

An ideal sample introduction system should provide droplet diameters $\leq 10\,\mu m$. In any case, droplets with diameters close to $25\,\mu m$ will never be totally desolvated. Note that it is necessary to have 1000 droplets with 1-μm diameter to obtain the equivalent of a single 10-μm diameter droplet in terms of amount of aerosol. In other words, a substantial amount of aerosol is contained in large droplets. Ideally, an aerosol should be quasi-monodisperse, as with the MDMI [387, 390, 831]. A sharp distribution of the signal intensity along the plasma axis would be observed. Practically, a large range of droplet diameters is obtained, leading to a broad-signal intensity distribution. Because of the strong influence of the carrier gas flow rate, a pneumatic nebulizer's ideal working range would be 0.5–1.2 L min^{-1}. With regard to plasma loading, because of the power range currently used (800–1600 W) a total amount of aqueous aerosol should be in the range 20–40 mg min^{-1}.

An ideal sample introduction system should provide droplet diameters $\leq 10\,\mu m$ with a total amount of aqueous aerosol being in the range 20–40 mg min^{-1}.

It will be seen later that this target is difficult to obtain. Current associations of pneumatic nebulizer and spray chamber will still provide a significant amount, that is 20–40% of partially desolvated droplets [698].

Use of organic solvents or mixture with water facilitates the obtaining of smaller droplets, but to the detriment of too large a loading.

The various parameters to characterize an aerosol are summarized in Table 2.5.

Table 2.5

Definitions of the parameters in order to characterize the mass and fineness of the aerosols for ICP

Parameter	Definition of the most often-used and identification of the acid origin
Liquid uptake rate, Q_l (mL min^{-1})	
Sauter mean diameter, $D_{3,2}$ (µm)	Corresponds to the surface mean diameter, $d_{sm} = \dfrac{\sum nd^3}{\sum nd^2}$, ($d$ is the diameter of an individual droplet)
Mass or volume mean diameter, $D_{4,3}$ or mass median equivalent diameters, MMED (µm)	This parameter is given by $d_{mm} = \dfrac{\sum nd^4}{\sum nd^3}$, and can be used, as the $D_{3,2}$, in order to evaluate the influence of a given variable on the aerosol drop size. It has the advantage against the former that it represents the mean of the distribution based on a property directly related with the emission signal, that is the aerosol mass or volume.
Median of the volume drop size distribution, D_{50} (µm)	This indicates that the 50% of the liquid aerosol volume is contained below this diameter. Its usefulness lies in the fact that it gives the central trend of the drop size distribution.
Volume percentage in droplets below 1.2 µm, $V_{1.2}$ (%)	Is referred to as the aerosol liquid volume percentage that is present in form of droplets, the diameter of which is lower than 1.2 µm.
Volume concentration, VC (%)	It is the percentage of the aerosol measurement volume that is occupied by liquid droplets. The measurement of this parameter at the exit of the spray chamber could be a relative indication of the influence of a given variable on the solution transported through it.
Volume flux of droplets (cm^3 s^{-1} cm^{-2})	Defined as the total volume of droplets passing a cross-sectional area in a given direction per unit of time. Also gives indication of the mass transport rate.
Total mass solvent transport rate, S_{tot}, (mg min^{-1} or µg s^{-1})	This variable is a direct reading of the solvent (liquid and vapour) that is introduced in the plasma base per unit of time.
Total mass analyte transport rate, W_{tot}, (normally µg s^{-1} or µg min^{-1})	Directly indicates the mass of analyte leaving the spray chamber per unit of time.

2.9 CHEMICAL RESISTANCE

Besides producing droplets of an adequate diameter and amount, the sample introduction system must be resistant to chemical attacks from acids used for sample storage and preparation, and to organic solvents.

Materials to be used for nebulizers or spray chambers are usually selected for their chemical resistance to reagents and their ease of processing. Note that chemical resistance is usually evaluated using continuous immersion, sometimes above room temperature, which is not the use of these materials as nebulizers or spray chambers. Brief exposures followed by washing are most of the time acceptable.

Glass is widely used either as a borosilicate glass, that is containing about 10% boron oxide, or as a silica glass, so-called quartz. Chemical resistance is very high to any acids or organic solvents, with the exception of hydrofluoric acid. The major limitation of glass is that glass blowing is the only way to obtain a required shape, with glass in a molten semiliquid state. Complex shapes and reproducibility are, therefore, highly challenging.

The alternative to glass is the use of polymers, which can be processed by extrusion, injection and moulding. The major difference between polytetrafluoroethylene (PTFE) and other polymers is that PTFE cannot be thermoformed or readily welded, while other polymers are melt processable.

Perfluoroalkoxy or PFA: Dyneon^{TM} (Dyneon LLC), Neoflon® (Daikin America Inc), Teflon® PFA (E.I. DuPont de Nemours Company). The PFA compounds have excellent thermal and chemical resistance and can be preferred to PTFE for ultrahigh purity. They can be used up to temperatures of 260 °C (melting point 305 °C), and they do not suffer from effects because of strong acids and strong oxidants, aromatic, aliphatic and chlorinated solvents.

Polychlorotrifluoroethylene or PCTFE: Kel-F® was originally the trademark of 3M Company (this trade name is now discontinued). The current trademark is Neoflon® (Daikin America, Inc.). PCTFE has a chemical structure similar to that of PTFE but with one chlorine atom substituted for a fluorine atom in the repeat group. PCTFE can be machined to precise dimensions. PCTFE has a melting point of 212 °C. It exhibits chemical resistance to most organic solvents, except THF, CCl_4 and chloroform.

Polyetheretherketone or PEEK, also referred to as polyketones, is a hard, yet slightly flexible polymer. PEEK melts at around 350 °C. PEEK shows poor resistance in concentrated sulfuric, nitric and hydrochloric mineral acids, although performance may be adequate for short-term use with these acids in very dilute form. Its resistance to hydrofluoric acid is very poor.

Polyimide (sometimes abbreviated as PI) is a polymer of imide monomers with a variety of trade names, for example Vespel® (E.I. DuPont de Nemours Company) and Duratron® (Quadrant DSM Engineering Plastic Products). Polyimide materials are characterized by a ring-shaped molecular structure containing nitrogen. They can be obtained with high purity and are relatively easy to machine. They do not melt. They have a poor chemical resistance to concentrated acids.

Polyphenylene sulfide or PPS, known as its trademark Ryton® (Chevron Phillips Chemical), is a resin known for dimensional stability and resistance to corrosive environment. Nonoxidizing, water-based acid, base, and salt solutions do not have a significantly different effect on PPS than water alone, except under very acidic conditions. Unfilled PPS or 40% glass PPS compounds are expected to be suitable for extensive exposure to reagents. Normally, extensive exposure to hydrochloric acid, hydrofluoric acid and nitric acid above 0.1% is not recommended by the manufacturer under immersion conditions. Use of PPS is not recommended for aqua regia. Kerosene is acceptable.

Polypropylene or PP (poly(1-methylethylene)) is a thermoplastic polymer that is rugged and unusually resistant to many chemical solvents, bases and acids. Polypropylene has a melting point of 160 °C. This contrasts with polyethylene, which has a much lower melting point. Polypropylene is most commonly used for plastic mouldings, where it is injected into a mould while molten, with the possibility of forming complex shapes. Polypropylene, whilst having excellent chemical resistance to a very wide range of chemicals, is attacked, however, by strongly oxidizing reagents, for example concentrated nitric and sulphuric acid. Use of PP is satisfactory for 30% hydrochloric acid, 40% hydrofluoric acid, 30% nitric acid, 30% sulphuric acid, but not for aqua regia. In the presence of certain organic solvents such as xylene, toluene and mineral oils, some swelling may occur with polypropylene at room temperature because of absorption.

Polytetrafluoroethylene or PTFE has different trade names: Teflon® (E.I. DuPont de Nemours Company, Neoflon®, Hyflon®). PTFE is a fully fluorinated polymer available in various unmodified and modified grades. PTFE is processed by compacting the powder under pressure at ambient or slightly higher than ambient temperatures. Teflon is inert to virtually all chemicals and can be moulded successfully from PFA Teflon.

Polyvinylidene difluoride, or PVDF (poly-1,1-difluoroethene) is a highly nonreactive and pure thermoplastic fluoropolymer. Trademarks are KYNAR® or HYLAR®. PVDF is a specialty plastic material in the fluoropolymer family; it is used generally in applications requiring the highest purity, strength and resistance to organic solvents and acids, including HF. Compared to other fluoropolymers, it has an easier melt process because of its relatively low melting point, and is easily fabricated into finished parts. It has a relatively low density (1.78) and low cost compared to the other fluoropolymers.

Some other features for sample introduction system design may also be of concern such as fast wash-out and equilibrium times and absence of memory effects between samples. These processes are related not only to the dead volume of the system, but also to the wettability of the material of the spray chamber. The wettability of a liquid is defined as the angle θ formed by the droplet tangent at the contact level and at the plane surface of a solid. The value of the angle depends on the interfacial liquid–gas, solid–liquid and solid–gas tensions (Figure 2.1). Low wettability is obtained for $0 < \theta < 90°$, while high

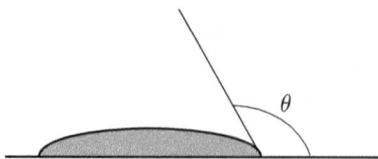

Figure 2.1 Schematic drawing of a droplet on a flat surface of a solid, with the angle θ, formed by the droplet tangent at the contact level and at the plane surface.

wettability is obtained for $90° < \theta < 180°$. Practically, this corresponds to either a droplet or a thin film at the surface of the material. Usually, polymers exhibit low wettability. PTFE- and PFA-made spray chambers may have an internal surface that is treated to change their molecular structure so as to improve wettability.

Commercially available nebulizers make use of borosilicate glass, quartz, PEEK, PTFE, PFA, polyimide, PVDF, alumina ceramic, stainless steel (with a Pt needle) or Ryton body with ruby tips. Commercially available spray chambers make use of borosilicate glass, PP, PFA and PTFE.

2.10 OTHER CONSTRAINTS IN SAMPLE INTRODUCTION SYSTEMS

Some other constraints may be important:

(i) no contamination from the sample introduction system (nebulizer, spray chamber and tubing)
(ii) tolerance to high levels of dissolved solids, particularly in emission spectrometry
(iii) possibility of injecting suspensions
(iv) no risk of nebulizer blocking

Another important specification is that the system should not drift and not be sensitive to the environment, for example temperature. For instance, an external jacket can be added to a spray chamber for the circulation of water at a constant temperature, either room temperature or a cooler one.

At last, setup, ease of use, maintenance, lifetime of the different components and cost should also be considered. Design of sample introduction systems is, therefore, not an easy task, and explains why only a limited number of companies are involved in this field.

– 3 –

Pneumatic Nebulizer Design

3.1 INTRODUCTION

The term *pneumatic* is defined as 'of or relating to or using air or a similar gas'. The word 'nebulizer' is derived from the Latin 'nebula' meaning mist and is defined as 'an instrument for converting a liquid into a fine spray'. Therefore, a pneumatic nebulizer is literally a device for converting a liquid into a fine spray that uses a gas as the driving force [807].

The choice of the nebulizer design in inductively coupled plasma (ICP) techniques for the analysis of liquid samples is critical. Table 3.1 highlights the complexity of the situation by comparing the conditions that must be fulfilled by an ideal nebulizer (see Chapter 2) for use in ICP–atomic emission spectrometry (ICP–AES) and ICP–mass spectrometry (ICP–MS) with the general characteristics of the systems currently used. From this table, it is puzzling to understand why pneumatic nebulizers have been and are still the most commonly used.

Because of the limitations, due to the constraints given in Table 3.1, a large effort has been devoted to develop more efficient and appropriate designs of pneumatic nebulizers. Despite this effort, the existing devices are not yet fully adapted to the plasma requirements. Thus, it will be indicated that these systems still suffer from some general drawbacks:

1. Aerosols produced by pneumatic nebulizers (i.e. primary aerosols) are, in most cases, too coarse. This problem involves that a second component, a spray chamber, for example must be used in order to further modify primary aerosols before introducing them into the plasma, so as to reduce their size.
2. Primary aerosols are normally fast and turbulent because of the use of a high-velocity gas stream to break-up the liquid bulk.
3. The aerosols are unstable. Both the aerosol drop number density and the aerosol total volume change as a function of time. This fact is due to local changes in the gas velocity and temporal changes in the effective liquid flow rate. Renebulization is one of the processes contributing to the unstable aerosol generation. Finally, the geometry of the liquid and gas interaction may also change during the aerosol production event.

Table 3.1

Comparison between the current situation and the desirable one with regard to the production of droplets by a pneumatic nebulizer

Characteristic of an ideal aerosol for ICP	Characteristics of generated aerosols
As fine as possible (Droplets with diameters <10 μm)	Too coarse (Droplets with diameters even >100 μm)
Must be as slow as possible	High-velocity aerosols
Drop velocity ∼ gas carrier velocity in the spray chamber (2 ms^{-1} for a double-pass spray chamber)	Drop velocity up to 80 ms^{-1}
All droplets with identical diameter (monodisperse)	Droplets with diameters ranging from several tens of nanometres up to hundreds of micrometres (very polydisperse)
Uniform droplet number density	Very heterogeneous in terms of spatial droplet number density
Uniform spatial droplet diameters	Coarse droplets are usually located at preferential aerosol locations
Similar characteristics irrespective of the sample composition	Physical properties (e.g. surface tension and viscosity) affect the aerosol characteristics

Despite the poor quality of the aerosol, pneumatic nebulizers are still the most common aerosol production devices for liquid sample introduction in ICP techniques for the following reasons:

1. Historically, the liquid sample introduction system in ICP was adapted from flame atomic absorption spectrometry, for which the aerosol was pneumatically generated.
2. Pneumatic aerosol production requires gas. As the plasma is produced from an argon gas, a logical approach is to energize an argon stream to generate aerosols.
3. Although with limitations, there is good knowledge about the pneumatic aerosol generation process. Consequently, within certain limits, the aerosol properties can be modified according to the requirements of analysis.
4. The versatility of the aerosol production process allows researchers to develop new designs adapted to special needs (i.e. high dissolved solid nebulizers, hydrofluoric acid (HF)-resistant nebulizers, micronebulizers).
5. Pneumatic nebulizers are usually robust and reliable.
6. So far, there is no other alternative showing a better compromise between quality of the results, nebulizer robustness and simplicity of operation.

This chapter is aimed at complementing the work already carried out by other authors [310, 553] in order to facilitate the understanding of the processes and the variables affecting the behaviour of a pneumatic nebulizer. The most common and novel nebulizers will be discussed in this part of the book.

3.2 MECHANISMS INVOLVED IN PNEUMATIC AEROSOL GENERATION

From a simplistic point of view, pneumatically generated aerosols are a consequence of the interaction between a high-velocity gas stream and a liquid solution. A fraction of the kinetic energy of the gas stream is transferred to the liquid bulk, thus yielding droplets. A thorough study of the pneumatic aerosol generation mechanism in the field of ICP nebulizers is difficult because it is an extremely fast process, on the order of several microseconds. In 1988, Sharp published an excellent review on the pneumatic aerosol generation principle based on knowledge already existing in engineering applications and demonstrated the complexity of the events occurring when the aerosol is produced through a liquid–gas interaction [226, 231].

> In the following paragraphs, the most salient points of the pneumatic nebulization mechanisms will be synthesized.

It can be indicated that the strict control of the pneumatic aerosol production is poor at least for the devices and operating conditions used in plasma spectrometry. This owes to the fact that most of the studies aimed at determining the mechanisms of aerosol production have been performed in fields such as engineering and aeronautics. In these areas, the nebulizers (or incorrectly called atomizers) used, as well as the liquid and gas flows–operated, are several orders of magnitude higher than those normally used in ICP–AES and ICP–MS [270]. In addition, it should be considered that there is not a single process responsible for droplet production. Another added difficulty is the fact that the only means for studying the aerosol production mechanism is the measurement of the aerosol characteristics and further interpretation of the results, as well as the development of models [687, 796]. However, once the droplets are produced, the aerosol properties are rapidly modified because of the appearance of other phenomena that occur just after the nebulization event, that is near the nebulizer tip. Because of this fast change, the experimental determination of the aerosol properties is, therefore, very difficult.

The pneumatic aerosol generation can be divided into two different processes: (i) generation of waves on the solution surface, and (ii) growing and break-up of these instabilities to finally produce droplets.

3.2.1 Wave generation

It can be assumed that the liquid surface of the column at the exit of the nebulizer sample capillary is flat. When a high-velocity gas stream comes in contact with it, only tangential forces are acting, resulting in low energy transfer. In fact, the velocity of the gas and liquid at the liquid surface is less than three per cent that of the gas stream [226]. Under these circumstances, the small fraction of energy that is transferred from the gas stream leads to the appearance of waves on the liquid surface. There are some forces that are opposite to the instability generation such as gravity and surface tension. Thus, the gas stream must have a given excess of energy to overcome these forces. The wavelength of the perturbations created on liquid surfaces under the conditions normally used in ICP pneumatic nebulizers is of the order of several tens of micrometres.

3.2.2 Wave growing and break-up

Once the waves are generated on the liquid surface, the gas–liquid interaction degree increases. Therefore, the gas interacts perpendicularly with each one of these waves. The magnitude of the force on a single wave depends directly on the gas–liquid relative velocity, the wave dimensions and the drag coefficient. Another important factor is the flow regime. A turbulent flow favours the penetration of the gas in the liquid bulk and, hence, promotes the generation and growing of the waves. This process is completed until the waves become instable, thus giving rise to the aerosol droplets. The destruction of the waves can be produced through three different mechanisms: (i) the direct stripping of the surface; (ii) the appearance of thin sheets or ligaments produced from the wave crests and (iii) disintegration of the wave. The process (i) is produced when the gas energy overcomes the surface tension of the sample and, as a result, an aerosol with very fine droplets is produced. The process (ii) also favours the production of small droplets. Meanwhile, the process (iii) is observed when the wave acceleration produced by the gas is high enough. According to this mechanism, the gas can penetrate the liquid bulk, thus producing chaotic instabilities and open-ended gas columns. As a result, large liquid masses can be detached, thus giving rise to coarse droplets.

3.2.3 Need for a supersonic gas velocity

The aerosol production depends on the difference in velocity of both the liquid and the gas streams. Therefore, an interesting point is the evaluation of the amount of energy or the gas velocity at the exit of pneumatic nebulizers. It has been indicated that the critical gas velocity required for aerosol generation is about $4.9 \, \text{m s}^{-1}$ [226]. Thus, supersonic gas streams are appropriate to generate fine droplets. However, it has been suggested that the maximum gas velocity at the tip of a converging nozzle (with geometry similar to that of many ICP pneumatic nebulizers) is the sonic velocity [226]. This fact owes to the compressibility of the gas [226]. Thus, in the acceleration region, the density decreases slower than the velocity increases, whereas once the sonic velocity is reached, the density decreases faster than the velocity increases.

   The pressure that must be applied to the argon stream to reach the sonic velocity at the converging nozzle throat is about 2 bar ($\sim$30 psig). Many of the pneumatic nebulizers require higher pressures to reach typical nebulizer gas flow rates (ca. 0.7–$1 \, \text{L min}^{-1}$). At pressures above the critical value, no further change in gas velocity is found. As the gas energy increases with the applied pressure, the mass of gas should increase. In fact, the increase of the pressure above the critical value yields a rise in the gas density. The nebulizer nozzle is thus acting as a mass flow controller, and a useful means of detecting any nebulizer blocking is to measure the back pressure through a manometer inserted from the instrument mass flow controller to the nebulizer. If a partial tip blocking is produced, the pressure required to keep the gas flow rate constant will increase.

To reach supersonic streams, a converging–diverging nozzle can be used (Figure 3.1). The converging region allows the velocity to increase, thus reaching the speed of sound, whereas the diverging one is used to further increase the gas velocity. Contrary to what has been mentioned above as Figure 3.2 indicates, the gas velocity at the exit of a pneumatic nebulizer is higher than the sonic velocity. This is not surprising, as already indicated, analytical nebulizers work very close or above the sonic point because of the nebulizer dimensions and typical liquid and gas flow rates [226]. Shock waves produced as the gas expands can be observed in Figure 3.2. According to this evidence, the gas velocity would be from 2 to 3 times higher than the sonic velocity (i.e. Mach 2–3).

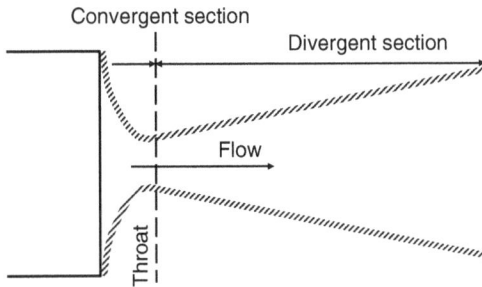

Figure 3.1 Scheme of a converging–diverging nozzle.

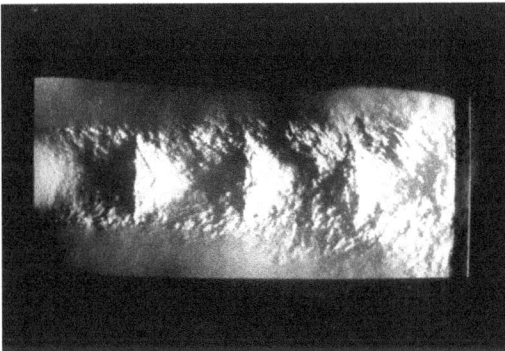

Figure 3.2 Shock waves generated at the exit of a pneumatic concentric nebulizer. (Courtesy of Meinhard Glass Products.)

Energy that is transferred from the gas stream leads to the appearance of waves on the liquid surface, of the order of several tens of micrometres. This facilitates the penetration of the gas in the liquid bulk and promotes the generation and growing of the waves. This process is completed until the waves become instable, thus giving rise to the aerosol droplets. Supersonic gas streams are appropriate to generate fine droplets, which requires a high backpressure, that is several bar.

3.2.4 Main pneumatic nebulizer designs used in ICP spectrometry

Aerosols for ICP–AES and ICP–MS are pneumatically produced according to different liquid- and gas-interaction geometries. As a consequence, there are many pneumatic nebulizers that have been developed [310]. Some pneumatic designs that have been described in detail are

a. concentric type either made of glass or alternative material (see Chapter 2);
b. cross-flow type that can be fixed or with adjustable tips;
c. high-solids nebulizers, that is nebulizers accepting solutions with high salt contents;
d. parallel path pneumatic nebulizer; and
e. high-pressure pneumatic nebulizers [553].

Note that the most commonly used nebulizer is still the concentric one.

Figure 3.3 shows simplified schemes of the pneumatic nebulizers most commonly used in ICP–AES. For concentric designs (Figure 3.3a), the gas causes the liquid break-up

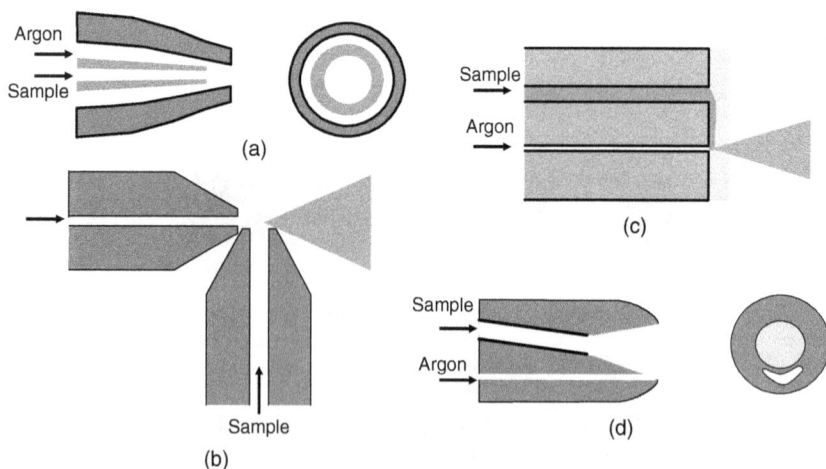

Figure 3.3 Geometrical principles of the aerosol generation for the most commonly used pneumatic nebulizers. (a) Concentric; (b) cross-flow; (c) high-solids nebulizer; (d) parallel-path nebulizer. (Taken from ref. [808].)

from the outermost zone of the liquid column that flows through a centred narrow capillary. Droplets are tangentially detached from the liquid surface. In the case of cross-flow systems (Figure 3.3b), and in contrast to the concentric design, the liquid stream literally impacts against the high-velocity gas stream, and droplets are released perpendicularly to the liquid-flowing direction. This principle is similar for the high-solids nebulizers (Figure 3.3c). Finally, the parallel path aerosol generation mechanism is partly like a concentric one (Figure 3.3d), although there are some recent modifications in which the liquid- and gas-interaction geometry has slightly changed.

3.2.5 Sample delivery

Because of their configuration, some pneumatic nebulizers are able to aspirate the liquid solution, the so-called free aspiration, without any external delivery device such as a pump. Because the gas stream is accelerated at the nebulizer tip, a reduction in the pressure at this particular nebulizer location is observed, called Venturi effect. In other words, a drop pressure is established between the liquid solution (at atmospheric pressure) and the nebulizer nozzle. Consequently, the liquid solution tends to flow from the high- to the low-pressure area, thus being aspirated.

The mathematical relationship developed by Hagen and Poiseuille helps to derive the liquid uptake rate as a function of some experimental variables and dimensions. According to these scientists (Eqn. 3.1), the volume flow rate of a homogeneous fluid passing through a tube (dV/dt) depends directly on the pressure drop (Δp) and on the fourth power of the tube radius (r) and depends inversely on the tube length (l) and fluid viscosity (η).

$$\frac{dV}{dt} = \frac{\pi r^4}{8\eta} \frac{\Delta p}{l} \tag{3.1}$$

Besides the properties of the liquid, there are many factors affecting the value of the liquid uptake rate. Among others we can consider (i) the inner diameter of the nebulizer capillary, (ii) the pressure drop at the nebulizer tip, (iii) the geometry of the nebulizer and (iv) the existence of small defaults on the nebulizer tip surface.

Free aspiration can only be used for concentric nebulizers. Uptake rate under free aspiration is then a characteristic of these nebulizers. Other pneumatic nebulizers have to make use of a delivery device, usually a peristaltic pump. Use of such a device eliminates the influence of the liquid properties such as viscosity, but introduces periodic noise in aerosol generation.

3.3 PNEUMATIC CONCENTRIC NEBULIZERS

3.3.1 Principle

The general characteristics of this nebulizer are (i) ease of operation, (ii) ability to aspirate the solution freely and (iii) possibility of manufacturing a particular nebulizer adapted to the sample nature.

Figure 3.4 shows a picture of a commonly used pneumatic concentric nebulizer. The liquid solution is introduced through a horizontal capillary, whereas the gas enters through a vertical conduction. At the nebulizer tip, the liquid flows concentrically to the gas. A fraction of the energy is transferred from the gas stream to the liquid bulk, thus producing the aerosol. This fraction of energy is rather low, because most of it is employed in the gas expansion and the droplet acceleration.

Figure 3.5 shows a picture of the tip of a pneumatic concentric nebulizer when only gas is flowing through it. The streamlines of the gas stream are highlighted in this image. It can be observed that the expansion of the gas is produced at a given distance from the tip exit. It is also observed that, at the exit of the sample capillary, there is a zone in which small gas turbulences appear. These turbulences correspond to the air that is being aspirated by the nebulizer. By taking an average value of the capillary's inner diameter

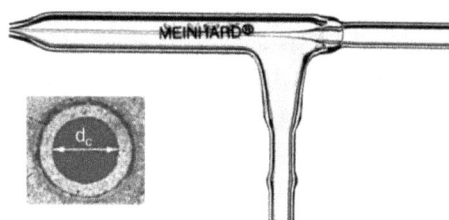

Figure 3.4 Picture of a pneumatic concentric nebulizer with a front view of the tip.

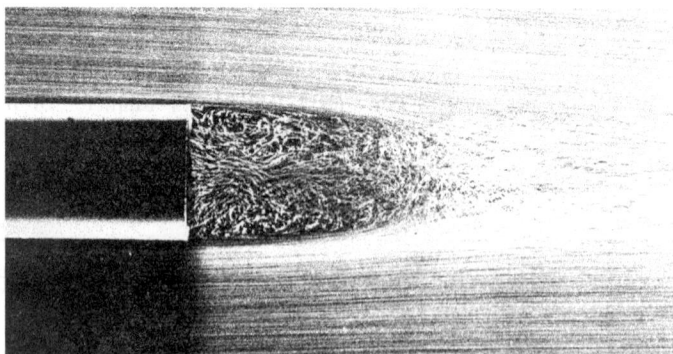

Figure 3.5 Picture of the gas dynamics at the exit of a pneumatic concentric nebulizer. (Courtesy of Meinhard Glass Products.)

close to 250 μm, which is representative for a conventional pneumatic concentric nebulizer, it can be verified that the turbulences aforementioned extend for about 750 μm (i.e. 3 times the inner diameter of the liquid capillary). If a liquid sample is present, the liquid core will remain unaltered for a longer distance because of the resistance offered by the surface tension. Indeed, it has been claimed that this distance extends to 5 times the inner diameter of the capillary [226]. Therefore, for the case studied in Figure 3.5, in the presence of liquid flowing through the capillary, the distance should be 1250 μm.

To produce efficiently fine droplets, the contact between the liquid and gas should be as efficient as possible. Figure 3.5 demonstrates that the concentric geometry is not the best choice for producing appropriate aerosols because an important fraction of the gas volume is not in direct contact with the liquid stream. Hence, part of the gas kinetic energy is not used and lost for droplet production. This energy is likely conserved in the gas expansion and droplet acceleration.

The liquid emerges from the centred capillary and, because of the action of the gas stream, moves towards the gas exit (Figure 3.6a). This effect causes the thickness of the liquid stream moving outwards from the nebulizer capillary to decrease [226]. This process is called prefilming and is one of the benefits of this kind of nebulizers. The fact that the liquid solution is prefilmed enhances the interaction of the liquid and gas streams. At a given liquid flow rate, the thickness of the solution on the front walls of the sample capillary slipping towards the gas stream is higher for thin capillaries than for coarser ones.

Figure 3.6b shows a picture of the nebulization process for an enlarged concentric pneumatic nebulizer that was built to observe the formation of the aerosol. In this particular design, a liquid capillary having a 10 mm i.d. was used. Although the dimensions of this nebulizer exceed the conventional ones by more than one order of magnitude, the aerosol generation pattern can be extrapolated. The gas velocity at the exit of the concentric nebulizer is high and causes the appearance of a gas vortex with a toroidal shape. As a consequence, the solution reaching the sample capillary end is spread out on the capillary walls and gives rise to a meniscus. A series of thin ligaments are formed from the contour of this meniscus (Figure 3.6). Each ligament collapses in fine droplets, thus giving rise to the aerosol. The length of the ligaments has been estimated to be of the same order as the inner diameter of the sample capillary. Unfortunately, the coalescence of these ligaments is a frequent phenomenon that promotes the generation of relatively coarse droplets.

Some concentric designs take advantage of the prefilming process to enhance the transfer of energy from the liquid to the gas stream [788]. In the case of the pneumatic extension nozzle (PEN), a 200-μm i.d. capillary is introduced into an argon pressurized chamber with an orifice in its final part (Figure 3.7). The liquid stream emerges from this capillary, and the gas flowing through the nebulizer causes the prefilming effect. At the exit of the nebulizer, the liquid jet is much thinner and hence the aerosol production becomes more efficient.

Prefilming effect is highly beneficial for aerosol production and can be enhanced with some nebulizer designs.

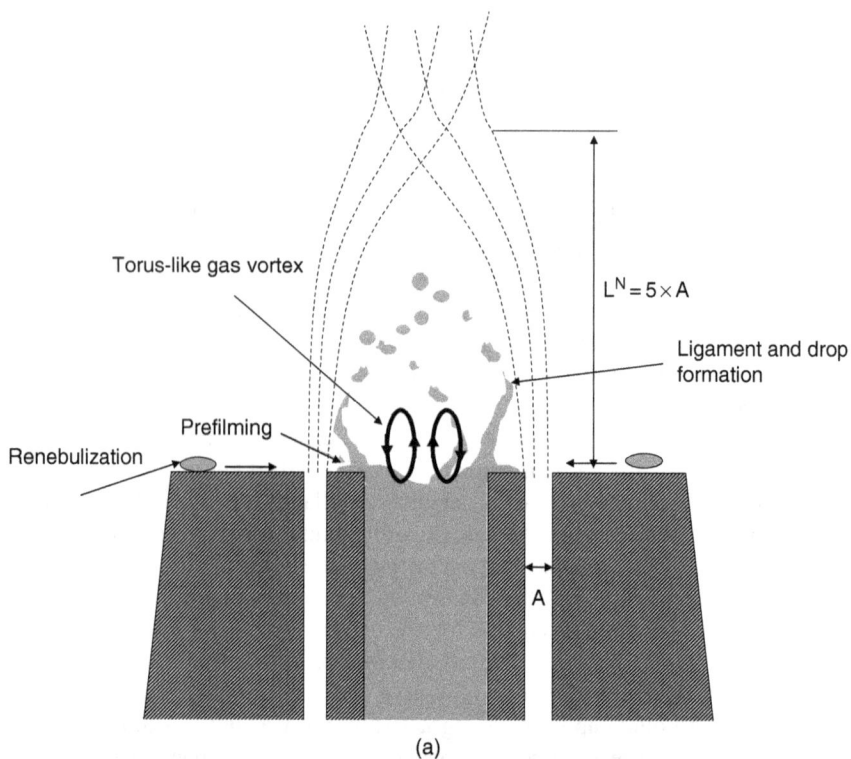

Torus-like gas vortex

$L^N = 5 \times A$

Ligament and drop formation

Prefilming

Renebulization

A

(a)

10 mm

(b)

Figure 3.6 (a) Schematic diagram of the processes taking place at the exit of a pneumatic concentric nebulizer (Adapted from ref. [226]) and high-velocity picture of the nebulization process for an enlarged pneumatic device (b).

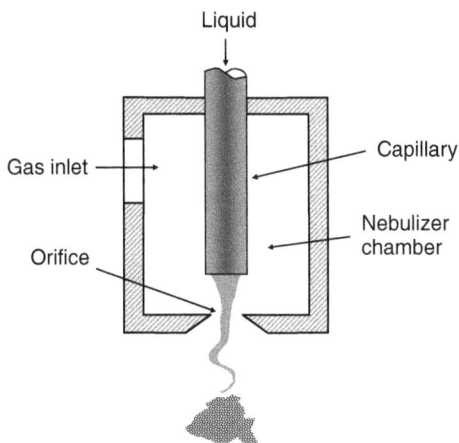

Figure 3.7 Scheme of the pneumatic extension nozzle (PEN). (Taken from ref. [788].)

3.3.2 Different designs

Several manufacturers have been involved in concentric nebulizer design and production. Although based on the principle described previously, some variations in the design may be observed between producers.

Various pneumatic concentric nebulizer designs are available from Meinhard Glass Products: the so-called type A, C and K nebulizers. Figure 3.8 shows the pictures of these three pneumatic concentric nebulizers. In the case of type A nebulizers (Figure 3.8 and Figure 3.9a), the sample capillary and the tip shell are coplanar. Furthermore, both the nozzle and the capillary have square and flat walls. The aerosol production starts outside the nebulizer. The type A was the first commercially available Meinhard nebulizer. For C and K nebulizers, the liquid capillary is recessed in the nozzle, which has some important implications. The solution suction effect is expected to be stronger for the type A arrangement because the gas pressure beyond the capillary is lower at the nebulizer exit. Note that the gas energy (and speed) is higher inside than outside the nebulizer. However, it has been claimed that the recess of a capillary in a nebulizer causes a 1.5–2 times increase in the free liquid uptake rate with respect to the situation found when the capillary is not recessed [349]. Initially, the capillary was recessed by about 0.8 mm, although nowadays, lower recess (i.e. 0.5–0.6 mm) is being used. Therefore, the liquid and gas mixing starts before these streams leave the nebulizer tip. The main difference between type C and K nebulizers is the shape of the final part of the sample capillary and the nebulizer tip. Thus, for the former system, the tip has been flame polished. Meanwhile, for nebulizers type K, the tip shape is similar to that of nebulizer type A (i.e. ground flat and square). This can make type K and C nebulizers more resistant to blocking than the type A ones when high salt content solutions are being used [127] and can also be efficiently used for the analysis of viscous samples [805].

A particularity of Glass Expansion nebulizers is that, unlike the Meinhard ones, the liquid capillary has thick walls and it is conically shaped. This can be observed from Figure 3.10. The corresponding commercial name is VitriCone^TM and several advantages can be

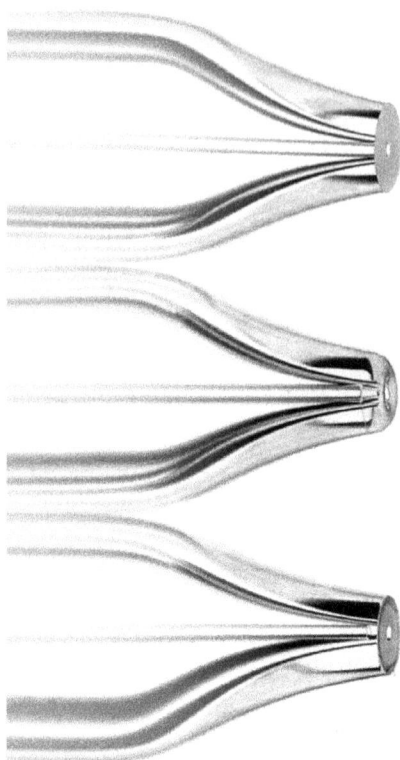

Figure 3.8 Picture of the tip of different Meinhard type pneumatic concentric nebulizers. From top to bottom, Types A, C and K. (Courtesy of Meinhard Glass Products.)

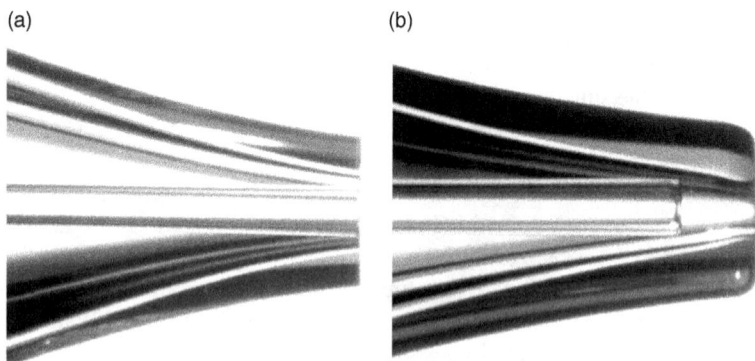

Figure 3.9 Magnified picture of the tips for pneumatic concentric nebulizers (a) type A and (b) type C. (Courtesy of Meinhard Glass Products.)

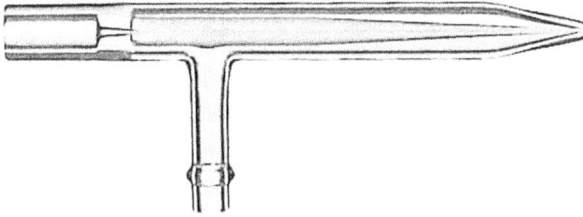

Figure 3.10 Picture of a pneumatic concentric nebulizer Glass Expansion type.

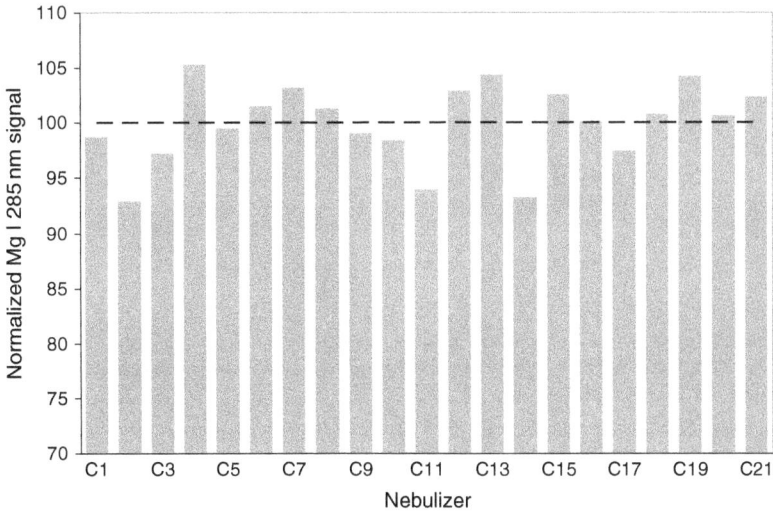

Figure 3.11 ICP–AES normalized emission signal for several pneumatic concentric Glass Expansion nebulizers (C. Dubuisson and J.M. Mermet, unpublished results).

mentioned: (i) higher robustness; (ii) higher resistance to the capillary vibration; (iii) less blocking problems, as the sample channel has a more uniform inner diameter and (iv) higher reproducibility in nebulizer construction and nebulizer-to-nebulizer performance. This last point is illustrated in Figure 3.11, in which the emission intensity is presented for a set of 21 different Glass Expansion nebulizers. It can be observed that, within a 10 per cent deviation, all the nebulizers provide similar ICP–AES sensitivities.

Pneumatic concentric nebulizers differ by the recess of the capillary tube with respect to the nebulizer tip, along with their thickness and their shape.

Special concentric nebulizers with Laval tips have been described. These devices are similar to glass concentric nebulizers, but a converging–diverging nozzle is adapted at the end of the nebulizer. This idea was first presented by Shuying and Yunlong in 1988 [232]. Figure 3.12a shows a drawing of special concentric nebulizers with Laval tips. As mentioned

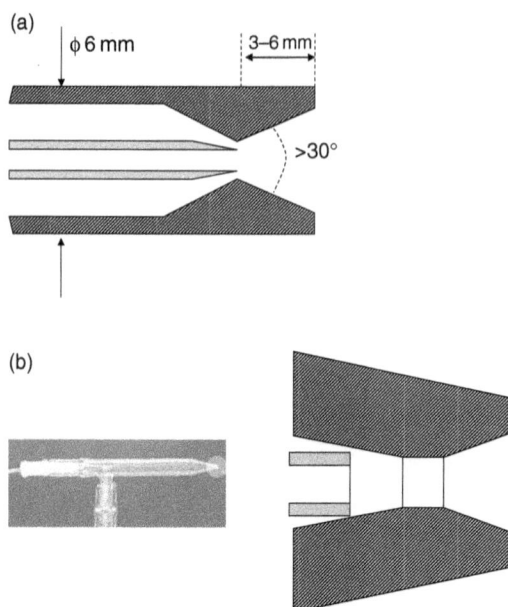

Figure 3.12 Pneumatic concentric nebulizer with the Laval nozzle (a) system initially developed by Shuying and Yunlong in 1988 [232]; (b) nebulizer commercialized by Epond (http://www.e-pond.biz).

before, in the case of this nebulizer, the gas velocity is higher than the speed of sound. This fact contributes to the generation of finer aerosols, thus enhancing the analytical ICP performance [232]. Furthermore, provided that there is fresh solution continuously wetting the inner walls of the nebulizer, this system is less prone to tip blocking. The experiments carried out with saturated sodium chloride solutions confirmed this fact. More recently, pneumatic concentric nebulizers with a Laval nozzle have been made commercially available (Figure 3.12b).

Demountable concentric nebulisers (DCNs) were described [592]. Thus, for example one of these devices can be constructed from fused silica tubing for the sample and a borosilicate glass serological pipette as the nebulizer external body [634]. This low-cost nebulizer is depicted in Figure 3.13. All the components can be easily purchased because they are commercially available.

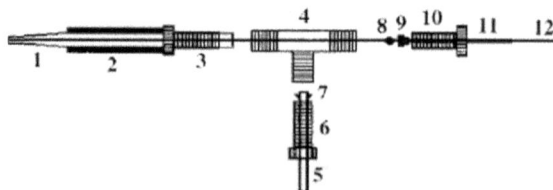

Figure 3.13 Laboratory-constructed concentric demountable nebulizer. (1) ~0.1 mL Pyrex serological pipet; (2) polyethylene tubing sleeve to fit standard spray chamber inlet; (3) PEEK nut; (4) PEEK tee; (5) Teflon argon supply tubing; (6) Tefzel nut; (7) Tefzel ferrule; (8) PTFE tape seal; (9) PEEK ferrule; (10) PEEK nut; (11) PEEK tubing; (12) fused-silica capillary tubing. (Taken from Ketterer and Hudson [634] with permission.)

One of the most important advantages of this nebulizer is that any part of it can be easily replaced. Thus, if the sample capillary is blocked by a salt deposit or a particle occlusion, the tubing can be easily removed and a new one can be adapted. An issue of this kind of nebulizer is the reproducibility of the analytical behaviour when it is assembled and disassembled several times. With regard to plasma performance, the demountable pneumatic concentric nebulizers provide similar figures of merit as conventional devices [762]. Another important advantage is that the price of this device is estimated at less than US$10 against the US$300–400 cost of classical glass pneumatic concentric nebulizer.

3.3.3   Possibility of free liquid uptake rate

As mentioned above, pneumatic concentric nebulizers are able to freely aspirate the solution. When a gas at high pressure flows through a bore and is introduced into a low-pressure gas zone, a stream is formed in the low-pressure gas zone. The gas at lower pressure is pushed away to the tip or orifice, thus leading to a current in the lower pressure gas zone. Some of the lower pressure gas is drawn into the higher pressure gas jet. This phenomenon causes significant suction that is responsible for the dragging of the liquid sample towards the gas stream, thus finally yielding the aerosol.

The sample capillary recess can modify the value of the pressure drop because a chamber is generated inside the nebulizer. The shape and size of this antechamber has a remarkable effect on the pressure drop. For example in the case of type C and K devices, the suction effect is stronger because of the appearance of the chamber inside the nebulizer. As reported in ref. [349], the portion of the nozzle extending beyond the capillary tip maintains some of the gas speed, initially gained by acceleration through the annulus, and therefore maintains lower gas pressure beyond the end of the capillary. This nozzle antechamber shields the emerging mist stream from immediate dissipation into the higher pressure of the atmosphere. When a capillary of a particular size is recessed, the natural liquid uptake rate will be 1.5–2 times greater than the rate observed when a similar capillary is not recessed.

The stability and reproducibility of the value of the free liquid uptake rate strongly depends on the dimensions of the liquid capillary. Thus, it has been observed that for a DCN, the free liquid uptake rate for a 180-µm i.d. capillary was $0.1 \, \text{mL min}^{-1}$ for a 1-L min^{-1} gas flow rate. However, the self-aspiration was not maintained if a gas slug was introduced. These problems were not observed when using capillaries with inner diameters above 250 µm [634].

Working under free-aspiration mode has several advantages as well as drawbacks. Aerosol generation pulsations are minimized when the peristaltic pump is removed. Furthermore, properly working under free aspiration mode ensures that the nebulizer is producing a good aerosol, because it is known that the best performance of concentric nebulizers is reached when they are operated at liquid flows close to the free uptake rate. However, if no external pumping device is used, the amount of liquid flow aspirated by the nebulizer per unit of time is a function of solution's physical properties such as the viscosity (Eqn. 3.1). Therefore, interferences are magnified if there are slight differences between the samples and the standards viscosity. Besides, the liquid and gas flow rates cannot be optimized separately because the former precludes the latter. Finally, the liquid flow rate depends on the capillary dimensions and its characteristics. The second factor is very important, and sometimes, it is observed that the sample aspiration can even cease.

> Although most concentric nebulizers can freely aspirate solutions, most instruments make use of a peristaltic pump to overcome aspiration problems related to viscosity. The delivery rate is usually adjusted to obtain a plateau in the signal, that is in the range 1–3 mL min^{-1}.

3.3.4   Critical dimensions

Taking into account the mechanism of pneumatic aerosol generation, it can be inferred that the variables affecting the aerosol characteristics are (i) the nebulizer dimensions that influence the gas velocity, (ii) the gas density after expansion, (iii) the liquid mass flow and (iv) the physical properties such as surface tension, viscosity or density.

With regard to nebulizer dimensions, there are two main considerations that have to be made in order to optimize the design for a pneumatic nebulizer, the liquid and gas interaction area and the distance beyond the liquid capillary along which a liquid stream is observed. Both directly affect the gas and liquid mixing geometry. The first one is related to the effective liquid and gas interaction area that, in turn, is referred to the area of aerosol production. The factors affecting the value of this parameter are the dimensions of the liquid vein and the gas velocity at the interaction region. The section considered in this case is only the zone in which the gas has a liquid propellant action. Once the gas loses the excess of energy, it is not able to detach droplets from the liquid vein. It is possible to calculate the length along which the gas is able to produce droplets as 5 times the inner diameter of the liquid capillary [226].

From the previous discussion, it is obvious that the performance of pneumatic nebulizers depends on the dimensions of the tip. The most relevant characteristics of these devices are (i) the gas exit cross-sectional area, (ii) the sample capillary inner diameter, (iii) the capillary wall thickness, (iv) the capillary recess and (v) the nebulizer shape. Figure 3.14 shows an outline of the tip of a pneumatic concentric nebulizer in which the critical dimensions are indicated.

At a given gas flow rate, the pressure required to apply to the gas stream is strongly correlated to the nebulizer annulus area. The lower this variable the higher the gas pressure. Hence, the gas has more kinetic energy to produce surface. Therefore, it may be expected that nebulizers with lower gas exit cross-section areas should generate finer aerosols. The dimensions of the sample capillary have an important effect on the liquid and gas interaction geometry. To produce fine droplets, both the inner diameter and the wall thickness should be rather low. In this way, the gas kinetic energy will be more efficiently used [382].

Table 3.2 summarizes representative values for the dimensions for several pneumatic concentric nebulizers available from Meinhard Glass Products and their range when several nebulizers of the same type were measured. At a first glance, it can be observed that the largest variability is for the annulus area.

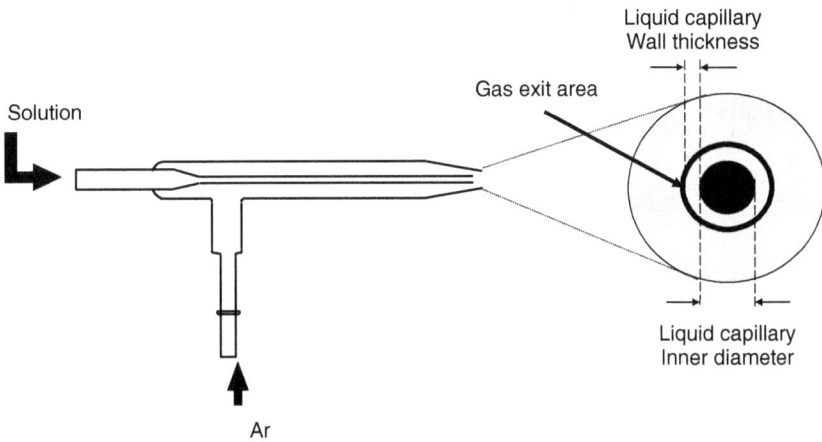

Figure 3.14 Scheme of a pneumatic concentric nebulizer tip together with its critical dimensions.

Table 3.2

Critical dimensions for some Meinhard Glass Products pneumatic concentric nebulizers

Type	Sample capillary inner diameter (μm)	Capillary wall thickness (μm)	Gas exit cross-sectional area (mm^2)	Recess of liquid capillary (μm)
TR-30-A0.5	150	70	0.07	
TR-30-A1	230–260	50–65	0.03–0.10	
TR-50-A1	200	45–60	0.03	
TR-30-A2	260–290	50–100	0.03–0.06	
TR-30-A3	320–330	45–70	0.08–0.13	
TR-50-C0.5	280	50	0.03	520
TR-50-C1	280–300	50	0.06–0.07	560–590
TR-50-C2	290	55	0.05	570
TR-50-C3	290	50	0.05	530
TR-30-K2	250–270	50–60	0.04–0.07	490–550

At this point, it is useful to indicate the meaning of the nomenclature used by a given manufacturer. When referring to one of these nebulizers, we normally use a reference as, for example that used by Meinhard Glass Products: TR-30-A1.

In this case, the two former characters refer to the material the nebulizer is made of. TR means that the nebulizer is made of glass. Following this, there is a number (20, 30, 50, . . .), which indicates the gas pressure in psig that is necessary to apply in order to reach a given gas flow rate. Type A and C nebulizers are calibrated at $1 \, L \, min^{-1}$, whereas for type K nebulizers, this number corresponds to the gas pressure when a 0.7-L min^{-1} gas flow rate is achieved. The geometry of the nebulizer is also given (A, C, K and more recently CK) and, finally, a number is included (0.5, 1, 2, . . .) that corresponds to the free liquid uptake rate when the nebulizer is operated at its nominal gas backpressure.

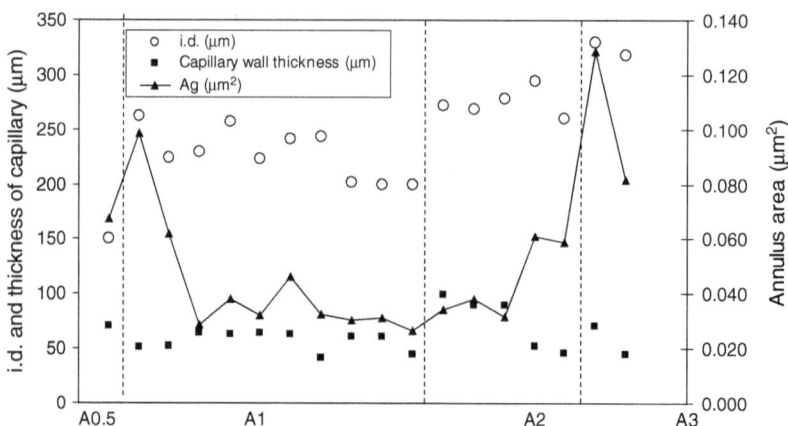

Figure 3.15 Critical dimensions of several glass pneumatic type A concentric nebulizers.

Another example of nomenclature is that used by Glass Expansion Ltd.: ARxx-yy-ZZnnn.

In this case, AR means that the gas used is argon. The characters xx, yy and nnn are the nominal argon pressure, flow rate and free liquid uptake rate, respectively, whereas ZZ refers to the nebulizer design.

The differences in dimensions can be evaluated in more detail by plotting them graphically. Figure 3.15 shows these data for 18 type A Meinhard nebulizers. The gas exit cross-section area, for a given family of nebulizers (e.g. type A1), may take dispersed values. However, the capillary dimensions vary from one nebulizer to another in a less remarkable way. Thus, for the nebulizers type A1, the inner diameter changes from 200 to 250 µm, whereas the wall thickness is virtually the same. Note that it has been demonstrated that the annulus area has a higher influence on the nebulizer performance than the capillary dimensions [302].

When comparing different nebulizer families, it can be observed that the average inner diameter of the sample capillary follows the increasing order: A0.5 < A1 < A2 < A3. This expected fact reveals that, at a given gas flow rate (e.g. $1\,L\,min^{-1}$) the higher the inner diameter of the sample capillary, the higher the free aspiration rate. However, if considering particular cases, it can be observed that there are some A1-type nebulizers having similar capillary dimensions and annulus areas as those for A2-type devices. However, the free uptake rate is different. Again, consideration must be given to the nebulizer tip shape and microscopic defaults to try to explain these observations.

From the data presented in Figure 3.15, it is possible to evaluate which nebulizer will be the most efficient in terms of aerosol production. As it has been previously mentioned, the effective liquid and gas interaction area value is a key issue in generating fine aerosols. It is possible to quantify this area by multiplying the length of maximum gas action by the perimeter of the gas jet. To evaluate the former parameter, the annulus width should be multiplied by 5. It has been claimed that this width takes typical values close to 20–30 µm [226]. For the nebulizers included in Figure 3.15, the annulus width has taken values ranging from about 20 to 70 µm. We could assume that the liquid and gas interaction is produced in the outermost zone of the sample capillary. The corresponding data are summarized in Figure 3.16. The gas–liquid interaction area is directly related with the efficiency of use of

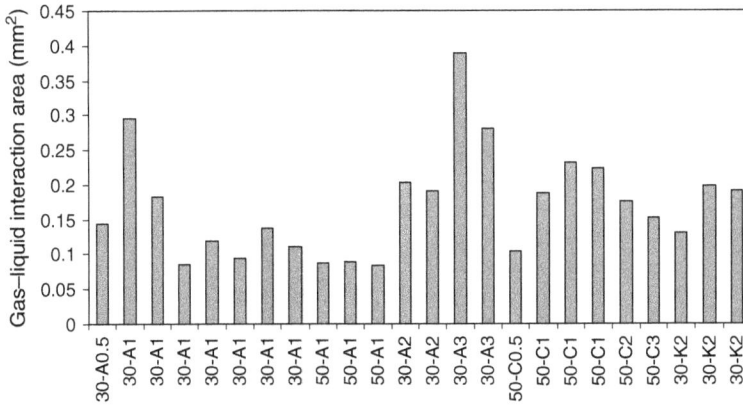

Figure 3.16 Effective gas–liquid interaction area for several pneumatic concentric nebulizers.

kinetic energy to produce droplets. This parameter may change significantly from one nebulizer to another theoretically identical nebulizer. Thus, for example in the case of nebulizers TR-30-A1, the interaction area may change by a factor of 3. This is a clear result of the differences in annulus width and capillary's inner diameter. As it may be observed from Figure 3.16, the type A nebulizers providing the highest value of the liquid and gas interaction area are those having the highest values of these dimensions (i.e. nebulizers A3). These data suggest that the gas kinetic energy is more efficiently used if the gas exit cross-section area and the inner diameter of the liquid capillary are high.

However, the gas energy should be taken into account in order to evaluate what nebulizer design would be the best in producing fine droplets. Note that there is a direct relationship between the gas area and the gas kinetic energy. As it is shown in Figure 3.16, the gas exit cross-sectional area for A3 nebulizers is, in general terms, higher than for A2 and A1 nebulizers. Therefore, the pressure that should be applied to maintain a given gas flow rate should follow the order: A3 < A2 < A1. In other words, there would be more kinetic energy available for aerosol production in the case of nebulizers A1 (except in one case) than for nebulizers A2 or A3. This fact would partially counteract the benefit of the enhanced gas–liquid interaction area mentioned before.

3.3.5 Renebulization

Renebulization is produced when a fraction of the solution is deposited on the walls of the nebulizer tip and subsequently renebulized (Figure 3.17).

This process has two different negative consequences on the performance of a pneumatic concentric nebulizer: (i) increase in the aerosol noise and (ii) tip blocking when high salt content solutions are introduced.

Figure 3.17 Picture of a pneumatic concentric nebulizer inserted into a spray chamber in which the droplet formation because of the renebulization is evidenced. (Courtesy of Burgener Research Inc.)

As it has been previously indicated, pneumatic nebulizers produce noisy aerosols. On the one hand, the liquid aerosol volume rate can vary as a result of processes such as renebulization. On the other hand, gas turbulences created at the exit of the nebulizer can also be responsible for changes in the aerosol droplet number density. The gas suction effect carries these droplets to the gas exit and, once they arrive to this nebulizer location, they are nebulized again.

This problem is especially important and unsolved in the case of pneumatic nebulizers. Renebulization depends on many factors such as the shape and material of the nebulizer, gas and liquid flow rates, the solution surface tension, the spray chamber design, among others. Figure 3.17 shows a droplet forming at the tip of a pneumatic concentric nebulizer when it is placed inside a double-pass spray chamber.

To try to minimize the renebulization of the solution, it is advisable to reduce the surfaces adjacent to the nebulizer gas exit. It has been indicated that in order to avoid renebulization, spray chambers without obstacles close to the nebulizer tip are the best situation (courtesy of Burgener Research Inc.). The nebulizer material also plays a drastic role, because it has been observed that for nebulizers made of wettable materials, the accumulation of coarse droplets that are suddenly nebulized and hence cause serious instability problems, is less frequent than in the case of nonwettable materials. Thus, for example nebulizers made of graphite are better candidates to overcome renebulization than glass or Teflon systems. With regard to the nebulizer shape, there are several choices that have been considered: (i) nebulizers with whiskers, (ii) devices with a minimum area inside the spray chamber and (iii) flat face nebulizers.

Figure 3.18 shows the situation found when a nebulizer with whiskers is placed inside the spray chamber. The droplets carried by the recirculating gas turbulences created inside the spray chamber impact against these protuberances. This solution grows, and it does not reach the nebulizer tip so that the problems of renebulization are virtually eliminated.

With a nebulizer having a flat surface (Figure 3.19), droplets that eventually deposit on the nebulizer cannot pass over a sharp surface, they grow and then they are lost as they fall down.

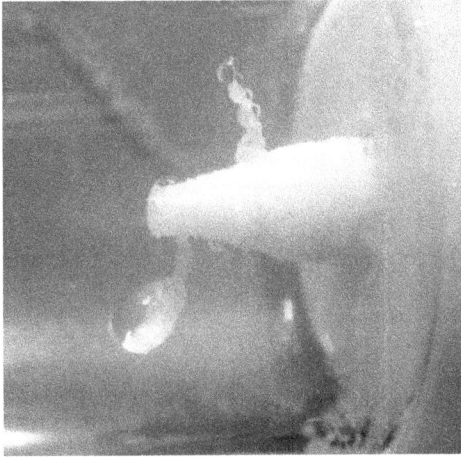

Figure 3.18 Pneumatic nebulizer with whiskers with the droplets formed on them. (Courtesy of Burgener Research Inc.)

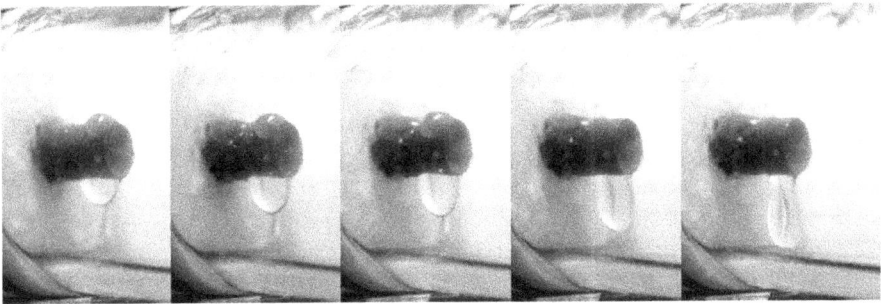

Figure 3.19 Pictures illustrating the formation of a droplet at the nebulizer tip and its further removal from the nozzle. (Courtesy of Burgener Research Inc.)

3.3.6 Nebulizer tip blocking

When working with high salt content solutions, pneumatic concentric nebulizers become blocked because of the solution renebulization. During salty solution nebulization, it has been claimed that the dry gas evaporates a fraction of the solvent, which causes a drop in the local temperature high enough to produce crystal formation at the gas annulus [226]. As soon as the annulus blocks, even partially, nebulization stops. Obviously, this effect would be more severe for nebulizers having a low gas exit cross-sectional area than for those with high values of this critical dimension. Therefore, among the nebulizers considered in Figure 3.15, the A3-type nebulizers would be less exposed to tip blockage.

Among the different approaches tested to reduce the significance of the blockage problems, the simplest one appears to be recessing the liquid capillary with respect to the nebulizer nozzle. This has been done in a number of nebulizers from Meinhard Glass Products (type C and K nebulizers) as well as from Glass Expansion Ltd. Figure 3.20 shows the pictures of the

(a)

(b)

(c)

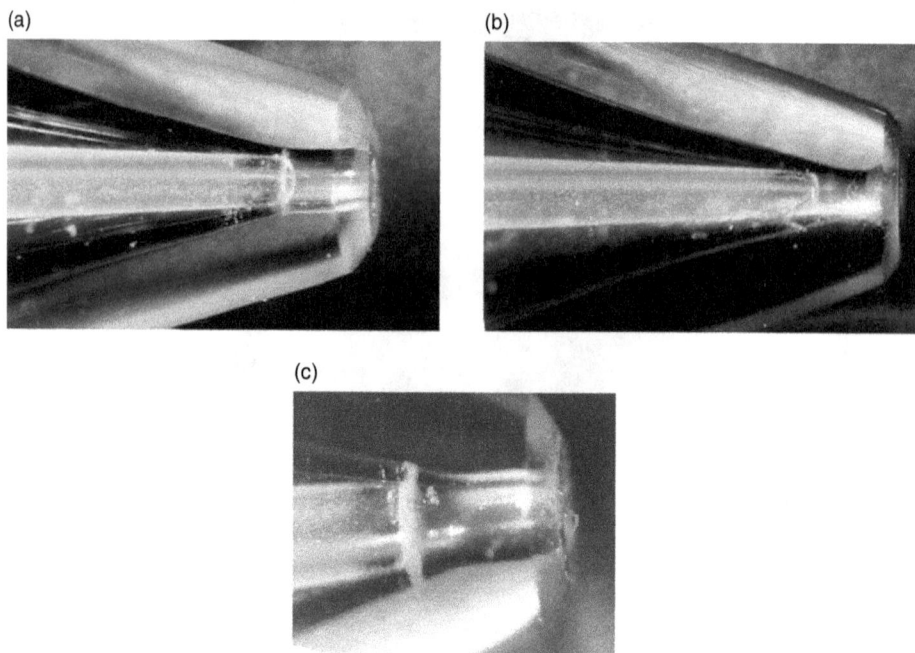

Figure 3.20 Pictures of the tips of several Glass Expansion nebulizers. (a) Conical nebulizer; (b) Sea spray nebulizer; (c) Slurry nebulizer.

tips of several Glass Expansion nebulizers. Table 3.3 in turn summarizes the critical dimensions of these nebulizers. For type C and K nebulizers, the capillary setback is of the order of 500–600 μm (Table 3.2), whereas for the Glass Expansion nebulizers, this dimension is about 2 times higher. In both groups of nebulizers, the salt deposit seems to be avoided, because the nebulizer tip is continuously refreshed with new solution. The different degree of recess affects the liquid and gas interaction efficiency. Thus, for the latter nebulizers (Table 3.3), because both streams interact for a higher length, the transfer of energy from the gas to the liquid stream must be more efficient than for the Meinhard Glass Products devices. This fact does not imply that the aerosols should be finer for nebulizers in Table 3.3, because there are other factors such as the value of the gas exit cross-sectional area and capillary dimensions that play a relevant role in terms of the production of fine droplets.

Table 3.3

Critical dimensions for some Glass Expansion pneumatic concentric nebulizers

Nebulizer	Sample capillary inner diameter (μm)	Capillary wall thickness (μm)	Gas exit cross-sectional area (mm^2)	Recess of liquid capillary (μm)
#1	220	50	0.014	800
#2	400	40	0.0104	1200
#3	200	100	0.015	920
#4	400	60	0.028	0
#5	500	30	0.014	1320

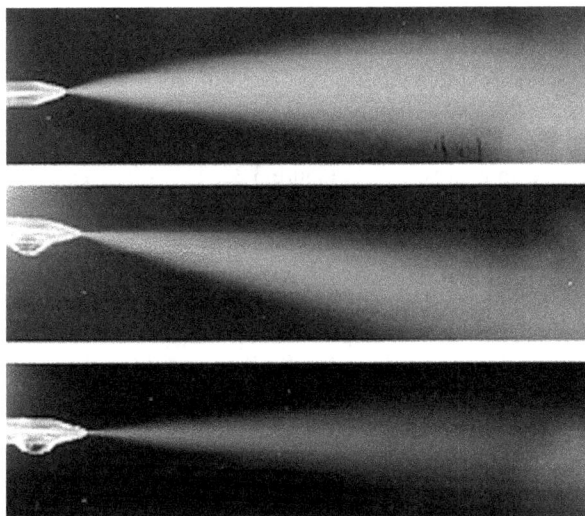

Figure 3.21 Pneumatic concentric nebulizers operating under the same liquid and gas flow rates with aqueous solutions.

If a drop forms at the nebulizer tip, then grows and finally falls down, the nebulization efficiency will be lower than 100 per cent. Figure 3.21 shows the pictures of the three pneumatic concentric nebulizers (as seen in Figure 3.20) working with water, under a given set of liquid and gas flow rates. For the nebulizer 1 (Figure 3.21) the solution is totally nebulized. However, for the other two nebulizers, only 43 and 32 per cent of the mass of liquid reaching them is turned out into aerosol [678]. By examining the differences in design of these three nebulizers, it can be observed that the recess for nebulizer 1 (a in Figure 3.20) is less conspicuous than for nebulizers 2 and 5 (Table 3.3) (b and c in Figure 3.20, respectively). These images would suggest that a fraction of the solution that is left between the capillary end and the nebulizer main orifice is nebulized inside the chamber. If droplets are generated in there, the likelihood for droplet collapse by impacts against the inner walls of the nebulizer tip should be rather high. This initial accumulation of liquid could act as an obstacle for the aerosol and, as a consequence, more droplets would impact, thus promoting the growing of a drop at the nebulizer exit. Therefore, the capillary setting back is beneficial, but if it is excessive, a detrimental effect can be produced. Another difference found among these three nebulizers is related with the value of the capillary's inner diameter. Thus, for nebulizer 1, a narrower sample capillary is used as compared with nebulizers 2 and 5. This fact could minimize the droplet production inside the nebulizer 1 because the solution velocity at the capillary exit is higher.

3.3.7 Aerosol drop characteristics

3.3.7.1 Influence of the gas and delivery rates on drop size distribution

All the aforementioned aspects have a direct effect on the characteristics of the aerosols produced by the nebulizers (see Section 2.8). In general, analysts are seeking nebulizers that are able to produce primary aerosols as fine as possible, because, as it will be later discussed, the finer the primary aerosols, the higher the mass of analyte delivered to the plasma and, hence, the higher the sensitivity. This cannot be considered as a rule, because for pneumatic nebulizers a 'fine' aerosol often means a 'fast' aerosol. As it will be discussed, in order to minimize losses inside the spray chamber, it is desirable to work with slow aerosols.

A complete drop size distribution curve obtained with a system based on the Fraunhofer diffraction of a laser beam is presented in Figure 3.22. The nebulizer used in this case was a Meinhard TR30-A-1 operated at 0.7 L min^{-1} gas flow rate and 1 mL min^{-1} liquid delivery rate.

The drop size distribution curves in band for pneumatic nebulizers is multimodal with several maxima indicating that there are preferential droplet diameters existing at higher frequencies (i.e. probabilities) than others. This can suggest the production of droplets through several mechanisms. For example coarsest droplets are expected to be produced following a mechanism known as Taylor instability, in which the waves created on the liquid surface in the early stages of the nebulization disintegrate as a result of the acceleration of their crests [226]. In contrast, the generation of small droplets may be due to the overcoming of the surface tension forces and the ligament or liquid sheet formation and disruption. Furthermore, there are other processes that can account for the generation of droplets by the appearance of maxima in this kind of curves such as the sample capillary oscillation at characteristic frequencies [507]. The position of the maxima on the drop size distributions in band depends on the nebulizer dimensions and the gas and liquid flow rates. Another point that adds complexity to the interpretation of the data corresponding to aerosol characterization is related with processes taking place from the aerosol production point to the measurement zone [231]. These processes (e.g. droplet recombination) are responsible for inconsistencies between some expected and obtained data.

Figure 3.23 shows the volume–drop size distribution curves in band for the primary aerosols generated by a pneumatic concentric nebulizer operated under different

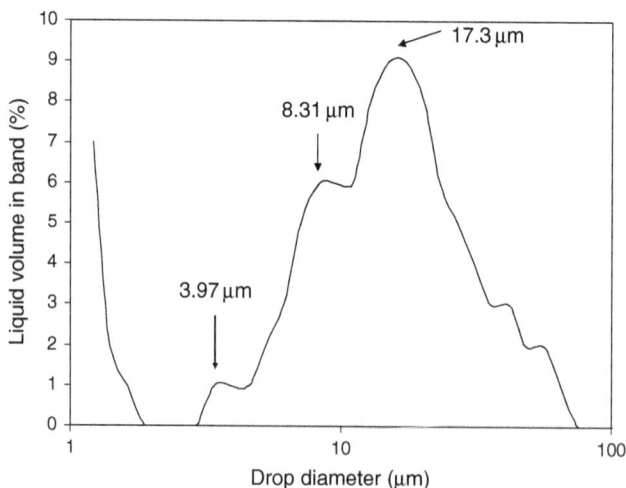

Figure 3.22 Typical complete volume drop size distribution curve for a pneumatic concentric Meinhard TR-30-A3 nebulizer.

(a)

(b)

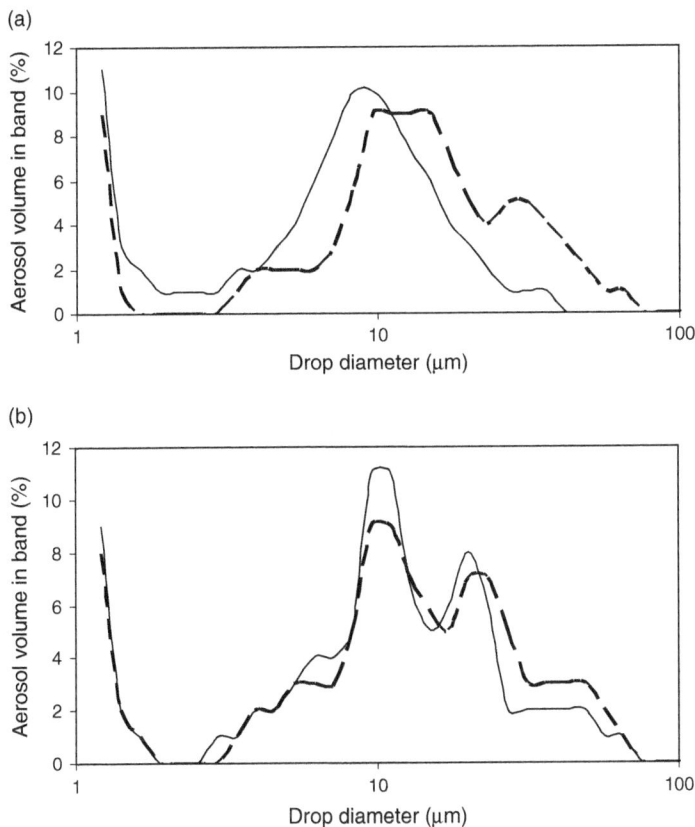

Figure 3.23 Aerosol volume drop size distribution curves in band for two different pneumatic concentric nebulizers operated under different conditions. (a) Liquid flow rate 1 mL min^{-1}; gas flow rate 0.6 L min^{-1} for the dotted line and 1 L min^{-1} for the continuous one. (b) Gas flow rate 0.7 L min^{-1}; liquid flow rate 0.6 mL min^{-1} for the continuous line and 1.4 mL min^{-1} for the dotted line.

conditions. This figure illustrates the influence of the gas and liquid flow rates on the aerosol characteristics. It is widely known that the increase in the gas flow rate is beneficial to generate fine droplets. This is a direct effect of the increase in the total amount of kinetic energy available to produce liquid surface. This trend is illustrated by a shift of the curve in Figure 3.23a towards the left (fine droplets) as the gas flow rate goes up. The increase in this variable leads to a rise in the percentage of droplets with diameters below 10 μm, which will be completely evaporated before the plasma observation zone [335]. A further effect of increasing this variable concerns the aerosol dispersion in terms of drop diameters. The maximum drop size produced by a nebulizer operated at a rather high gas flow rate is lower than that found when the nebulizer is operated at low values of this variable. Finally, Figure 3.23a reveals that the shape of the curve changes when modifying the gas flow rate. Thus, at high nebulizer gas flows, the drop size distribution shows smooth maxima and is more monodisperse than if the nebulizer is operated at low gas flow rates.

The trends with the gas flow rate are general for concentric nebulizers. For example for a demountable nebulizer, it has been found that the percentage of aerosol droplets whose diameters fall below $10\,\mu m$ is 20 and 70 per cent as the gas flow rate increases from 0.4 to $1.0\,L\,min^{-1}$ [762].

With regard to the liquid flow rate (Figure 3.23b), it can be mentioned that if a fine aerosol is needed, the liquid flow rate should be lowered. The reason for this is that the liquid to kinetic energy ratio is lowered as the former variable is increased. This is detrimental from the point of view of fine droplets generation. However, as it will be shown later, if the liquid flow rate decreases below a given value, the aerosol production becomes instable and coarse droplets are again generated [226]. Therefore, the liquid flow rate should be neither too low nor too high in order to produce aerosols as fine and stable as possible. It is commonly accepted that pneumatic concentric nebulizers perform the best when the liquid flow rate approaches to its free liquid uptake rate. Figure 3.24 shows that an increase in the liquid flow rate (controlled by a peristaltic pump) leads to a concomitant growth of the median of the volume drop size distribution (D_{50}), thus indicating that coarse aerosols are being produced. In this figure, the value of the D_{50} obtained under free aspiration mode is also included. It can be observed that the best performance is observed when the system is switched at delivery flow rates close to the free uptake rate.

> Globally speaking, it can be concluded that the effect of the gas flow rate on the aerosol characteristics is more noticeable than the influence of the liquid delivery rate.

Thus, it has been observed that for concentric nebulizers, the $D_{3,2}$ increases by just an average factor of 20 per cent as the liquid flow rate grows 5 times (i.e. from 0.4 to $2.0\,mL\,min^{-1}$) [436].

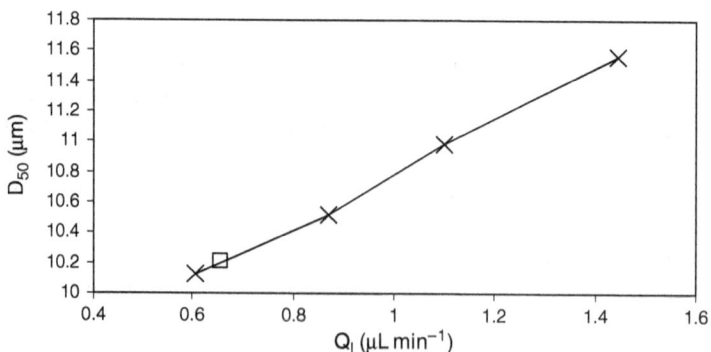

Figure 3.24 Median of the volume drop size distribution versus the liquid flow rate for the aerosols produced by a pneumatic concentric nebulizer. Gas flow rate: $0.7\,L\,min^{-1}$. The square on this figure indicates the D_{50} obtained when the nebulizer is operated under free aspiration mode.

An evaluation of the effect of the different nebulizer critical dimensions on the characteristics of aerosols can illustrate the complexity of the situation. Figure 3.25 shows the D_{50} values of the aerosols produced by several A-type pneumatic concentric nebulizers. The nebulizers are operated under the same liquid and gas flow rates. It is found that for nebulizers having the same design, the aerosols produced are actually quite different in size. D_{50} changes by about 3 μm. It is also interesting to notice that this is not a regular trend with regard to the effect of the gas exit cross-sectional area. More interesting is the fact that for several nebulizers having similar values of this dimension, the aerosol produced is different. Therefore, there should be another nebulizer dimension able to explain this result. The next nebulizer dimension that can be considered is the sample capillary inner diameter [382]. The correlation between D_{50} and this second critical dimension is quite good as it is demonstrated in Figure 3.25 (middle). However, there are some nebulizers that produce aerosols that are too coarse according to the capillary inner diameter only. The third dimension that can be considered is the capillary walls' thickness. However, as Figure 3.25 (bottom) demonstrates, this variable does not explain the results that are found. In fact, for two nebulizers having different capillary wall thickness, quite similar aerosols are generated, whereas for nebulizers with similar values of this dimension, the aerosol characteristics are rather different. Therefore, there should be additional nebulizer tip characteristics (shape, microscopic defaults or surface status) that have to be studied to understand why unexpected data are obtained.

A way of performing systematic studies concerning the effect of critical dimensions on the aerosol characteristics is to use a demountable pneumatic concentric nebulizer. Thus, with two demountable nebulizers, it has been verified that the nebulizer having the lowest inner diameter of the sample capillary and gas exit cross-sectional area produces the finest primary aerosols [762]. Although no more-detailed investigations have been reported, demountable nebulizers can provide a lot of information about nebulizer construction.

3.3.7.2 *Spatial distribution and velocity*

When a pneumatic concentric nebulizer is used, the aerosol produced has a conical shape. The dimensions (i.e. angle) of this cone are very important because they preclude the transport of solution towards the plasma. Obviously, this will depend on the spray chamber design. Thus, for example if a double-pass spray chamber is used, wide aerosol cones will favour the impact droplet losses, whereas narrow aerosols will promote the transport of solution through the spray chamber. Figure 3.26 shows the measured dimensions for three different pneumatic concentric nebulizers. The wideness of the cones depends on the nebulizer used. Thus, at a 15-mm distance from the nebulizer tip, the cone diameters range approximately 8–13 mm. This difference in aerosol dimensions induces remarkable changes in the mass of solution delivered to the plasma.

The aerosol velocity also determines the analyte transport rate. At low droplet velocities, the aerosol losses caused by gravitational settling are noticeable whereas at higher velocities the inertial impact losses become more significant. Thus, in order to optimize the nebulizer performance, this point should be taken into account. Increasing the gas flow rate is beneficial from the point of view of production of fine droplets, but it causes an enhancement

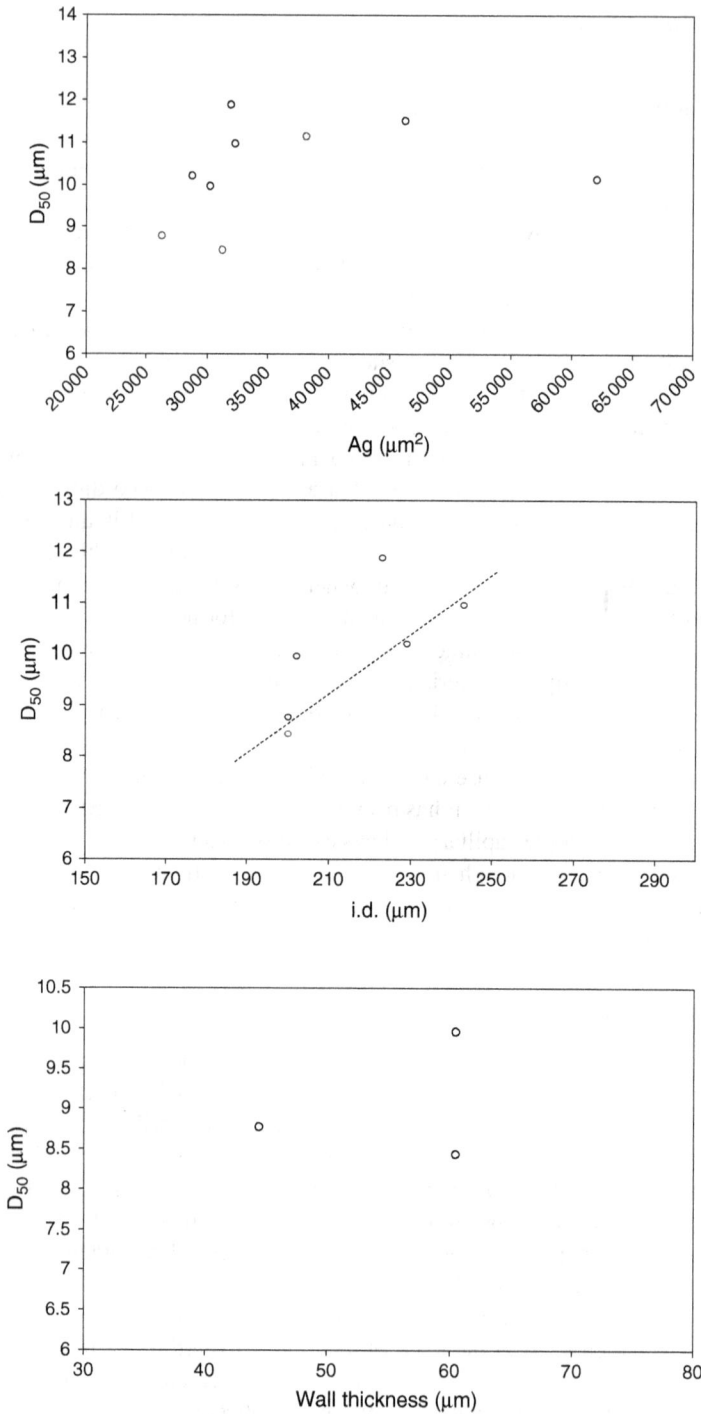

Figure 3.25 Median of the volume drop size distribution of primary aerosols produced by several type A glass pneumatic concentric nebulizers, (*top*) effect of the gas exit cross-sectional area, (*middle*) effect of the sample capillary inner diameter, (*bottom*) effect of capillary walls thickness.

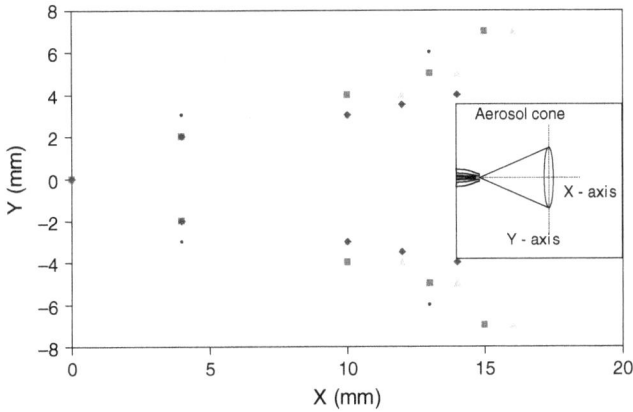

Figure 3.26 Dimensions of the aerosol cones produced by three different pneumatic concentric nebulizers. (Adapted from ref. [678].)

in the inertial droplet deposition inside the spray chamber. According to the measurements of aerosol droplet velocities, it has been found that both axial and radial velocities increase with increasing the nebulizer gas flow rate. Furthermore, the velocity distributions (i.e. the plotting of the number of droplets versus the velocity) become narrower as the gas flow rate goes up. Figure 3.27 shows an example of velocity distributions. From this figure, it can be observed that the aerosol cone for the considered nebulizer is not symmetric, because the radial velocity profiles are not symmetrically distributed. This result is a clear indication of small defaults on the nebulizer tip such as variations in the gas annulus width.

A possible solution to minimize inertial impact losses would be to decrease the nebulizer gas flow rate down to values that produce acceptable aerosols (not too coarse) and simultaneously to introduce a sheathing gas in order to compensate for the loss in droplet-carrying capability [846].

Not only the line of sight average diameters is important but also the spatial distribution is important in order to study a given pneumatic nebulizer. For concentric nebulizers, coarse droplets are mainly located at the edges of the aerosol cone, whereas finest ones are found at the innermost cone area [553]. This trend is quite general for concentric nebulizers as it has been recently demonstrated for micronebulizers [750]. This fact implies that to remove the aerosol coarse droplets, it is advisable to introduce the aerosol into tubes as in the case of double-pass spray chambers.

3.4 CROSS-FLOW NEBULIZERS

In a cross-flow nebulizer, the liquid and gas exits are perpendicularly mounted on a polymer (e.g. PTFE) body [22]. Cross-flow nebulizers can be either fixed or demountable, with either fixed or adjustable tips in the latter case.

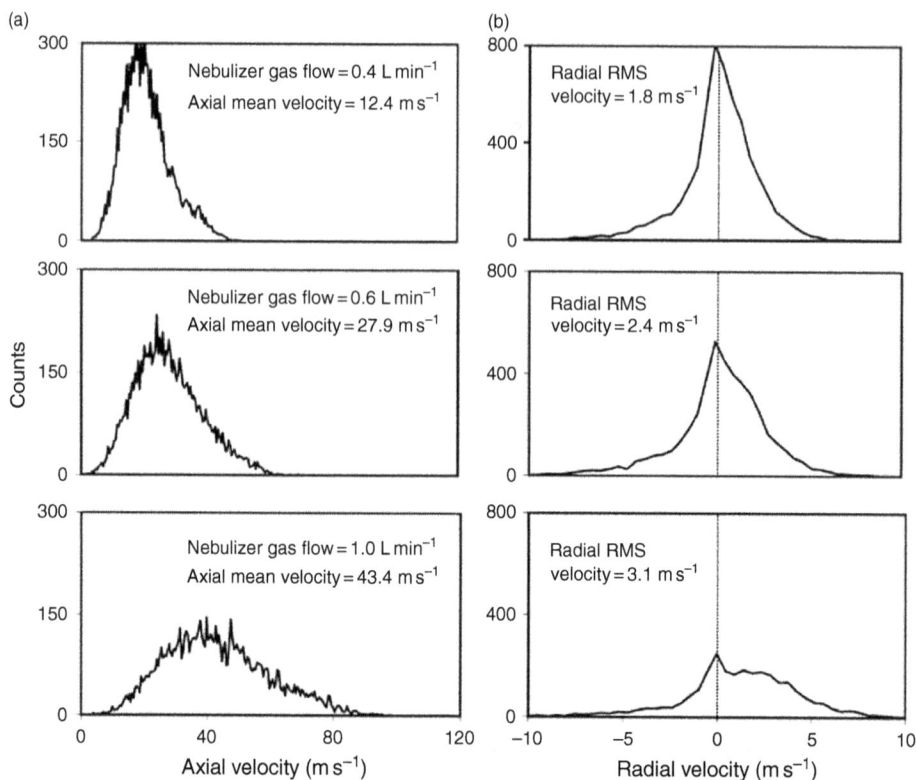

Figure 3.27 Axial (a) and radial (b) droplet velocity distributions for different nebulizer gas flow rates obtained with a demountable pneumatic concentric nebulizer. Liquid flow rate: $\sim 0.5\,mL\,min^{-1}$. Aerosol was probed at 15 mm from the nebulizer tip. (Taken from ref. [762] with permission.)

A cross-flow nebulizer, commonly supplied as a default device to commercially available ICP–AES systems, makes use of two demountable ruby nozzles, the positions of which is preadjusted in a polymer body (Figure 3.28). When using adjustable tips, the relative position of these tips can be regulated until optimum aerosol production. However, this step is tedious and critical and the nebulizer performance at long term can be degraded because the position of the tips can gradually change [158]. In contrast, fixed cross-flow nebulizers do not allow modification in the relative position of the two tips, thereby providing better precisions [50]. However, if one of the exits of a fixed cross-flow nebulizer is blocked, the complete nebulizer must be replaced.

The energy-transfer efficiency in cross-flow nebulizers is rather low. A way of enhancing the performance of these nebulizers is to decrease the gas exit cross-sectional area. Concomitantly, the nebulizer can be made more resistant to the tip blocking by increasing the inner diameter of the liquid inlet. However, this solution gives rise to a noisier aerosol because of the disproportion created between the liquid and the gas exit inner diameters [50].

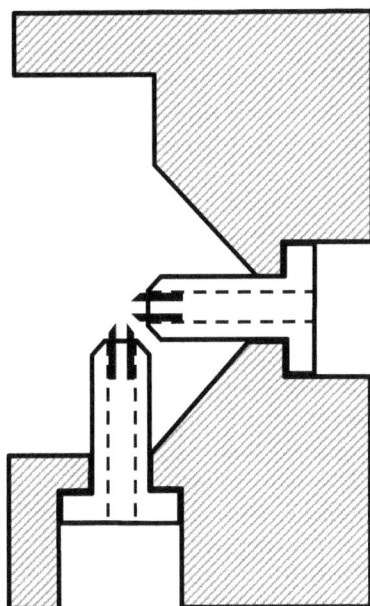

Figure 3.28 Drawing of a demountable cross-flow nebulizer.

This kind of nebulizer is very popular mainly because of its robustness and resistance to tip blocking as compared with glass concentric nebulizers. They have been widely used in atomic absorption spectrometry (AAS). This fact has been checked by demonstrating that the increase in the noise because of the sample introduction system induced by the presence of salts is more significant for a concentric than for a cross-flow nebulizer [611]. Furthermore, this nebulizer is also able to work with some slurries as it has been observed for cements or gypsum [741]. Unfortunately, few studies concerning the nebulizer improvement have appeared, and the most recent studies are only applications in which this nebulizer is used.

In cross-flow nebulizers, the aerosol is horizontally generated as a rising solution is exposed to the gas stream. The relative position of the nebulizer tips is one of the most critical point of this nebulizer. Thus, the best performance of the nebulizer is found when the gas exit is placed from the edge of the liquid one to half-way between the centre line and the liquid boundary [226]. This is likely because the transfer of kinetic energy from the gas to the liquid stream is more efficiently produced because of the longer interaction area between these two jets.

Cross-flow nebulizers are less sensitive to changes in the chemical nature of the analyte than other designs such as ultrasonic (not considered in the present book), as it has been proved for several selenium species [619].

A recent advance consists of the integration of two solution lines in the same cross-flow nebulizer by keeping a single gas orifice [851]. In this way, it is possible to perform online standard additions and internal standardization procedures in a rapid and efficient way. Two different configurations have been evaluated: in the first case, the three conductions (i.e. two for the solutions and one for the gas) are placed orthogonal to each other

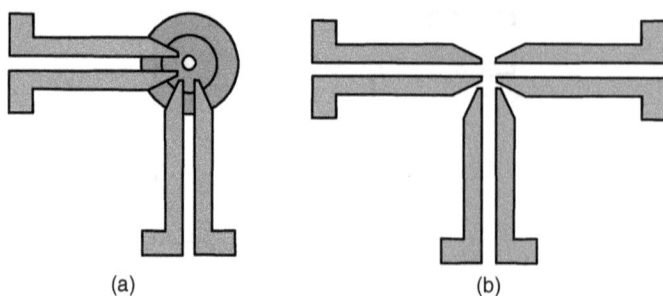

(a)   (b)

Figure 3.29 Scheme of the cross-flow nebulizers used for mitigating matrix effects. (a) Trihedral nebulizer; (b) T-shaped cross-flow nebulizer. 1, gas inlet; 2 and 3, liquid inlets. (Taken from Bauer and Broekaert [851], with permission.)

(trihedral cross-flow nebulizers, ThCFN), whereas in the second case, the gas capillary is placed orthogonal with respect to the liquid capillaries (T-shaped cross-flow nebulizers, T-CFN). The aerosols produced from each one of the liquid exits are mixed at the nebulizer tip. Figure 3.29 shows the geometry of these two nebulizers. With regard to the operation of these systems, it was found that both cross-flow devices were able to nebulize efficiently at gas flows above $0.8 \, \text{L min}^{-1}$. Meanwhile, a conventional cross-flow design provided good analytical figures of merit at flow rates of the order of $0.6–0.7 \, \text{L min}^{-1}$.

3.5 HIGH-SOLIDS NEBULIZERS

So-called high-solids nebulizers are nebulizers in which the interaction between the solution orifice and the gas–liquid have been modified to have a high tolerance to high total dissolved solids solutions (e.g. up to 20 per cent), with the possibilities of having slurries. It is usually obtained by increasing the size of the sample capillary bore.

The first nebulizer of this kind was developed by Babington in 1969. In this case, a hollow sphere with an orifice for the gas stream was placed under a capillary. The liquid solution exiting the capillary was deposited on the sphere. The aerosol was finally generated when the liquid sample reached the orifice [13, 14]. Several modifications have appeared in the literature making use of impact surfaces [58] or of orifices of different dimensions [261]. A problem of the initial Babington design is that the sample consumption is too high because a big fraction of the solution is lost as it does not come in contact with the gas exit.

This nebulizer has been extensively used for the analysis of high salt content solutions and slurries. There are several versions of the high-solids nebulizers, but all of them are based on the production of aerosol according to the Babington principle. The only difference lies in the way in which the liquid solution is delivered to the gas orifice. Among high-solids nebulizers, the V-Groove is the most commonly used [41]. This device is also made of several materials such as glass, plastic or ceramic [827]. In this case, the sample is pumped

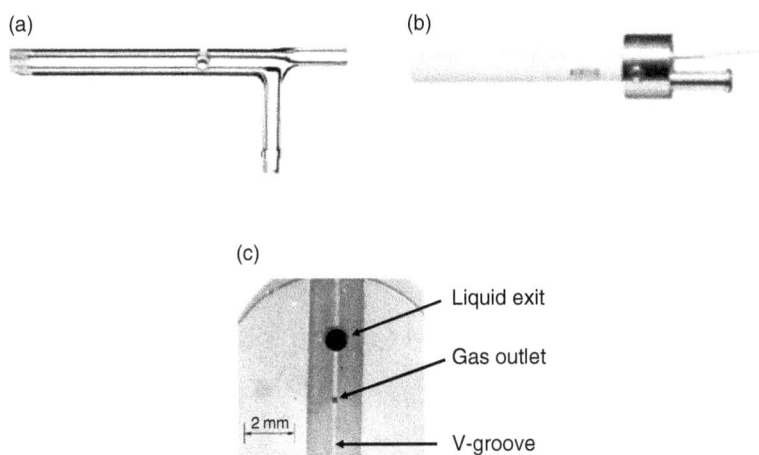

Figure 3.30 Babington V-Groove pneumatic nebulizers made of glass, (a) (Taken from www.laboratorynetwork.com) PTFE, (b) (Taken from www.parma.co.kr) and (c) enlarged photograph of the tip of a V-groove nebulizer.

typically through an 800-μm i.d. capillary, whereas the gas flows by a 100-μm bore i.d. (Figure 3.30). These dimensions require a gas backpressure close to 3.5 bar for reaching a 1-L min^{-1} gas flow rate [772].

A modified Babington nebulizer has been introduced [712]. This nebulizer is entirely made of PTFE, and it has a hood in order to reduce the problems associated with the renebulization of the droplets deposited on the nebulizer tip, once the solution is nebulized. It has been observed that the signal stability found with a conventional V-Groove nebulizer was poor with ICP–AES signals ranging from 100 to 1500 cps for a 1-μg ml^{-1} Mn solution. These fluctuations can be accounted for by the changes in the nebulized solution volume because of the renebulization. Then, by shielding the nebulizer nozzle, this unwanted process can be minimized and the signal stability improved significantly. Figure 3.31 depicts this new version of the Babington V-Groove nebulizer. Comparatively speaking, the hooded nebulizer provided higher ICP–AES emission signals and similar signal-to-background ratios as compared with a conventional unshielded V-Groove nebulizer.

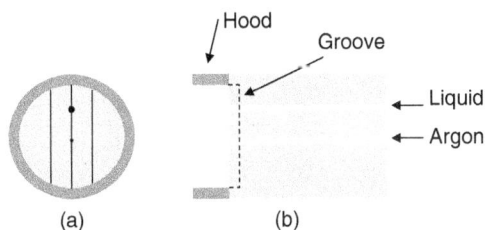

Figure 3.31 Drawing of the shielded V-Groove nebulizer. (a) Front view and (b) lateral view (Adapted from ref. [712].)

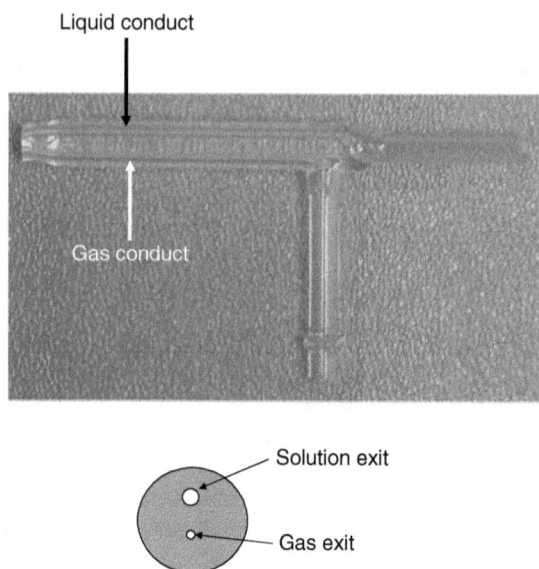

Figure 3.32 Picture of the 'excentric' nebulizer (Taken from www.cpiinternational.com) and front view of its tip.

Another commercially available high-solids nebulizer is the so-called excentric nebulizer (www.cpiinternational.com). This nebulizer is similar in design to the Babington, but it does not have any groove, and its tip is rather flat (Figure 3.32). Thus, the liquid solution exits through the corresponding hole, and it is led to the gas orifice by the action of gravity or by the recirculating gas stream created at the nebulizer tip. In this case, the nebulizer orientation has a significant effect on the characteristics of the aerosol generation. So far, the nebulizer has been successfully applied to the determination of some metals present in textile samples [850].

3.6 PARALLEL-PATH NEBULIZER

3.6.1 Principle

> The parallel-path nebulizer (PPN) is a new nebulization concept issued in 1995 in which the liquid and gas interaction is produced when the high-velocity gas stream tangentially enters in contact with the liquid one [442].

In the PPN, both streams are aligned close enough to create instabilities on the solution surface, thus providing an aerosol. The proximity of both streams is the key point of the operation with this nebulizer. The recirculation of gas at the exit of the nebulizer is beneficial to draw the liquid sample into the gas stream (Figure 3.33), thus producing

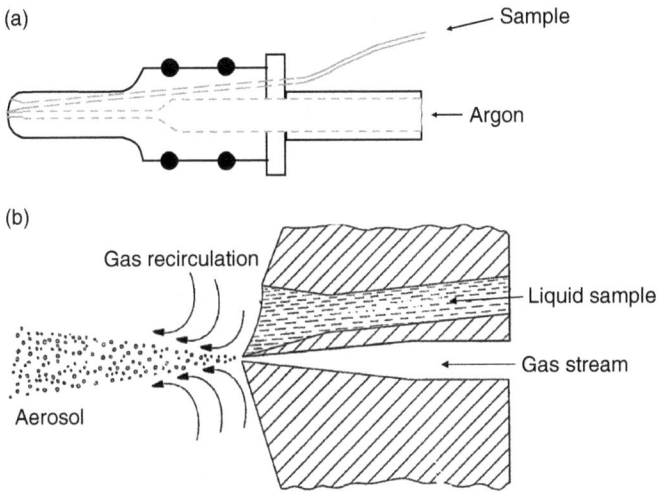

Figure 3.33 Drawing of the parallel-path pneumatic nebulizer, PPN (a) and detail of its tip and the nebulization process (b). (Adapted from ref. [442].)

the aerosol. The solution surface tension is responsible for the generation of a continuous sample passage from the liquid conduction to the gas exit. Because this stream does not need to be constrained, large orifices can be used to deliver the sample to the gas exit. This fact makes the nebulizer to be more difficult to block when particles or salts are present in the sample. In fact, some of these nebulizers have liquid passage inner diameters that allow the nebulization of slurries containing 100 μm particles.

Figure 3.33 shows a scheme of the PPN as well as a detailed view of its nozzle. The liquid can be drawn to the gas exit through a constrained path. To achieve 100 per cent nebulization yield, it is important to work at liquid flow rates below the flow that can be efficiently nebulized by the gas stream. At higher liquid flows, a fraction of the solution falls without being transformed into aerosol. In contrast, if the liquid flow rate is too low, a continuous mist is not produced. As it can be observed from Figure 3.33, the liquid path's inner diameter increases at the end of the nebulizer tip. As a result, any eventual nebulizer plugging will occur before the nebulizer tip, thus facilitating the system regeneration by replacement of the blocked liquid capillary. The maximum diameter of the liquid exit corresponds to the diameter of the drop formed at the exit of the nebulizer before it falls down in absence of nebulizer gas [442]. If the liquid exit has a lower diameter than the experimentally determined one, the surface tension of the liquid will maintain the liquid bulk and a drop will form at the exit of the nebulizer. The gas dynamics will draw the liquid towards the gas exit and an aerosol will be formed.

The gas exit of a PPN may have a much lower size than the liquid one. It has been claimed that the position of the gas outlet is not critical [442]. Thus, as the gas orifice is moved outside or inside the nebulizer, the only change is related to the point of aerosol generation. Nonetheless, the gas exit position can modify the direction of aerosol movement. This can have an important effect on the amount of aerosol transported through the spray chamber, because a bad aerosol orientation can induce a significant increase in the aerosol inertial losses.

Unlike concentric and other pneumatic nebulizers, no solution suction is produced in the case of the PPN. This nebulizer can be made of glass, polymer or metal. PTFE nebulizers are also available. These nebulizers are chemically inert and nonwetting. This fact makes the salt deposits not to form because droplets of solution do not stay at the nebulizer tip. As a consequence, it is not necessary to humidify the argon stream to prevent this problem.

3.6.2   Critical dimensions

Table 3.4 summarizes some of the characteristics of the PPN. For all the nebulizers except for the AriMist®, the sample capillary has a higher inner diameter than for pneumatic concentric nebulizers, thus making the formers more resistant to blocking when salts or particles are present in the sample. However, the gas orifice can be easily blocked if the argon stream carries particles. In fact, this has been considered as one of the most important sources of PPN blocking (www.burgener.com). As it can also be observed from this table, the gas backpressure required to reach a gas flow rate usually employed in optical emission spectrometry ranges from 30 to 45 psig. This means that, in order to reach the optimal rates for ICP–MS, the pressure of the gas line must raise up to 50 psig approximately. Because two PPNs of the same kind supply different gas flow rates at a given pressure, the optimal operating gas pressure should be determined each time a new PPN is used. Finally, Table 3.4 provides the data corresponding to the range of liquid flow rate that can be operated for each nebulizer. Capillaries with low inner diameter allow working at low liquid flow rates. Indeed, nebulizers with high liquid exit inner diameters are not appropriate to work at low liquid flow rates, likely because a stable liquid stream towards the gas exit is not formed for these devices. Thus, it is not surprising that the AriMist is a device specially conceived to work for Capillary Electrophoresis applications. In contrast, if high liquid flow rates should be

Table 3.4

Representative dimensions and characteristics of several parallel-path pneumatic nebulizers (PPN) (Adapted from: http://www.burgener.com.)

Model	Body material	Sample capillary i.d. (μm)	Psig operating pressure at $0.8\,L\,min^{-1}$	Sample flow at ($\mu L\,min^{-1}$)
PEEK Mira Mist	PEEK	425	35–45	200–2500
Teflon Mira Mist	PTFE	425	35–45	200–2500
T 2100	PTFE	750	30–40	500–2500
Ari Mist	PEEK	225	35–45	50–1000
TJA Trace	PTFE	425–800	35–45	800–2500
T2002	PTFE	425–800	35–45	800–2500

operated, it is necessary to use capillaries with high inner diameters in order to avoid the problem of solution dropping mentioned before.

The aerosols generated with the PPN emerge at a given angle with respect to the nebulizer body. Hence, the nebulizer position must be optimized with respect to the spray chamber, because the analytical results vary as a function of the nebulizer orientation. This effect is especially important in the case of using cyclonic spray chambers. It is, therefore, important to rotate the nebulizer inside the spray chamber in order to find its optimal position.

Because of drag forces, the velocity of a gas stream flowing through a conduction decreases as it moves towards the tubing walls. Furthermore, the effective gas flow is about twice higher at the centre of the conduction than at its walls. For the nebulizers shown so far, the liquid and gas interaction takes place at the edge of the sample capillaries where the gas stream has lost an important fraction of its energy. It might be then more efficient to expose the liquid solution to the gas action at the point of maximum energy. In the so-called enhanced parallel-path nebulizer (EPPN) (Figure 3.34), a spout is machined at the gas exit (Figure 3.35). As it is represented by the dark grey zones appearing along the gas conduction in this figure, the gas energy is up to 100 times higher at the centre than in the outermost zone of the conduction. The spout delivers the liquid solution at this point, and, hence, the aerosols produced are finer for this nebulizer than for a conventional PPN. Figure 3.35 shows a front view of one of these designs. The main features of this nebulizer are (i) production of slightly finer aerosols than a concentric device, (ii) slightly higher analyte transport efficiencies than other conventional pneumatic nebulizers [820] and (iii) good ICP signal reproducibility and repeatability.

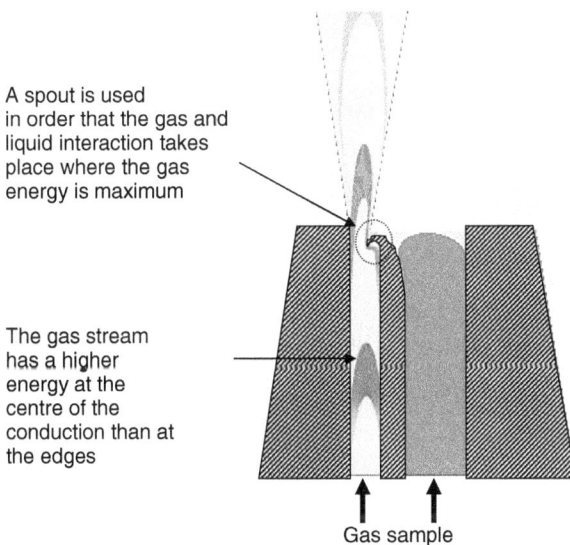

A spout is used in order that the gas and liquid interaction takes place where the gas energy is maximum

The gas stream has a higher energy at the centre of the conduction than at the edges

Gas sample

Figure 3.34 Scheme of the Enhanced Parallel-Path Nebulizer (EPPN).

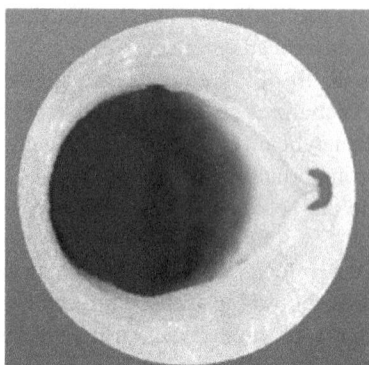

Figure 3.35 Front view of a Enhanced Parallel-Path Nebulizer (EPPN).

3.7 COMPARISON OF THE DIFFERENT CONVENTIONAL PNEUMATIC NEBULIZERS

Analytical results obtained in ICP–AES depend critically on the characteristics of the aerosols produced by the nebulizer as well as on the spray chamber design [488, 745]. For a given set of operating conditions, the results finally obtained by a nebulizer are a consequence of its critical dimensions. Table 3.5 establishes a comparison in terms of design of the different nebulizers mentioned in this chapter. To obtain the gas–liquid interaction area, the guidelines given by Sharp [226] have been followed. Thus, for a pneumatic nebulizer, the gas action extends up to 5 times the diameter of the gas exit (L). This length should be multiplied by the liquid–gas interaction perimeter. For a concentric nebulizer, L is the capillary nozzle gap (typically 10–30 μm) and this magnitude must be multiplied by the capillary internal perimeter.

Cross-flow nebulizers have liquid and gas interaction areas slightly superior to those estimated for concentric nebulizers. However, they are less efficient in the production of fine droplets. This is likely because of the stagnation of a fraction of the solution at the rear of the liquid capillary. At this position, the gas stream has lost an important fraction of its initial kinetic energy, and, hence, coarse droplets are produced.

Table 3.5

Comparison between the dimensions of some pneumatic nebulizers

Nebulizer	Gas cross-sectional area (mm^2)	Liquid inner diameter (μm)	Gas–liquid interaction area (mm^2)
Concentric	$2 \times 10^{-2} – 10^{-1}$	~ 150–400	1.25×10^{-1}(*)
Cross flow	$4 \times 10^{-3} – 10^{-1}$	~ 500	$\sim 2 \times 10^{-1}$(*)
V-Groove	$\sim 8 \times 10^{-3} – 3 \times 10^{-2}$	~ 500–800	$\sim 1.6 \times 10^{-1}$(*)
MiraMist®	~ 0.01	425	$\sim 6 \times 10^{-2}$

(*) representative values taken from ref. [226].

For the MiraMist® (Table 3.5), it is difficult to determine the interaction area because the geometry of the gas exit is not circular. In this case, L could be the diameter of the gas orifice (considering it as circular) whereas the liquid–gas interaction perimeter would be the gas outlet diameter. From the data presented in Table 3.5, this nebulizer would be the device with the lowest value of the liquid–gas interaction area. This cannot be considered as a rule, because the values of L and liquid–gas perimeter for concentric nebulizers take very disperse values depending on the particular design. Therefore, by recalculating the interaction area, considering the wide range of critical dimensions for concentric nebulizers, it could be indicated that the interaction area would range from 2×10^{-2} to $1.9 \times 10^{-1}\,mm^2$. In other words, there would have some concentric nebulizers having lower interaction areas than the MiraMist.

The differences in terms of nebulizer dimensions have an important effect on the optimum operating conditions. An example is given in Table 3.6, in which the optimum conditions are summarized for different pneumatic nebulizers. Because of the increased gas exit cross-sectional area, cross-flow systems require lower backpressures than concentric ones. With regard to Babington nebulizers, it has been found that the liquid flow rate required for obtaining the best ICP–AES limits of detection is less than 2 times lower than for a concentric device [610]. This fact likely demonstrates the deterioration in aerosol production observed for this kind of pneumatic nebulizers.

As mentioned previously, cross-flow nebulizers are tolerant to dissolved salts. In fact, the ICP–AES matrix effects caused by solutions containing ammonium sulphate, sodium chloride and sodium tetraborate were more pronounced for a concentric nebulizer than for a cross-flow design [611].

If prefilming-based nebulizers are used such as the PEN [788], it was found that finer aerosols were obtained as compared with conventional pneumatic concentric nebulizers. Thus, the fraction of liquid aerosol volume contained in droplets with diameters below $10\,\mu m$ was 2 times higher for the former nebulizer. This, of course, led to lower ICP–AES detection limits.

The Babington-type nebulizers used to provide worse ICP analytical figures of merit than concentric devices. Table 3.7 illustrates this for a set of four different matrices. Similar results have been found in the literature when comparing between a Meinhard TR-30-A3 [610] or TR-50-C1 [830] and a V-Groove Babington nebulizer. The degradation in the analytical figures of merit when using the latter nebulizer is due to the production of coarser droplets than for a concentric nebulizer. Thus, for an 8 per cent sodium chloride solution, reported $D_{3,2}$ values for primary aerosols are about 5 and $8\,\mu m$ for a Seaspray (Glass Expansion) and a

Table 3.6

Optimum conditions for the different pneumatic nebulizers*

Nebulizer	Gas backpressure (bar)	Liquid flow rate (mL min^{-1})
Concentric	0.9	1.5
Cross Flow	1.9	1.2
Babington		1.0
Concentric		2.4

*The conditions have been optimized in terms of ICP–AES sensitivity.

Table 3.7

ICP–AES limits of detection obtained by a V-Groove nebulizer divided by those found
for a concentric sea spray type nebulizer (Glass Expansion) (Adapted from ref. [678].)

Element	Water	2% NaCl	4% NaCl	6% NaCl
Ba	1	0.833	1.875	1.75
Cd	1.475	1.636	1.542	1.833
Cr	1.396	1.948	1.481	2.059
Cu	1.314	1.024	2.479	2.091
Mg	1.5	1.667		2
Mn	1.636	2.222	2.777	2.454
Ni	1.773	1.983	2.254	1.496

V-Groove nebulizer, respectively [678]. This leads to an analyte transport rate about 2 times
higher for the concentric nebulizer. A comparison established between a V-Groove nebulizer
and a Meinhard C-type nebulizer is also in agreement with these findings. Thus, in this case,
the solvent transport rate as well as the analyte transport efficiencies are 2 times higher for
the Meinhard nebulizer than for the V-Groove [830]. Furthermore, the cone angle is higher
for the V-Groove than for a concentric nebulizer [678].

   In a comparative study [854] six different nebulizers have been tested. The accumulated
volume drop size distribution curves are presented in Figure 3.36. The nebulizers considered in
this figure have been randomly selected. As previously mentioned, the nebulizer's critical
dimensions play a very important role on its performance. Nonetheless, these results can be
considered as those potentially obtained by a customer who would buy a nebulizer of each type.
First, it can be observed that concentric nebulizers are the best in terms of aerosol fineness. As
previously indicated, concentric designs are among the most efficient systems for transferring
energy from the gas to the liquid stream. However, the nebulizer based on the Laval nozzle used
for these experiments has provided the coarsest aerosols. Cross-flow, V-groove and parallel-
path nebulizers provide quite similar aerosols.

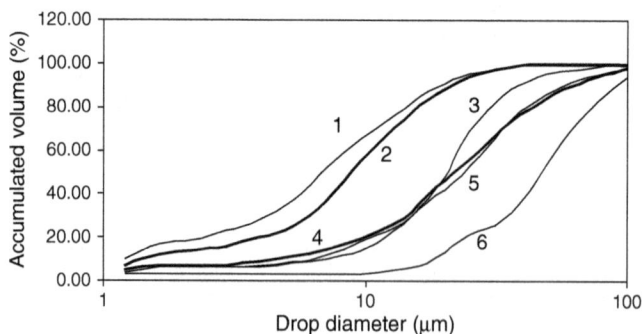

Figure 3.36 Cumulative volume drop size distributions of the primary aerosols produced by several
pneumatic nebulizers. (1) PFA concentric nebulizer; (2) glass concentric; (3) Cross-flow;
(4) V-groove; (5) Parallel-Path Nebulizer; (6) Laval-based nebulizer. Gas liquid flow rate:
$1 \, L \, min^{-1}$, liquid flow rate: $0.8 \, mL \, min^{-1}$.

3.8 PNEUMATIC MICRONEBULIZERS

So-called micronebulizers are nebulizers optimized to work at a solution delivery rate below 200–300 µL min^{-1}, usually in the range 20–100 µL min^{-1}. One major modification is the reduction of the liquid-capillary inner diameter.

Nebulizer miniaturization has been a subject of great interest in the last 15 years. This research area is especially interesting in those situations in which the sample volume or flow are the limiting factors such as forensic, biological or separation ICP hyphenation. Table 3.8 shows examples of situations in which the amount of sample available is very low [738].

The nebulizers mentioned in previous sections are operated at typical sample flow rates of 0.5–2 mL min^{-1}. This makes it necessary to have a sample volume available for the analysis ranging from about 1 to 10 mL. One of the most popular and simple proposed solutions for microsample analysis has been to decrease the liquid flow rate down to 10–300 µL min^{-1}. Because with conventional pneumatic nebulizers, working under these conditions leads to a dramatic loss of sensitivity and an increase in the washout times, new nebulizers have been purposely developed. These are the so-called micronebulizers.

The characteristic critical dimensions for micronebulizers are summarized in Table 3.9. In general terms, pneumatic micronebulizers have reduced critical dimensions (i.e. liquid capillary inner diameter, gas exit cross-sectional area) as compared with conventional devices. The final result of the modifications performed on micronebulizers is that the liquid and gas interaction becomes more efficient for the formers, thus favouring the generation of finer primary aerosols.

In addition to the nebulizers shown in Table 3.9, there are other devices that have been used for the introduction of liquid microsamples in ICP techniques, such as the Babington, Glass Frit, Microultrasonic nebulizers, the Monodisperse Dried Microparticulate Injector and the Electrospray Nebulizer. The principles of operation and the results reached by these additional micronebulizers can be found in the literature [553].

Micronebulizers are especially efficient when they are operated at low liquid flow rates. Under these conditions, conventional nebulizers described in previous sections are not able to provide satisfactory results. This is because the aerosol generation process becomes erratic as the nebulizer is operated at delivery rates much lower than the free

Table 3.8

Available sample volumes for determination of various metals (Taken from ref. [738].)

Types of analysis	Available sample volume, mass or flow rate	Analytes
Cells	20 µL	Na, Mg, K, Fe, Cu, Zn, Cd, Se
Suspended nanoparticles	100 µL	Fe
Brain	1 mg	Fe, Ca, K
Metalloproteins	50 µL	Fe, Cu, Zn
Dust	5 mg	B, Mg, Si, Mn, Sr, Zn
CE Speciation	nL min^{-1}	As, Se, Fe, Cr

Table 3.9

Dimensions of pneumatic nebulizers used for working with low (below 100–200 μL min⁻¹ range) liquid flow rates. Comparison with conventional nebulizers

Nebulizer	Gas exit cross-sectional area (mm²)	Liquid capillary inner diameter (μm)	Liquid capillary wall thickness (μm)	Gas back pressure at 1 L min⁻¹ argon (psig)	Nebulizer dead volume (μL)
Conventional nebulizers (optimum for liquid flow rates ~0.5–1.0 mL min⁻¹)					
Concentric nebulizer	~0.028	400	60	30–40	~100
Cross-flow nebulizer	0.02	500	200	30–40	
Parallel-path nebulizer	0.03	425	—	30–40	
Micronebulizers (suitable for liquid flor rates < 100–200 μL min⁻¹)					
High-Efficiency Nebulizer (HEN)	0.007–0.008	80–100	30	150	
MicroMist (MMN)	0.018	140	50	50	
PFA Nebulizer (PFAN)	0.021	270		40	
Microconcentric Nebulizer (MCN)	0.017	100	30	50	
Parallel-path micronebulizer	~0.015	75	—	90–110	0.64
Sonic-Spray Nebulizer	0.019	150	50	72	
Oscillating Capillary Nebulizer	—	50	50	120–200	
Demountable concentric nebulizer DCN	0.032	95	50		
Direct-Injection Nebulizer (DIN)		60	30	45/70*	<1,2 pl
Direct-Injection High-Efficiency Nebulizer (DIHEN)	0.0094	104	20	155	10–55
Large Bore Direct-Injection High-Efficiency Nebulizer (LB-DIHEN)	0.0371	318	16	36	
Demountable DIHEN	0.005–0.008	40–100	5–21	20–70#	1.2–11

* Pressures required to reach 0.25 and 0.6 L min⁻¹ gas flow rates, respectively.
# Gas flow rate: 0.2 L min⁻¹.

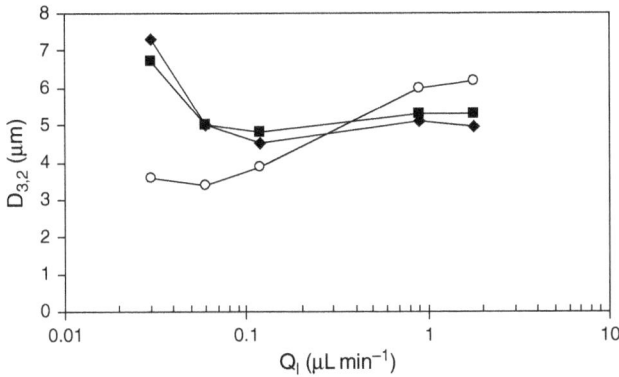

Figure 3.37 Variation of the aerosol surface mean diameter ($D_{3,2}$) with the liquid flow rate (Q_1) for three different pneumatic nebulizers: conventional concentric nebulizer type A (A), conventional concentric nebulizer type K (B) and Micromist (C). Nebulizer gas flow rate, Q_g: $0.8 \, L \, min^{-1}$.

aspiration rate. This fact is illustrated in Figure 3.37, where the Sauter mean diameter ($D_{3,2}$) is plotted versus the liquid flow rate for two conventional pneumatic concentric nebulizers and a micronebulizer. At low liquid flow rates, conventional nebulizers provide coarser aerosols (i.e. higher $D_{3,2}$ values) than the micronebulizer. This result evidences that the performance of a given pneumatic concentric nebulizer is best as it is operated at liquid flow rates close to the free aspiration rate [726].

Another fact that demonstrates the highly efficient nebulizer performance is that these devices supply ICP sensitivities higher than or similar to those encountered with conventional nebulizers operated at delivery flow rates more than one order of magnitude higher. This benefit was corroborated in the earliest studies on micronebulizers (see, e.g. [388]).

As Table 3.9 indicates there are many nebulizers especially devoted to work at low liquid flow rates. The micronebulizers mentioned in this section can be classified as follows:

a. The High-Efficiency Nebulizer (HEN)
b. The MicroConcentric Nebulizer (MCN)
c. The MicroMist nebulizer (MMN)
d. The PFA micronebulizer (PFAN)
e. Demountable concentric micronebulizers
f. High-efficiency cross-flow micronebulizer (HECFMN)
g. Parallel-path micronebulizer (PPMN)
h. Sonic-spray nebulizer (SSN)
i. Oscillating-capillary nebulizer (OCN)
j. High-solids micronebulizer (HSM)
k. Direct-Injection Nebulizers (DIN, either fixed or demountable Direct Injection High-Efficiency Nebulizer, DIHEN and dDIHEN, respectively, Vulkan DIN)

3.8.1 High-Efficiency Nebulizer (HEN)

The HEN [592, 366] is made entirely of glass and is an adapted version of the Meinhard® A-type pneumatic concentric nebulizer. Figure 3.38 compares the tips of these two nebulizers. The data presented in Table 3.9 are confirmed; the sample capillary is clearly

Figure 3.38 Pictures comparing the nozzle of the HEN with that of an A-type pneumatic concentric nebulizer. (Courtesy of Meinhard Glass Products.)

narrower for the micronebulizer. For this reason, with the HEN, even clean aqueous solutions must be filtered to avoid tip blockage caused by the presence of fibres or small particles. Furthermore, it must be carefully handled, because the capillary of this nebulizer can be easily broken when cleaning the HEN.

Because of the low cross-sectional area of the gas exit for the HEN, the gas pressure should be rather high (Table 3.9). As a consequence, it is necessary to use an external additional gas cylinder and high-pressure adapters and lines for the gas stream.

3.8.2 Microconcentric Nebulizer (MCN)

The MCN consists of a polyamide narrow capillary adapted to a tee-shaped plastic body (Figure 3.39a). A sapphire adapter is placed at the nebulizer tip. As can be observed from the picture shown in Figure 3.39b the liquid capillary ends outside the nebulizer. This fact has several drawbacks: (i) the aerosol is generated at the exit of the nebulizer where the gas stream has lost a fraction of its kinetic energy; (ii) the capillary tip can deteriorate in the long term and, consequently, aerosol production is degraded and (iii) the position of the polyamide capillary with respect to the nebulizer nozzle is a critical variable, and small changes in it can lead to noticeable modifications in the nebulizer performance.

In ICP–AES, the MCN gives rise to limits of detection similar to or slightly higher than those calculated for conventional nebulizers operated at liquid flow rates more than 10 times higher [478]. This nebulizer provides higher ICP–MS sensitivities than conventional pneumatic nebulizers operated at nearly the same rates. In agreement with these results, at a given liquid and nebulizer gas flow rate, the MCN leads to higher oxide ratios $(UO^+:U^+)$ than the cross-flow nebulizer. These results arise because of the higher mass of solution transported to the plasma in the case of the micronebulizer.

The MCN appears to show a high tolerance to high-dissolved solids [554]. Thus, for samples having salinity up to 3.5 per cent, no blockage problems were observed [528]. Because it is entirely made of rugged polymeric materials, the MCN shows high tolerance to HF. In contrast, the MCN can be more sensitive to changes in the sample

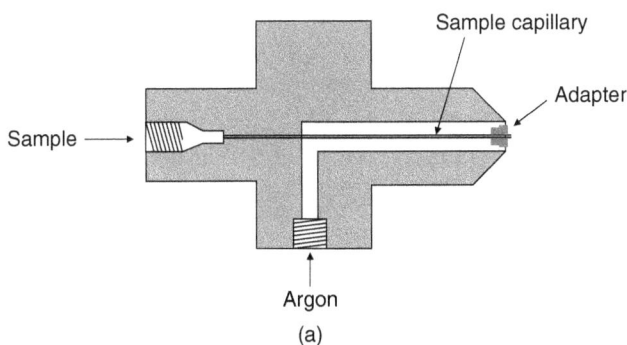

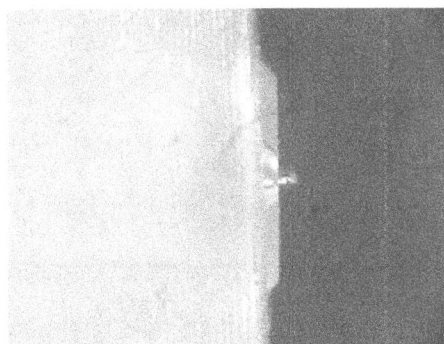

Figure 3.39 Scheme of the MicroConcentric Nebulizer (a) and enlarged picture of its tip (b). (Taken from ref. [828].)

matrix than a conventional pneumatic concentric device. De Wit and Blust demonstrated that even for diluted salt solutions, the signal stability was significantly poorer for a MCN operated at $100\,\mu L\,min^{-1}$ than for a conventional pneumatic nebulizer at $1.6\,mL\,min^{-1}$. The analytical results depend on the particular MCN considered. Thus, the precision [562] as well as the intensity of the matrix effects [529] depend on the MCN used.

3.8.3 MicroMist nebulizer (MMN)

The MMN is entirely made of glass and in this case, the liquid capillary is recessed with respect to the nebulizer tip (Figure 3.40). This fact confers to the MMN the ability to work with high salt content solutions without suffering from nebulizer tip blockage. Furthermore, the outer wall of the inner capillary is a tapered, ground glass piece. As a result the MMN is more robust than other glass micronebulizers.

(a)

(b)

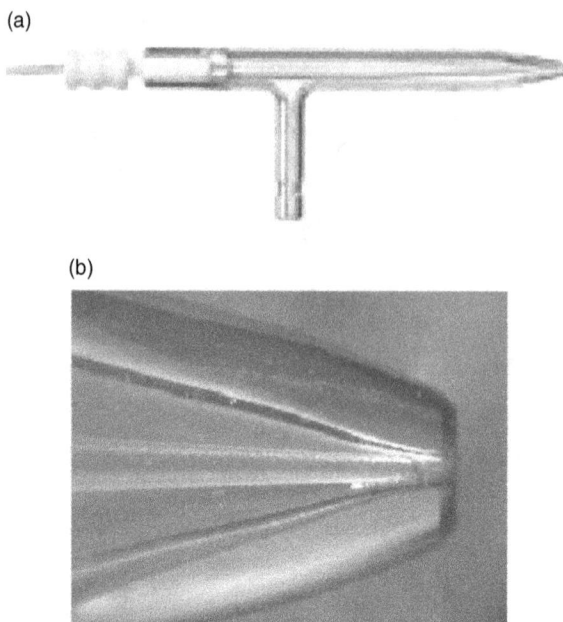

Figure 3.40 Picture of the MicroMist Nebulizer (a) and enlarged picture of its tip (b). (Taken from ref. [828].)

The MMN also suffers from nebulizer-to-nebulizer dimensional irreproducibility. Thus, the free liquid uptake rate can increase, decrease or reach a maximum with the gas flow rate depending on the MMN used [733]. Furthermore, the absolute values of the liquid free uptake rate also depend on the particular MMN. Thus, for three nominally identical nebulizers and at $Q_g = 1.05 \, L \, min^{-1}$, the liquid flow rates were roughly 20, 140 and 180 $\mu L \, min^{-1}$ [733].

3.8.4 PFA micronebulizer (PFAN)

In this case, the material used for construction is PFA (tetrafluoroethylene–per-fluoroalkyl vinyl ether copolymer). The PFAN sample tubing (Figure 3.41) is much more recessed than in the case of the MMN (i.e. 6 mm and about 1 mm, respectively). Therefore, with this micronebulizer, the liquid stream emerges through the capillary and is deposited on the inner walls of the nebulizer. A liquid prefilming is produced because of the action of the gas and hence the thickness of the liquid vein attached to the inner walls of the nebulizer decreases [270]. As a result, a closer interaction between streams takes place and promotes the generation of fine aerosols at relatively low gas back-pressures [742].

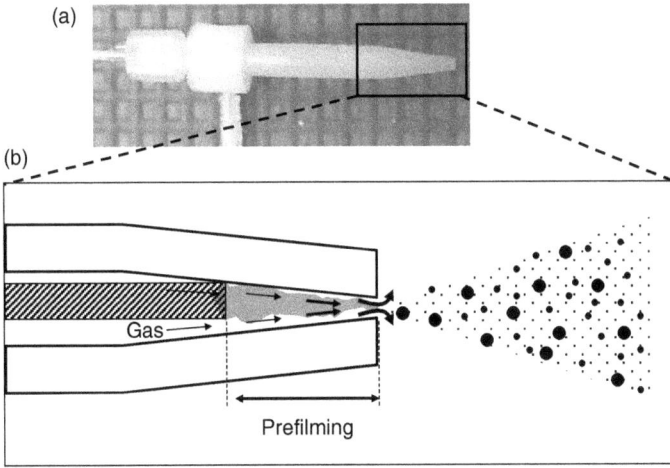

Figure 3.41 Photograph of the PFA Nebulizer (a) and enlarged picture of its tip (b). (Taken from ref. [828].)

3.8.5 Demountable concentric micronebulizers

In the so-called Concentric Capillary Nebulizer (CCN) [630], a stainless steel body is used to adapt a polyetheretherketone (PEEK) tube containing the sample capillary (Figure 3.42). A second body fits the nebulizer to a conventional spray chamber. A Teflon ferrule is used to prevent the system from gas leakage. Characterization of

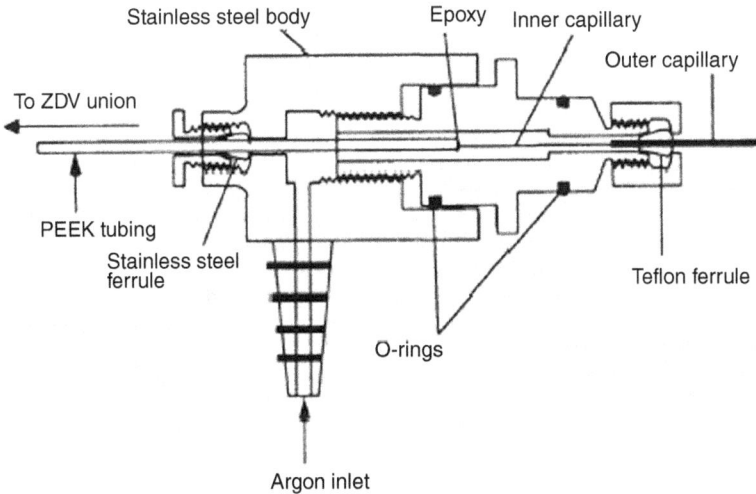

Figure 3.42 Scheme of a demountable pneumatic concentric micronebulizer. (Taken from ref. [630] with permission.)

the aerosols reveals that most of the droplets produced by the CCN had diameters below 10 μm. Under optimum conditions, tertiary aerosols for the CCN are finer and less dispersed than those found for conventional pneumatic nebulizers.

Another DCN has been recently characterized [762]. The nozzle of the DCN is made from a 100 μL surgical pipette and the solution is driven to it by means of fused-silica tubing having 95 or 190 μm inner diameters. This assembly is fitted by means of a brass tee. According to O'Brien et al., a decrease in the sample capillary inner diameter gave rise to finer primary aerosols.

3.8.6　High-efficiency cross-flow micronebulizer (HECFMN)

Concentric micronebulizers have several common drawbacks: (i) low tolerance to dissolved solids, (ii) fragility and (iii) all of them have a suction created in the sample uptake tubing that degrades the separation efficiency in techniques such as capillary electrophoresis or μ-high-performance liquid chromatography (HPLC). These problems can be overcome by using a micro cross-flow nebulizer [642]. The nebulizer body is made of PTFE, whereas a fused-silica capillary is used for the sample and a PEEK nozzle is employed for the argon stream (Figure 3.43). This capillary has reduced dimensions with respect to that used in conventional cross-flow nebulizers (i.e. 75–150 μm i.d. and 375 μm o.d.). The alignment of the capillaries can be performed by means of a U-shaped cut performed at the end of the gas exit capillary. The ICP–MS limits of detection provided by the HECFMN at 50 μL min^{-1} are up to 4 times lower than those calculated for a conventional nebulizer operated at 1 mL min^{-1}.

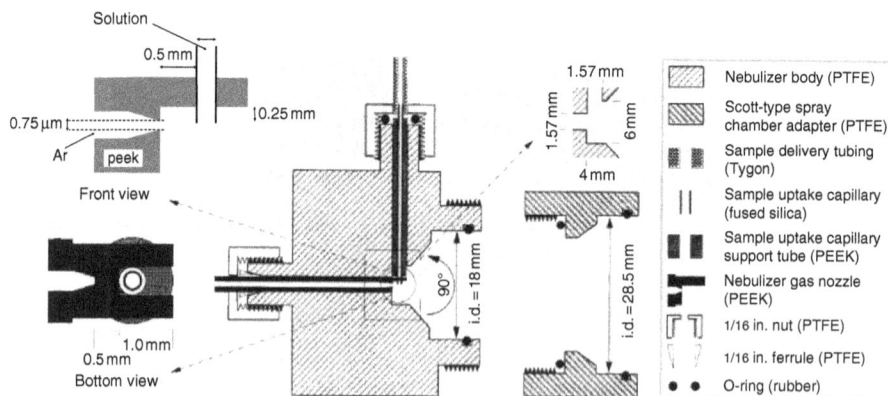

Figure 3.43 Schematic of the high-efficiency cross-flow micronebulizer, HECFMN, (Reprinted with permission from ref. [642].)

3.8.7 Parallel-Path Micronebulizer (PPMN)

The parallel-path micronebulizer (MiraMist®) has been used for the introduction of liquid samples in plasma spectrometry [739]. This device is based on the above mentioned aerosol generation principle (Section 3.6), but it has reduced dimensions.

As its precursor, the PPMN is made completely of PTFE and has a relatively large conduit for the sample that permits the fitting of a 350 µm o.d. glass capillary to work at liquid flow rates in the order of several microlitres per minute. As can be seen in Table 3.9, the gas backpressure is higher than that available in most plasma instruments. Therefore, it is necessary to use an extra gas line when this nebulizer is being operated. Because the liquid and gas outlets are separated, this nebulizer has no suction [773]. For that reason, it is compatible with CE–ICP coupling. According to the manufacturer, the PPMN does not suffer from blockage when working with high salt content solutions or slurries.

The position of the nebulizer inside the spray chamber can induce up to twofold changes in sensitivity [802]. Furthermore, comparison of different PPMNs provides relative signal changes as high as 100 per cent.

3.8.8 Sonic-Spray Nebulizer (SSN)

The Sonic-Spray Nebulizer (SSN) [554] has a cavity with a final orifice (Figure 3.44a). At the end of a 150 µm o.d., 50 µm i.d. silica capillary is placed in the middle of this orifice. With the SSN, absolute sensitivities for ICP–AES are improved more than fivefold with respect to conventional pneumatic nebulizers [600] what appears to be due to an increase in the nebulization efficiency.

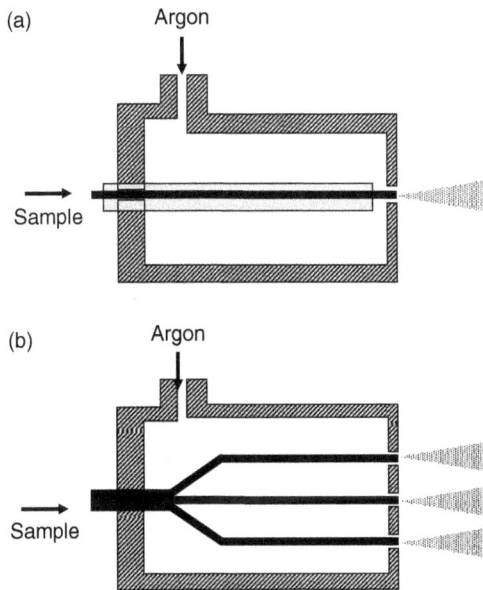

Figure 3.44 Design of the Sonic-Spray Nebulizer (a) and the Multimicrospray Nebulizer (b). (Taken from ref. [828].)

An enhanced version of the SSN is the Multimicrospray Nebulizer (MMSN) [596, 600]. The sample solution is divided into three streams (Figure 3.44b). Each one of the three capillaries employed is centred with three respective gas exit orifices. Thus, there are three aerosol-generation points (i.e. 'nebulization units') behaving like three micronebulizers. As a result, the gas energy is more efficiently employed in the aerosol generation. Indeed, it has been demonstrated that the analyte transport efficiencies and sensitivities reached by the MMSN are up to 2 times higher than those provided by a conventional pneumatic concentric nebulizer and a SSN.

3.8.9 Oscillating-Capillary Nebulizer (OCN)

The OCN [463, 697] consists of two coaxially mounted silica capillaries. Samples travel along the central capillary (40–50 μm i.d., 105–150 μm o.d.), whereas gas flows through the area left between this capillary and an external capillary (250 μm i.d., 350–510 μm o.d.). The gas stream induces liquid capillary oscillations. A longitudinal standing wave appears along the liquid capillary, which is partially responsible for the aerosol generation. Droplets are also produced as a result of the liquid and high-velocity gas stream interaction. This device is also demountable and if damage is produced on any component, it can be easily replaced. The analyte transport efficiencies reached by the OCN approach to 100 per cent because it produces fine aerosols. To obtain the best results with this nebulizer, sample tubes with lower inner diameters and wall thickness are required.

3.8.10 High-Solids MicroNebulizer (HSMN)

The HSMN is a nebulizer made of glass. In this case, the liquid sample emerges through an upper conduction and moves downwards until it reaches the gas exit orifice, thus giving rise to the aerosol. Once the solution is atomized, it impacts against a glass surface placed in front of the nebulizer tip (Figure 3.45a). As a result of the configuration of the tip, the aerosol is deviated

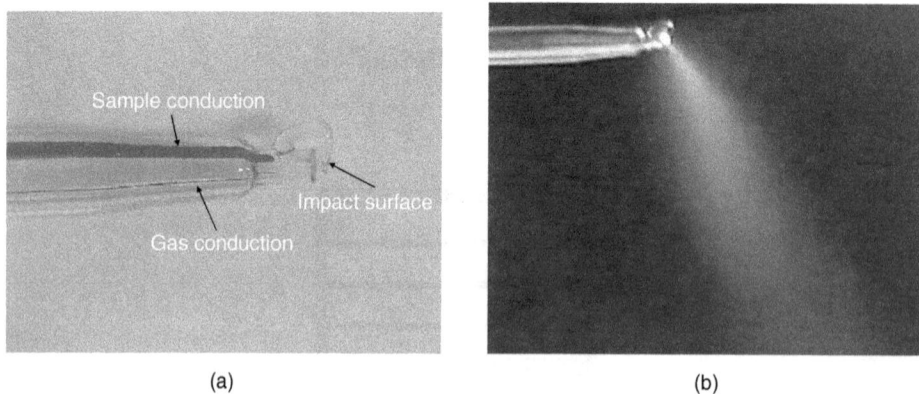

(a)   (b)

Figure 3.45 Picture of the High-Solids MicroNebulizer tip (a) and HSMN working (b).

as it is shown in Figure 3.45b. Consequently, the nebulizer position inside the chamber is a critical factor. When the HSMN is used together with conventional spray chambers, a big fraction of the aerosol impacts against the chamber inner walls and is lost [742].

3.8.11 Direct-Injection Nebulizers

Direct-injection nebulizers are micronebulizers that do not make use of a spray chamber to filter the aerosol. Their tip is located at the base of the plasma so that the aerosol is directly and totally injected into it. There is then no need for an injector.

The micronebulizers previously mentioned are normally used in conjunction with a spray chamber (see Chapter 4) or desolvation systems (see Chapter 5). As will be discussed later, these aerosol transport devices suffer from several drawbacks: (i) existence of memory effects, (ii) intensification of matrix effects, (iii) increase of signal noise, (iv) removal of a high proportion of the analyte nebulized with subsequent loss of sensitivity, (v) waste generation and (vi) postcolumn broadening effects when separation methods are coupled to ICP techniques.

In the case of direct-injection nebulizers, primary aerosol is introduced in the plasma base [120, 185]. In this way, the totality of the analyte delivered to the system is transported to the plasma. However, there are two conditions that must be fulfilled: (i) the liquid flow rate should be lower than the maximum amount that can be accepted by the plasma under a set of given conditions and (ii) the aerosols should be fine enough to ensure an efficient analyte excitation/ionization in the short time it expends in the plasma.

3.8.11.1 Direct-Injection Nebulizer (DIN)

In the DIN, an external ceramic/stainless tube is adapted into the torch and a narrow (i.e. about $120\,\mu m$ o.d., $60\,\mu m$ i.d.) [528, 529] sample capillary is inserted into this ceramic body. The gas flows through the annulus left between both tubes (i.e. typically $15\,\mu m$ wide). The sample solution is injected into the line through a computer-controlled six-port valve and delivered to the nebulizer by a gas displacement pump. The operation of this pumping system requires high pressures (ca., 40–50 bar) to reach liquid flow rates in the range 50–$100\,\mu L\,min^{-1}$. In this way, excellent signal stabilities are obtained. The gas is supplied by means of a cylinder. Due to the low gas flow rates used with this nebulizer (ca., 0.2–$0.5\,L\,min^{-1}$), an additional $\sim0.3\,L\,min^{-1}$ argon stream is introduced in order to efficiently inject the aerosol into the plasma central channel.

There are several variables that influence the behaviour of the DIN such as the dimensions of the external tube [542] and the position of the capillary end with respect to the main nebulizer body [312]. Another important variable is the distance between the nebulizer and the plasma base. The DIN tip is normally placed about 1 mm below the torch central tube to produce a nebulizer nozzle–initial plasma radiation zone gap close to 4 mm. However, it has been demonstrated that the aerosols produced by the DIN at 4 mm from its tip are generally finer than those measured just at the exit of the nebulizer [371]. The DIN is a system that should be operated at low liquid delivery rates. An increase in this variable causes a linear raise of the ICP–AES and ICP–MS signal up to a given value (close to $120–160\,\mu L\,min^{-1}$) and then decreases for higher values of this variable [120, 312].

As it has been mentioned for other micronebulizers, primary aerosols for the DIN are finer than those generated by conventional pneumatic nebulizers [312, 323]. As for other low-sample consumption devices, this result owes to the reduced sample capillary dimensions (Table 3.9), which promotes an efficient interaction between the gas and liquid streams during the nebulization event. Spatially speaking, as for other pneumatic concentric nebulizers, for the DIN the coarsest droplets were preferentially found at the aerosol cone edges, whereas the finest ones were located at the cone centre [565]. An additional consequence of the reduced dimensions of the DIN capillary is that the wash-out times are shortened with respect to a conventional sample introduction system. This is especially interesting when determining elements that are difficult to rinse, such as boron [769], or for the analysis of very low liquid sample volumes. In fact, it has been observed that with the DIN it is possible to obtain steady signals by injecting sample volumes as low as $100\,\mu L$ [421, 659].

Direct-injection nebulizers are prone to tip blocking. To mitigate this problem, the sheathing gas flow rate can be increased and the liquid capillary can be extended 0.5 mm past the end of the nebulizer tip [315]. Another problem that can be observed with the DIN is the tip overheating, which is especially significant when the high-frequency (hf) power increases above 1.3 kW [346].

The low inner volume of the DIN makes it very useful as an interface between separation techniques and ICP. Thus, the postcolumn band broadening is minimized for an HPLC–ICP–MS association [430]. If the mobile phase contains organic solvents such as methanol, ICP operating conditions should be reoptimized, because they are different to those for the analysis of plain water samples [688]. Because it does not show any significant suction, the DIN is also suitable for CE–ICP coupling [418, 758]. To accomplish this, the CE capillary can be placed inside the fused-silica nebulizer capillary. The make-up liquid stream used to establish the continuous electrical contact is pumped through the nebulizer capillary. By doing so, the nebulization and the separation processes can be optimized separately. With this assembly, it is observed that liquid flow rates below $10–15\,\mu L\,min^{-1}$ induce signal pulsing [542]. However, by performing some modifications on the DIN design, such as reducing the gas exit cross-sectional area and recessing the capillary about 1 mm with respect to the nebulizer body, it is possible to work properly at liquid flow rates ranging from 1 to $7\,\mu L\,min^{-1}$. By further extending the liquid capillary 1 mm beyond the nebulizer tip, the system is suitable to work at liquid flow rates ranging from 9 to $15\,\mu L\,min^{-1}$ [542].

3.8.11.2 *Direct-Injection High-Efficiency Nebulizer (DIHEN)*

A relatively low-cost version of the DIN is the so-called Direct-Injection High-Efficiency Nebulizer (DIHEN) [507]. The DIHEN is entirely made of glass or quartz (Q-DIHEN). This nebulizer is similar to a HEN (Table 3.9), but it is longer (200 mm instead of the 75 mm for the HEN) in order to be easily fitted into the torch (Figure 3.46). A support tube

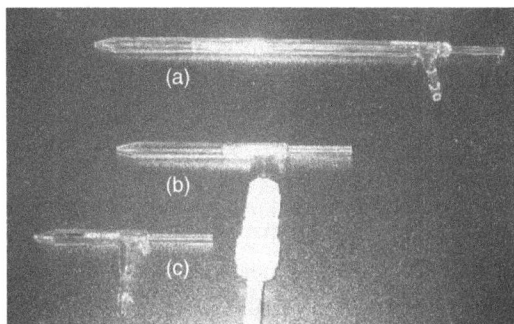

Figure 3.46 Pictures of the DIHEN (a), TL-HEN (b) and HEN (c).

is used for the sample capillary in order to reduce the capillary damage caused by the oscillations induced by the gas stream, thus enhancing the nebulizer robustness (Figure 3.47). Unlike the DIN, the DIHEN can be used with a peristaltic pump. Nonetheless, this can cause severe variations in the signal with time due to the pulses [694, 696], and a syringe pump is advisable when the spray chamber is removed from the sample introduction system [665]. The dead volume of the DIHEN (i.e. $55\,\mu L$) is much larger than that for the DIN (below $1\,\mu L$). Nonetheless, by inserting a PTFE tube, the inner volume can be reduced to 10–$15\,\mu L$ [689, 718]. The early studies performed using

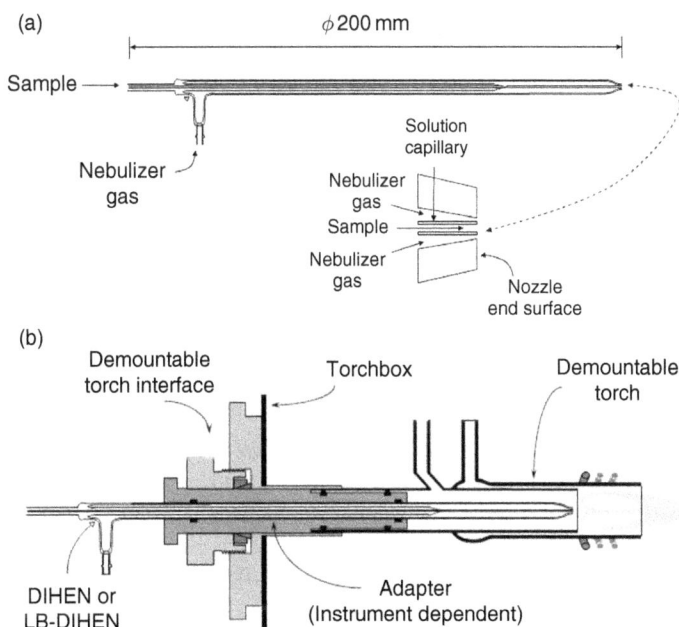

Figure 3.47 Scheme of the DIHEN (Courtesy of A. Montaser.)

ICP–MS with the DIHEN revealed that the optimum signals were obtained at high hf power values (i.e. 1.4–1.5 kW) and low nebulizer gas flow rates (i.e. 0.16–0.25 L min^{-1}). In fact, it was found that the signal increased steeply by either increasing the former variable or decreasing the latter one [507].

The critical dimensions vary from one DIHEN to another. For this reason, the results found for four different nebulizers reveal that the gas backpressure for a gas flow rate of 0.3 L min^{-1} range from 3.4 to 4.4 bar [716]. As a result, the optimum gas flow rates are also different according to the particular DIHEN considered [608]. This fact has a direct effect on the maximum ICP–MS signal attainable with two of these nebulizers [555].

There are several observations that can explain why the DIHEN is not widely used for routine analysis despite its advantages:

1. In general terms, the analytical figures of merit are not as good as expected. This is due to two main reasons: first, coarse aerosols, with drop diameters above 30 μm [652, 750], are introduced into the plasma and, second, the rotational movement of the aerosol [665] is responsible for the radial motion of droplets, which leads to a dispersion of the aerosol across the torch [779]. Only about 30 per cent of the aerosol generated by the DIHEN is introduced into the plasma central channel and thus contributes efficiently to the analytical signal. The aerosol spread has been attributed to the large cone of the DIHEN aerosol, that is at 50 mm from the nebulizer tip, the aerosol cone diameter was around 30 mm [665].

   Because of differences in aerosol drop size and velocity, the number of surviving droplets after the plasma observation zone is much higher for the DIHEN than for a nebulizer–spray chamber combination [664, 779]. Because of all these facts, the optimum hf power when using the DIHEN must be high (i.e. about 1.5 kW) [507].
2. The DIHEN is costly and care should be taken in order to prevent the melting of the nebulizer tip, which can easily melt because of the proximity of the nebulizer tip to the plasma [689, 727] and/or the very low gas flow rate used that reduces the gas cooling action [652]. A recommendation to protect the nebulizer tip from excessive heating is to increase the intermediate gas flow up to 1.2 L min^{-1} [668]. The tip melting can be more problematic during the plasma ignition step. Therefore, it is also recommended to use high values of both the coolant and the auxiliary gas streams (i.e. 20 and 4 L min^{-1}, respectively) and, once the plasma has been ignited, to decrease these variables to the values used for the signal measurement (i.e. 15 and 2 L min^{-1}, respectively) [727].
3. The DIHEN is prone to tip blockage because of the narrow capillary used [689]. Thus, it has been indicated that the DIHEN tip becomes clogged when a one per cent sodium chloride solution was analysed [668].
4. If some organic solvents such as ethanol are nebulized at liquid flow rates from 25 to 100 μL min^{-1}, a carbon deposit is observed at the nebulizer tip [705]. This produces an asymmetric aerosol with a subsequent loss in sensitivity.

To overcome the tip blocking, a new version of the DIHEN has been developed; the so-called large bore direct-injection nebulizer (LB-DIHEN) [593]. This is a modified version of the DIHEN in which the sample capillary and the gas annulus area have been enlarged. This modification gives rise to an increase in the aerosol mean diameter and a decrease in the drop velocity. As a result, the LB-DIHEN provides lower ICP–MS sensitivities [771] and more severe matrix effects [649] than the DIHEN. Furthermore, the precision reached with the LB-DIHEN is worse than that afforded by a pneumatic nebulizer coupled to a spray chamber [593]. The LB-DIHEN is suitable for the analysis of high salt content solutions [751] and slurries, and for this reason it has been applied to the analysis of biological samples [607] having a high concentration in cell agglomerations with diameters ranging from 16 to 18 μm. To improve the liquid and gas generation, while preventing the nebulizer tip blockage, the inner diameter of the sample capillary has been reduced from 320 to 205 μm [751].

An advantage of the DIHEN is related with its low inner volume. Hence, it can be successfully applied for interfacing chromatographic separation techniques and ICP spectrochemistry [655]. With a further reduction of its inner volume through adapting capillaries inside the DIHEN liquid sample tubing, it can used for micro and nano HPLC–ICP–MS coupling. When compared with other low sample consumption systems, the DIHEN has provided better peak resolution [677, 817]. This nebulizer has also been used as a CE–ICP–MS interface. The main advantages over an interface consisting of a cross-flow nebulizer coupled to a double-pass spray chamber are [582] (i) the peaks are sharper and more symmetrical and (2) provided that the amount of buffer is lower, lower background levels are obtained. An interesting application combined anodic stripping voltammetry (ASV) with ICP–MS, wherein the DIHEN served as an efficient interface [708]. In the case of the SF–MS spectrometer equipped with a torch shield, the DIHEN provided absolute sensitivities about sevenfold higher than a PFAN [756]. Meanwhile, with a modified guard electrode [648], the sensitivity found for the DIHEN with the new configuration was sixfold higher than that for the initial design.

Because of the fragility of the DIHEN and in order to reduce its cost, a demountable Direct-Injection Nebulizer has been developed [650]. The sample capillary and the nebulizer shield are adjusted by means of a PEEK piece (Figure 3.48). In this way, if the nebulizer is melted, it is easy to replace the damaged part. Because it has mobile parts, the nebulization process can be optimized by adjusting the position of the sample capillary at 0.1–0.2 mm beyond the nozzle surface [658, 695]. Under these circumstances, better ICP–MS analytical figures of merit can be obtained compared to the conventional DIHEN [650]. The reasons given for these results are the reduced dimensions of the demountable direct-injection nebulizer with respect to the classical one (Table 3.9). More recently, an adjustment vernier has been used in order to precisely modify the position of the liquid capillary [771]. In this case, the nebulizer dead volume is just 11 μL and it is not necessary to use any special capillary to further lower it. Because of its optimized design, ICP–MS sensitivities obtained under optimum conditions are about 2.4 times higher for the demountable DIHEN than for the classical one [771].

3.8.11.3 *Vulkan Direct-Injection Nebulizer*

A new version of direct-injection nebulizer is commercially available [http://www.geicp.com] and has been recently characterized for use in ICP–AES [844]. This nebulizer is similar to the DIHEN, in the sense that it is fitted to the torch by means of a special adapter and its tip is positioned at 2–3 mm from the plasma base.

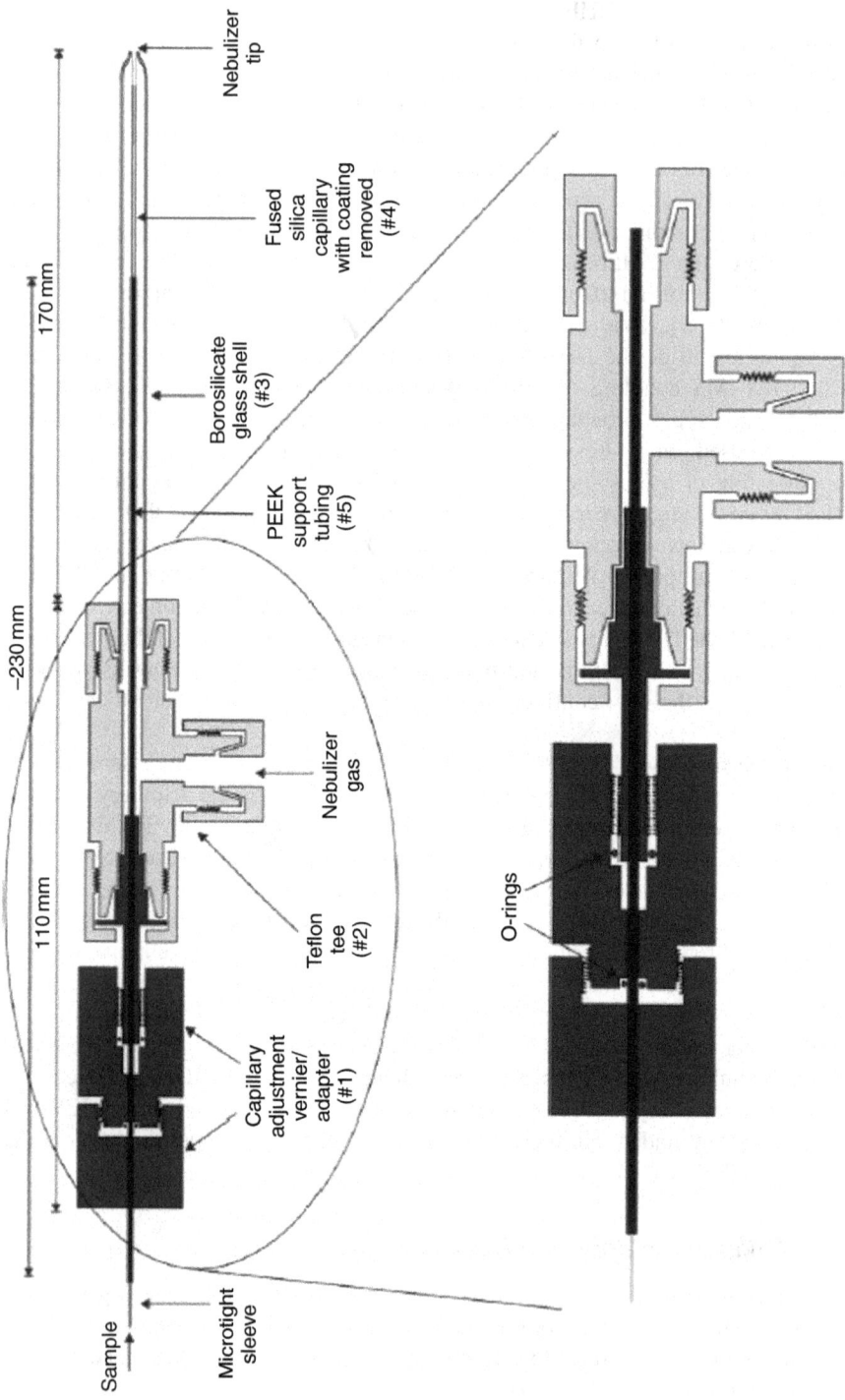

Figure 3.48 Demountable DIHEN. (Taken from ref. [771] with permission.)

When the Vulkan DIN is operated at a 90-μL min^{-1} delivery flow rate, it has been observed that the ICP–AES signal enhances 2–3 times with respect to that found in the case of a MCN coupled to a cyclonic spray chamber. This was found for lines having E_{sum} values (i.e. sum of ionization and excitation potentials) lower than 3 eV. On the contrary, when the E_{sum} of the lines increased, the sensitivities found for the MCN were 1–4 times higher than those measured for the Vulkan DIN. Further studies led to the conclusion that in the case of the Vulkan DIN, the energy transfer efficiency from the induction region to the species was poor. This fact was in agreement with the observation that the plasma ionization temperature was 300–1100 K lower for the Vulkan DIN than for a MCN coupled to a cyclonic spray chamber. Ion number density and MgII/MgI ratios were lower for the former nebulizer [844]. These facts led to the existence of nonspectroscopic matrix effects for this direct-injection nebulizer [797].

3.9 COMPARISON OF MICRONEBULIZERS

As a result of their critical dimensions (Table 3.9), the liquid and gas interaction takes place more efficiently, and finer primary aerosols are generated for the micronebulizers than for the conventional ones [576]. By examining some pneumatic concentric micro-nebulizers, it can be concluded that, in general, the gas backpressure follows the decreasing order: HEN > MMN > PFAN. The kinetic energy available for aerosol generation is higher for the HEN than for the remaining micronebulizers, leading to finer primary aerosols. It can be observed that, for given gas and liquid flow rates, the HEN generates the finest primary aerosols among the devices tested (i.e. the drop size distribution curves are shifted towards the left with respect to the remaining nebulizers). Approximately 89 per cent of the aerosol liquid volume generated by the HEN is contained in droplets having diameters less than 10 μm [388, 465]. Figure 3.49 shows the accumulated volume

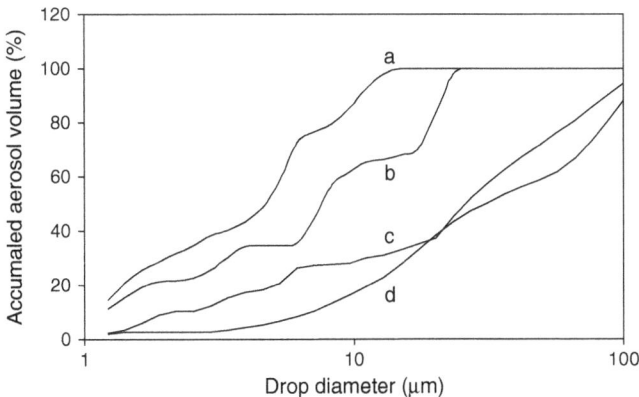

Figure 3.49 Volume drop size distribution curves for the aerosols generated by several micro nebulizers. (a) High-Efficiency Nebulizer, (b) PFA micronebulizer, (c) MicroMist, (d) High-Solids MicroNebulizer.

drop size distributions of the aerosols produced by four different micronebulizers. The results found for the PFAN are highly interesting. It can be seen that the gas backpressure is lower than for the MMN. Therefore, the amount of kinetic energy required to produce the aerosol is higher for MMN. Nonetheless, coarser primary aerosols are reported for the MMN than for the PFAN. This result is explained by taking into account the aforementioned liquid prefilming. The HSMN, in turn, produces the coarsest aerosols among the systems considered in Figure 3.49, likely because of the action of the impact surface.

As a result of the reduced inner diameters of the liquid capillaries, micronebulizers usually have reduced free liquid uptake rates than conventional devices. This fact is illustrated in Figure 3.50, in which this parameter is summarized for a set of six different micronebulizers. As it can be verified, these systems aspirate aqueous solutions at rates from one to two orders of magnitude lower than the conventional pneumatic nebulizers.

In comparison to other pneumatic nebulizers, there is a variability of results from one HEN nebulizer to another. Thus, at $50\,\mu L\,min^{-1}$ liquid flow rate and $0.7\,L\,min^{-1}$ gas flow rate, the medians of the primary aerosol volume drop size distribution were 2.8 and 3.9 μm for two similar HEN designs.

The DIHEN has recently been compared with the Vulkan DIN nebulizer [844]. The differences in performance are due to their different critical dimensions and design. In summary, the main differences observed between these two nebulizers are: (i) in the case of the Vulkan DIN, the liquid capillary is recessed with respect to the main body (Figure 3.51), whereas for the DIHEN, the capillary tip is located at the nebulizer tip and (ii) liquid capillary inner diameter and wall thickness as well as gas cross-sectional area are higher for the former device (Table 3.10). As a result of the critical dimensions, the liquid and gas interaction takes place less efficiently for the Vulkan than for the DIHEN. Hence, aerosols are coarser and the distributions are broader for the former than for the latter design (Figure 3.52a). However, these discrepancies in aerosol characteristics are not

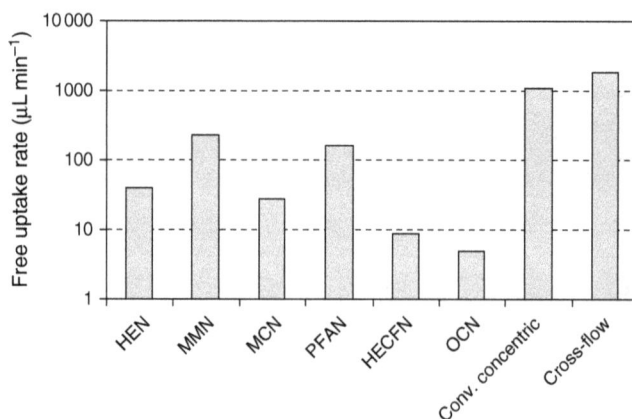

Figure 3.50 Free liquid uptake rates measured for different pneumatic micronebulizers operated at nebulizer gas flow rates ranging from 0.7 to $1\,L\,min^{-1}$.

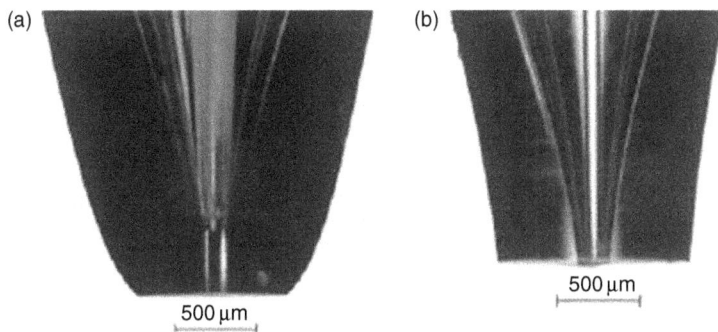

Figure 3.51 Magnified pictures of the tip of a Vulkan DIN nebulizer (a) and a DIHEN (b). (Taken from ref. [844] with permission.)

Table 3.10

Comparison of critical dimensions for the DIHEN and the Vulkan
direct-injection nebulizer (Taken from ref. [844].)

Dimension	Vulkan DIN	DIHEN
Solution capillary i.d. (μm)	125	104
Solution capillary o.d. (μm)	250	144
Gas exit cross-sectional area (mm^2)	0.0126	0.0094

related to similar differences in terms of ICP–MS analytical figures of merit. Thus, The Vulkan gives rise to lower sensitivities than the DIHEN, but for some isotopes (e.g. ^{24}Mg$^+$, ^{103}Rh$^+$, ^{111}Cd$^+$ and ^{140}Ce$^+$) lower limits of detection are reported for the former nebulizer. The reason for these findings appears to be related with the fact that although the aerosols are finer for DIHEN, their velocities are lower for Vulkan DIN (Figure 3.52b). Furthermore, the aerosols can be more confined within the plasma central channel for Vulkan because of the capillary recess, as it has been anticipated for conventional pneumatic concentric nebulizers [349]. With regard to the signal precision, it has been recently claimed that the Vulkan DIN provides higher white noise than the DIHEN [818].

Direct-injection nebulizers are costly and fragile, and require specific plasma-operating parameters (high power, unusually low carrier gas flow rate), which explains why they are not routinely used, although they exhibit some advantages. In marked contrast, micronebulizers associated with a spray chamber, possibly with a simplified design (see Chapter 4), can be routinely used, without losses in sensitivity and repeatability.

It should be noted that most of the pneumatic nebulizers described in this chapter have an external normalized diameter of 6 mm. Interchangeability between nebulizers is then easy, depending on the type of solutions.

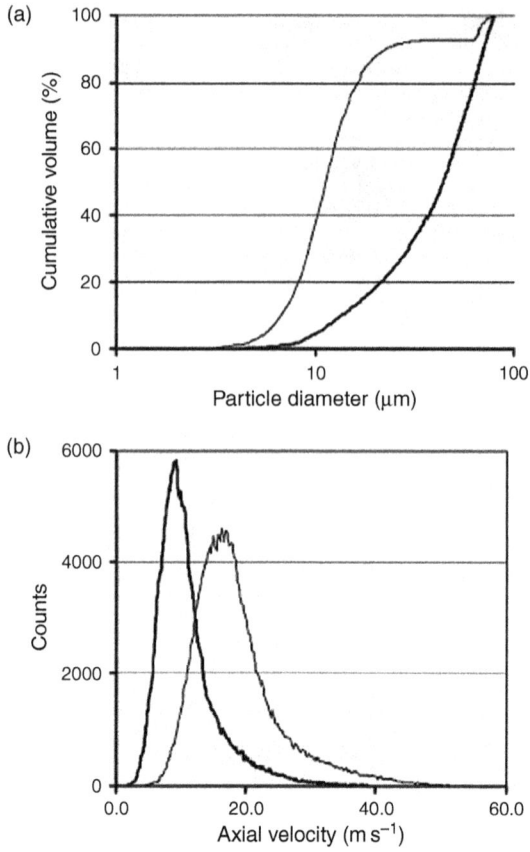

Figure 3.52 Cumulative volume drop size distribution curves (a) and axial velocity drop size distributions (b) for the aerosols generated by a DIHEN and a Vulkan DIN. (Taken from ref. [818] with permission.)

– 4 –

Spray Chamber Design

4.1 INTRODUCTION

Aerosols pneumatically generated have characteristics that prevent their direct introduction into the plasma (see Chapter 3). Therefore, it is necessary to use an additional component to adapt the aerosol properties to the plasma requirements. Usually, the nebulizer is fitted to a spray chamber. Although in some instances (e.g. hydride generation ICP determinations) a spray chamber is used as a gas–liquid separation system [459, 462], its main function is to remove coarse droplets from primary aerosol [231]. Unfortunately, the majority of the useful droplets generated by the nebulizer are lost in this second component of a typical liquid sample introduction system. Obviously, the mass of solution lost in the spray chamber and the signal are functions of the quality of primary aerosol in terms of fineness [115], velocity and shape. These characteristics are given by the ability of a nebulizer to produce fine droplets as well as by the nebulization conditions. Nevertheless, it is considered that the spray chamber rather than the nebulizer determines the mass and characteristics of the aerosol introduced into the plasma. For the currently used spray chambers, the maximum attainable analyte transport efficiencies at conventional liquid flow rates (i.e. on the order of $1\,\text{mL}\,\text{min}^{-1}$) are about 2–4%. Therefore, from the early stages of the ICP development, it was evidenced that the spray chamber design had to be optimized [46].

The signal finally obtained in ICP–AES and ICP–MS depends on the mass of the analyte and solvent reaching the plasma per unit of time. Both magnitudes are functions of the particular nebulizer–spray chamber combination. Therefore, these parameters are good indicators of the signal potentially produced by a given sample introduction system [553].

Analyte mass transport rate (see Chapter 2), W_{tot}, is defined as the mass of the analyte leaving the spray chamber per unit of time. W_{tot} is measured by means of the so-called direct method. In this case, the sample (i.e. a stock solution usually with a $100\,\mu\text{g/mL}$ analyte concentration) is trapped on one or two glass fibre filters adapted at the exit of the spray chamber [187]. The system is operated under controlled conditions for a given time ($10\,\text{min}$ used to be sufficient). Then the filters are washed, normally with a hot nitric acid solution. The total solution volume is led to a given volume (e.g. $50\,\text{mL}$) and it is analysed by a suitable technique, thus giving rise to the analyte concentration in the new solution.

Provided that the total volume is known, it is possible to determine W_{tot} by dividing the calculated analyte mass by the trapping time. Analyte transport efficiency, in turn, is defined according to:

$$\varepsilon_n = 100 \frac{W_{tot}}{Q_1 C}, \qquad (4.1)$$

where Q_1 is the nebulizer liquid flow rate and C is the analyte concentration in the stock solution.

Similar parameters can be defined to quantify the solvent plasma load. The total mass of solvent transport rate leaving the spray chamber (S_{tot}) can be measured by means of a direct method [355]. In this case, a U tube full with silica gel is adapted at the exit of the chamber. The system is run for a given period of time. S_{tot} is obtained by weighing the tube before and after the experiment and dividing the resulting magnitude by the trapping time. An indirect method in which the drain is weighed can also be employed for S_{tot} determination [81]. Similarly, solvent transport efficiency is determined by dividing S_{tot} by the liquid mass flow rate and multiplying it by 100.

4.2 AEROSOL TRANSPORT PHENOMENA

As mentioned in the previous section, primary aerosols are modified along their passage through the spray chamber. The phenomena taking place from the aerosol production until the introduction of the remaining droplets into the plasma are known as aerosol transport phenomena. The intensity of production of each one of these processes depends on the spray chamber design, the aerosol characteristics (e.g. fineness, velocity, shape) and the solution physical properties.

The final results of the aerosol transport phenomena are (i) the removal of a high proportion of coarse droplets, (ii) the reduction of the total aerosol mass or liquid volume, (iii) the decrease in the aerosol droplet velocity and (iv) the reduction of the polydispersion degree in terms of aerosol droplet diameters. Furthermore, a modification of the charge equilibrium and the establishment of the thermal equilibrium are also produced [231]. The aerosol reaching the plasma is known as *tertiary aerosol* and the sensitivity finally obtained depends mainly on the properties of this aerosol.

The main aerosol transport phenomena occurring inside a spray chamber used for ICP techniques are evaporation, coagulation or coalescence and impacts caused by droplet inertia, gravitational settling and turbulences.

Figure 4.1 illustrates all these processes occurring simultaneously inside a double-pass spray chamber. In this figure, the influence of these processes on the aerosol

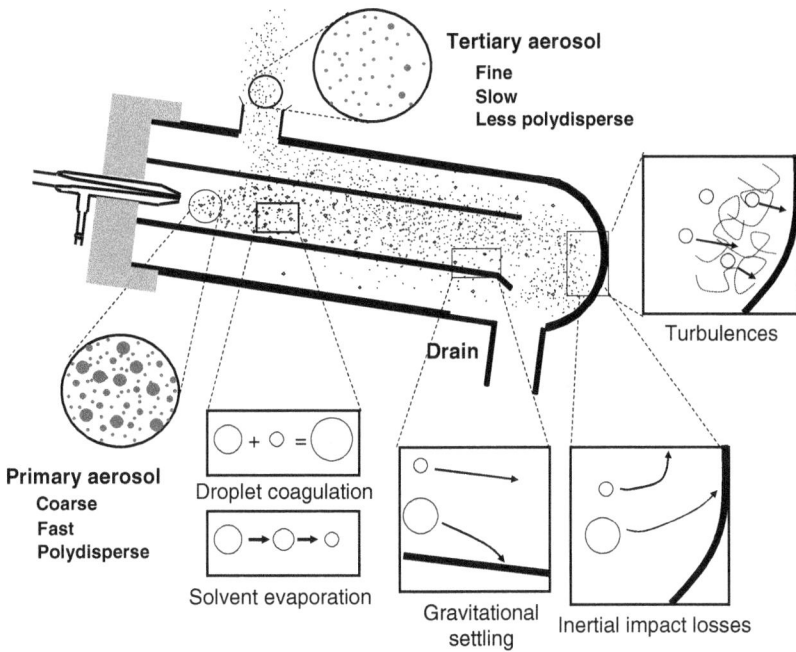

Figure 4.1 Scheme illustrating the aerosol transport phenomena and their consequences in a sample introduction system consisting of a concentric nebulizer coupled to a double-pass spray chamber.

characteristics can also be seen. The present section is aimed at describing briefly each one of these phenomena and its effects on the aerosol characteristics.

4.2.1 Droplet evaporation

Droplet evaporation inside a spray chamber precludes the net mass of solvent reaching the plasma, thus influencing its thermal properties. In addition, the droplet diameters decrease as the solvent is evaporated, hence increasing the likelihood for the droplet transport through the chamber. Solvent evaporation is favoured because of two reasons: firstly, the liquid surface in aerosol phase can be up to about four orders of magnitude higher than in non-dispersed phase. For example, measurements with a laser diffraction system indicate that for a pneumatic concentric nebulizer operated at typically $1\,\mathrm{mL}$ min^{-1} and $1\,\mathrm{L\,min}^{-1}$ liquid and gas flow rates, respectively, the total aerosol liquid (AL) surface is about $3\,\mathrm{m^2\,min}^{-1}$ (i.e. a surface increase factor close to $50\,000$). Secondly, the increased vapour pressure caused by the curvature of droplets favours the solvent evaporation.

Droplet evaporation is expected to be mainly produced just after the aerosol generation at the nebulizer nozzle. At this location, the gas is not saturated in solvent and the relative gas–liquid velocities are very different. As a consequence, the droplet environment is continuously renewed with dry gas, which in turn promotes further solvent evaporation. When both streams move at the same velocity, this process becomes less significant.

The variation of the diameter of a droplet with time is given by [101]:

$$\frac{dD}{dt} = \frac{4D_v M}{R\rho d}\left(\frac{P_\infty}{T_\infty} - \frac{P_d}{T_d}\right),$$

(4.2)

where D_v is the solvent diffusion coefficient, M is its molecular weight, R is the gas constant, ρ is the solvent density, P_∞ and P_d are the partial pressure of the solvent in the carrier gas and the droplet surface, respectively, and T_∞ and T_d are the temperatures in the carrier gas and droplet surface, respectively. Usually, the partial pressure in the carrier gas is lower than at the droplet surface; as a result, the term into parenthesis is negative and drop diameter decreases with time.

By integrating eqn. 4.2, it is possible to find a relationship between the drop diameter and the time.

$$(d)_t^2 - (d)_0^2 = \frac{8D_v Mt}{R\rho}\left(\frac{P_\infty}{T_\infty} - \frac{P_d}{T_d}\right),$$

(4.3)

where $(d)_t$ is the drop diameter at a given time t and $(d)_0$ is the initial drop diameter.

It is known that fine droplets evaporate faster than coarse ones. This is a consequence of the following reasoning: if considering the change in drop area (dA/dt) caused by evaporation instead of the variation in drop diameter in eqn. 4.2 (i.e. $dA/dd = 2\pi D$), eqn. 4.2 is transformed into:

$$\frac{dA}{dt} = \frac{8\pi MD_v}{R\rho}\left(\frac{P_\infty}{T_\infty} - \frac{P_d}{T_d}\right).$$

(4.4)

For a given set of pressure and temperature values, dA/dt is independent of the drop diameter. Therefore, the mass of solvent evaporated is proportionally higher for small than for big droplets. This phenomenon is called the Kelvin effect, which consists of an increase in the solvent vapour pressure observed when switching from a flat surface (solution into a beaker) to a curved surface (aerosol droplets). In fact, it has been indicated that only the aerosol finest droplets contribute to the solvent evaporation inside the spray chamber [231]. The increased vapour pressure induced by the droplet curvature would explain why a small enough droplet would evaporate spontaneously in a solvent-saturated environment.

In eqn. 4.3 it has not been considered that some droplets of an aerosol generated from a pure solvent will not evaporate completely. The diameter of these droplets will be given by the Thompson–Gibbs or Kelvin equation,

$$\frac{P}{P_s} = \exp\left[\frac{4\sigma M}{\rho RTD}\right].$$

(4.5)

By taking this fact into account, eqn. 4.3 is transformed into [57]:

$$(D)_t^3 = (D)_0^3 - \left[48 D_v M^2 P_s \sigma (\rho RT)^{-2}\right] t \tag{4.6}$$

or

$$(D)_t^3 = (D)_0^3 - Et, \tag{4.7}$$

where E is the so-called evaporation factor. The most important assumptions made throughout this reasoning are that the aerosol is under isothermal conditions and that the aerosol flow regime is laminar [213]. It is then interesting to note that the droplet evaporation velocity may increase up to a maximum value that would correspond to the situation of an aerosol being produced into a vacuum medium. However, evaporation velocity depends on the solvent vapour diffusion. As the solvent is evaporated, it cannot leave the droplet environment as fast as required for further evaporation. In addition, solvent evaporation causes a decrease in the temperature at the drop surface. From eqn. 4.7, it can also be calculated that the time required for the complete evaporation of a 2-μm droplet diameter is of the order of several seconds [57]. Note that the time of production of other phenomena such as droplet inertial impacts is on the millisecond time scale.

Equation 4.7 allows modelling the change of the complete drop size distribution caused by solvent evaporation. If the aerosol is measured by means of a laser Fraunhofer diffraction system, the drop size distribution is obtained as a histogram with a total of 31 bars. This corresponds to the 31 rings of the sizer detector. In this curve, the percentage in band is plotted versus the drop diameter. If eqn. 4.7 is applied to each drop diameter interval, the x-axis of the plot represents the variation of the diameter with time. By considering a single-drop diameter range (i), the evolution of the volume of the droplets can be calculated according to

$$V_i = \frac{1}{6} \pi \left[(D_0)_i^3 - Et\right], \tag{4.8}$$

where $(D_0)_i$ and V_i are the initial mean droplet diameter and volume for a given size range, respectively. For a given $(D_0)_i$, the evolution of V_i with time can be calculated by means of eqn. 4.8. The non-evaporated (NE) aerosol fraction is obtained by dividing V_i at a given time by the initial one $(V_0)_i$.

The NE values can be multiplied by the percentage of AL volume contained in each one of the diameter ranges and by the liquid flow rate to calculate the AL flow rate contained in each size range at a given time.

$$AL = NE \times Q_1 \times (\% \text{ in band}) \tag{4.9}$$

$$AL = \left[1 - \frac{Et}{(D_0)_i^3}\right] Q_1 \, (\% \text{ in band}). \tag{4.10}$$

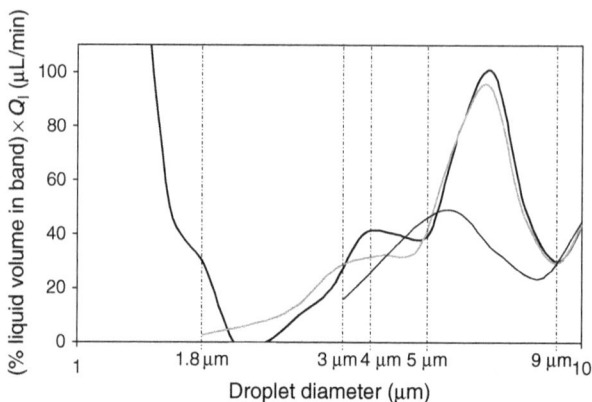

Figure 4.2 Evolution of the aerosol drop size distribution with time by taking into account that the only phenomenon taking place is solvent evaporation. Nebulizer: pneumatic concentric; liquid flow rate: 1 mL min^{-1}; gas flow rate: 1 L min^{-1}. Coarse black line: aerosol at the exit of the nebulizer; grey line: aerosol 1 s after its formation; fine black line: aerosol 10 s after the aerosol formation.

This is a means of estimating what would be the contribution of evaporation to the modification of the aerosol characteristics. Figure 4.2 shows the volume drop size distribution curve for the primary aerosol produced by a pneumatic concentric nebulizer operated under typical ICP–MS conditions. To get better resolution for small droplets, scale in Figure 4.2 only extends up to 10 μm. The calculated curves are also shown. Three different situations have been considered: 0 s, corresponding to the primary aerosol, 1 and 10 s after the aerosol generation. One second after the aerosol production, the theoretical calculations indicate that the droplets with diameters lower than 1.8 μm have completely evaporated. This result is in agreement with published calculations, which indicate that only droplets whose diameters fall below 2 μm contribute efficiently to the solvent evaporation [57, 37]. By comparing the curves at 0 and 1 s, it can be observed that the fraction of droplets with diameters from about 2 to 3 μm is higher for the latter case than for the former one. Meanwhile, the opposite is found for droplets with diameters from 4 to 5 μm. With regard to fine droplets, the highest population of droplets with diameters under 1.2 μm are obviously found when the aerosol has just been generated. Finally, both curves are virtually identical for diameters above 5 μm.

It can be concluded that it is very difficult to predict what will be the effect of the solvent evaporation on the fineness of the aerosols. Thus, for example, for the data presented in Figure 4.2, the median of the volume drop size distributions (D_{50}) are 6.2 and 8.9 μm just after the aerosol generation and 1 s after its production, respectively. This fact would indicate that globally speaking, the aerosol becomes coarser as the solvent evaporates from the droplets surfaces because finest droplets have disappeared [37, 231]. Therefore, the coarse aerosol could produce a higher ICP signal than the fine one, provided that the initially finest droplets have lost their solvent and have transformed into analyte particles virtually dry.

Ten seconds after the aerosol generation, the calculations indicate that all the droplets have diameters higher than $3\,\mu m$ (Figure 4.2). In this case, the prediction reveals that solvent evaporation induces a noticeable drop in the AL mass (or volume) contained in droplets with diameters going from 5 to $9\,\mu m$. This period of time has been considered because it fits with the estimated gas residence time at a gas flow rate of $0.7\,L\,min^{-1}$ inside a $100\,cm^3$ double-pass spray chamber (i.e. $9\,s$). According to some calculations, a droplet can spend inside a cyclonic spray chamber a time as long as $50\,s$ before abandoning it [724]. This time is enough to evaporate a $9\,\mu m$ droplet.

By taking into account the previous discussion on evaporation and the value of the maximum diameter of a droplet that can be completely vaporized in the plasma (i.e. about $12\,\mu m$) [335], it would be possible to conclude that in order to reach a 100% aerosol transport efficiency, it would be necessary to use nebulizers able to produce aerosols with droplets under roughly $9\,\mu m$. This value is approximate, because of the assumptions inherent to the droplet evaporation calculations. Furthermore, depending on the liquid flow rate and temperature, the argon can become saturated, thus avoiding complete evaporation.

By integration of the curves considered in Figure 4.2, the mass of water contained in the aerosol and its evolution with time can be estimated. For a pneumatic concentric nebulizer operated at a liquid flow rate of $1\,mL\,min^{-1}$, the amount of water evaporated $1\,s$ after the aerosol production would range from 140 to $300\,mg\,min^{-1}$. However, as theoretical thermodynamic calculations reveal, the mass of water that can evaporate at $20°C$ is about $20\,mg\,L^{-1}$ of argon [231, 388]. Therefore, it can be concluded that at this liquid flow rate, the gas stream becomes saturated with water in less than a second. According to our calculations, under the conditions mentioned before, the saturation of the argon stream would occur in a period of time close to $30\,ms$. This result is in good agreement with early calculations in which it was demonstrated that the 90% of the change in droplet size occurs in a time range from 1 to $100\,ms$ [37]. The change in drop size distribution caused by evaporation $30\,ms$ after the aerosol generation can be calculated. To perform this new calculation, eqn. 4.7 can be applied to each drop diameter range provided by the laser diffraction sizer. The initial and final volume for each drop size can be calculated. Then the percentage of AL volume can be obtained by means of:

$$\% \text{ change in aerosol liquid volume} = \frac{V_0}{V_{30\,ms}} = \frac{D_0^3}{D_{30\,ms}^3}, \qquad (4.11)$$

where D_0 is the initial drop diameter and $D_{30\,ms}$ is the calculated drop diameter $30\,ms$ later. Figure 4.3 shows the proportion in liquid volume that would remain in the aerosol stream $30\,ms$ after the aerosol production. As expected, the percentage of liquid volume is lower for small droplets than for coarse ones. However, the predicted change in the liquid aerosol volume contained in droplets with $1.2\,\mu m$ diameter is just 20%. Meanwhile, for droplets with diameters higher than $3.7\,\mu m$ approximately, this change is lower than 1%.

In summary, it can be indicated that evaporation from droplets is an important phenomenon, but just after the aerosol production, at least at $20°C$. All theoretical calculations indicate that evaporation is responsible for only a slight change in aerosol properties.

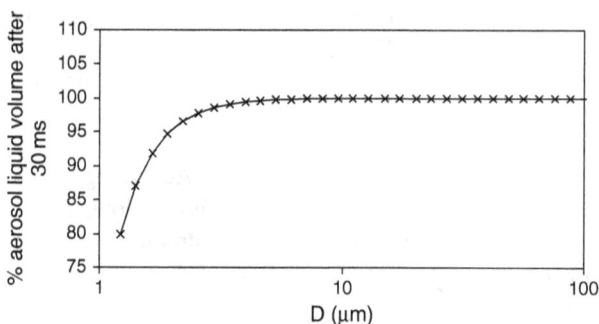

Figure 4.3 Percentage liquid volume remaining in the aerosol versus the drop diameter 30 ms after the aerosol production by a pneumatic nebulizer operated at $1 \, L \, min^{-1}$ and $1 \, mL \, min^{-1}$ gas and liquid flow rate, respectively.

There are other situations in which the aerosol evaporation plays a role of capital importance in terms of aerosol modification such as the work at high temperature (e.g. desolvation systems) and at low liquid flow rates (e.g. low sample consumption systems) also considered in this book (see Chapter 5).

Under the conditions normally used with a conventional liquid sample introduction system, it can be indicated that evaporation remains negligible, whereas aerosol transport phenomena may be the main responsible for the modification of aerosol characteristics. These processes are considered in the following sections.

4.2.2 Droplet coagulation

Droplet coagulation or coalescence consists of the droplet fusion as they collide, leading to an increase in droplet diameter.

This process slows the solvent evaporation, because when two droplets combine, the overall surface decreases. There are several causes of coalescence: if it is produced because of the Brownian particle movement, the process is called thermal coalescence, whereas if it is due to causes such as droplet inertia, repulsion electrical forces or gravity, the process is called cinematic coalescence. The latter phenomenon can be due to three reasons [231]: (i) the existence of velocity gradients at the exit of the nebulizer, (ii) mixing of small particles entrained in the turbulent eddies and (iii) different aerosol drop diameters and differences in droplet acceleration.

Coalescence is a complex phenomenon and its final result depends on the diameters and velocities of the droplets involved in this process. Furthermore, the impact angle and the particle diameter ratio are the key parameters in determining the magnitude of the coagulation. Thus, for droplets with diameter ratios of 1:1 and 1:2, coagulation is not observed, instead the phenomenon produced consists of reflex disjunction. In the case of droplets with 1:3 diameter ratio, coalescence occurs. Depending on the velocities of the droplets, their combination can be either favoured (for droplet velocities above $1\,\mathrm{m\,s^{-1}}$) or avoided (for droplet velocities below $1\,\mathrm{m\,s^{-1}}$) as the drop diameter decreases [15]. Therefore, if the kinetic energy of the collision is higher than the loss in energy caused by the reduction in liquid surface, the droplets can suffer from separation. The drop coalescence can be complete or partial. When a small droplet collides with a large one, two processes are possible: combination and initial combination + detachment. This latter process would be referred as partial coalescence and it involves the transfer of a net solvent mass between the two droplets, hence, the droplets diameters vary with respect to the original ones [296]. The droplet separation is an extremely rapid process. It is estimated that the time elapsed from droplet coagulation to separation takes values from 10^{-7} to $10^{-8}\,\mathrm{s}$ for 1–10-μm droplets diameters [734]. This is illustrated in Figure 4.4 in which two 15-μm diameter droplets initially collide and are separated $1.5 \times 10^{-10}\,\mathrm{s}$ later. In this case, the mass transfer can be neglected, since this period of time is too short. The simulation presented in Figure 4.4 is based on molecular dynamics that take into account the Newton attractive forces and electrical repulsive forces acting among different solvent molecules [734].

It is also interesting to notice that once droplets collide, the resulting liquid bulk is deformed and a liquid bridge can be established. The magnitude of this liquid bridge

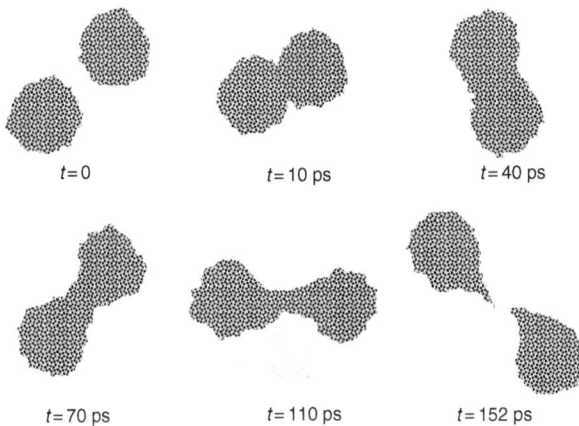

Figure 4.4 Simulation of the evolution of two droplets having 2500 water molecules, each one suffering from partial coagulation. (Taken from ref. [734], with permission.)

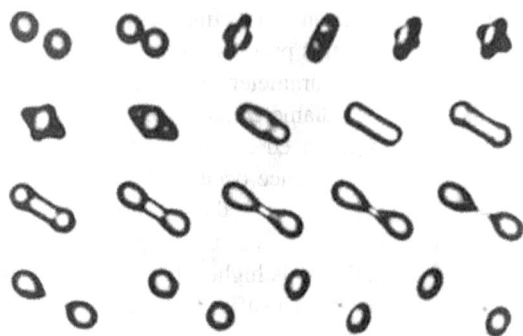

Figure 4.5 Sequence illustrating the reflex disjection suffered by two droplets approaching at a 1.5 m s^{-1} velocity. (Taken from http://parkscrapbook.us/drops with permission.)

depends on the drop velocity and the angle of interaction between the coagulating droplets. There are other processes that can be produced when two droplets are approaching at high velocities: they are the disruption and the drop rebound (Figure 4.5).

Coalescence has a direct effect on droplet diameters. In fact, drop size increases as a result of this phenomenon. For monodisperse aerosols, the variation of diameter with time $d(t)$ is given by:

$$d(t) = d_0(1 + N_0Kt)^{1/3}, \tag{4.12}$$

where d_0 is the initial diameter, N_0 is the initial particle number density and K is the coagulation coefficient.

Equation 4.12 can be applied to determine the change in drop diameter with time. The results are presented in Figure 4.6 for selected droplet diameters. The drop diameter at a given time has been calculated by assuming that the aerosol is monodisperse within each drop size range supplied by the sizer (a system based on the Fraunhofer diffraction of a laser beam). For most cases, and as eqn. 4.12 indicates, the drop diameter increases with time as a result of droplet coagulation. The x-axis scale has been extended up to 10 s, which is the estimated gas residence time in a typical double-pass spray chamber at a 0.7 L min^{-1} flow rate.

It is also interesting to notice that the higher the drop diameter the less pronounced the droplet growth through a coalescence mechanism. This is likely due to the fact that the droplet number density (N_0 in eqn. 4.12) is lower for coarse than for fine droplets. As it may be seen from Figure 4.6, two different delivery flow rates are considered. At 1.5 mL min^{-1} (Figure 4.6a), the increase of drop size is more significant than at 0.5 mL min^{-1} (Figure 4.6b). Thus, for droplets with diameters of 1.44 μm, the coalescence causes a rise in their diameters by a factor of 2 and 1.5 at sample delivery rates of 1.5 and 0.5 mL min^{-1}, respectively. Again, this fact is due to the higher initial droplet number density.

Coagulation caused by differences in droplet velocity is significant mainly near the nebulizer nozzle. At this location, the difference in velocity of droplets is higher than at other locations inside the spray chamber [389]. Besides, coalescence is intimately related

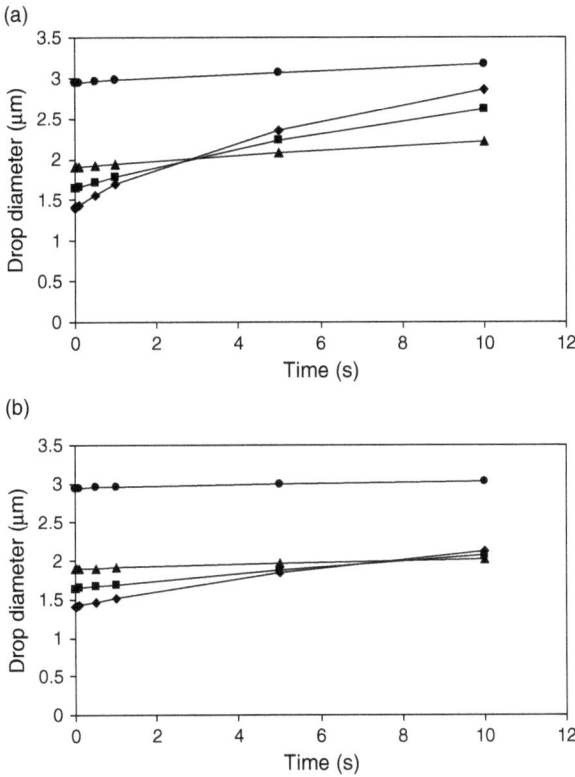

Figure 4.6 Change in the drop diameter with time caused by droplet coagulation for several initial droplet diameters at $1.5 \, \text{mL min}^{-1}$ (a) and $0.5 \, \text{mL min}^{-1}$ (b). Nebulizer: pneumatic concentric. Gas flow rate: $0.7 \, \text{L min}^{-1}$. Initial drop diameters (μm): ($\blacklozenge$) 1.41, ($\blacksquare$) 1.64, ($\blacktriangle$) 1.9, ($\bullet$) 2.95.

with the droplet number density. At the exit of the nebulizer tip, the droplet number density for a $1.5 \, \text{mL min}^{-1}$ liquid flow rate has been calculated to be close to 3×10^{14} particles m^{-3} [231]. These calculations were performed assuming that all droplets had a diameter of $10 \, \mu m$. However, aerosols produced by pneumatic concentric nebulizers are polydisperse and a high number of small droplets can be found. If we consider the measurements performed by a laser diffraction sizer and that the probed aerosol volume is a cone of $10 \, \text{mm}$ length (the diameter of the laser beam) and $8 \, \text{mm}$ width (see Figure 3.26, Chapter 3), the droplet number density in the aerosol volume delimitated by the sizer laser beam at a liquid flow rate of $1.5 \, \text{mL min}^{-1}$ is about 5×10^{17} particles/(m^3 min). This presumably will produce a huge number of droplet collisions.

To illustrate the importance of the droplet coalescence, the number of collisions between droplets can be calculated. For a pneumatic nebulizer and at a $1.5 \, \text{mL min}^{-1}$ liquid flow rate, the number of collisions between droplets at the exit of the nebulizer (i.e. in a cylinder whose diameter and length are $100 \, \mu m$ and $0.1 \, \text{mm}$, respectively) is about $1020 \, \text{s}^{-1} \text{m}^{-3}$, which indicates the enormous significance of the coalescence process [231].

The number of collisions can be calculated in a different way, depending on the coagulation mechanism considered. If the coagulation is mainly produced by the existence of velocity gradients in the aerosol jet boundary, the number of collisions (N) is given by

$$N = \frac{32}{3} n^2 \Gamma a^3, \qquad (4.13)$$

where n is the number of droplets per volume unit, Γ is the velocity gradient and a is the particle radius. Equation 4.13 indicates that the number of collisions between droplets varies with the square of the aerosol droplet density. This is also accomplished for the other two droplet coagulation mechanisms [231]. According to Sharp, at 1.5 mL min^{-1} the number of collisions for 10 μm droplets is 4×10^{19}. Thus, at a lower liquid flow rate (e.g. 0.5 mL min^{-1}), N should be about three times lower.

Droplet coalescence can be the reason to explain why an increase in the liquid flow rate causes a slight modification in the mass of solution delivered to the plasma [436]. As a result, the analyte transport efficiency decreases. Measured transport efficiencies for a pneumatic concentric nebulizer coupled to a double-pass spray chamber are 2 and 0.7% as the liquid flow rate goes from 0.6 to 1.2 mL min^{-1}. The explanation is simple: as the liquid flow rate increases, the drop number density goes up, thus promoting droplet coalescence. Therefore, coarse droplets are formed, which have less likelihood to escape the spray chamber. Impact losses hence increase.

Because of the high aerosol density, droplet coalescence is a significant phenomenon in the spray chamber. Coarse droplets that result of coalescence will not escape the spray chamber through impact losses.

4.2.3 Droplet impacts

To remove the coarsest aerosol droplets, surfaces strategically placed along the aerosol path in the spray chambers will be used so that droplets will impact to them and, therefore, will be eliminated. Additionally, complex aerosol trajectories may be imposed.

The impact of droplets has been a process used in the ICP sample introduction systems to promote the transport of just the smallest aerosol droplets to the plasma. The impact losses can be (i) inertial, (ii) gravitational and (iii) caused by turbulences.

It has been demonstrated that a droplet with a diameter of 1 μm is able to easily accommodate its velocity to that of the gas stream just at the exit of the nebulizer.

However, droplets having diameters of 10 and 100 μm require travel for about 20–150 mm to equal their velocities to that of the gas under the same conditions [231]. Note that, when they are produced, these three types of particles would have the same velocity. Traditionally, either spherical or flat impact surfaces have been placed inside the spray chamber. The efficiency of these impactors can be evaluated from the Stokes number (Stk), which is the relation between the particle stopping distance and the impact surface characteristic dimension:

$$\text{Stk} = \frac{\rho d^2 U C_c}{9\eta_g w},\tag{4.14}$$

where U is the mean velocity of the aerosol stream at the exit of the nebulizer, w is the characteristic dimension of the impact surface (i.e. the bead diameter in the case of a spherical surface), C_c is the correction Cunninghan factor, d is the drop diameter, ρ the density of the drops and η_g is the gas viscosity.

Because of the fact that a droplet will be completely vaporized in the plasma if this diameter is lower than 12 μm, an impact surface should be placed at a distance not farther than 20 mm from the nebulizer tip to be used efficiently. If the surface is placed closer to the nebulizer, a fraction of droplets with diameters ranging from 2 to 10 μm will be lost, thus causing a reduction in the sensitivity potentially achievable.

A droplet with a 10 μm diameter only needs 300 μs to reach a surface placed at 15 mm from the nebulizer exit. Meanwhile, this time is 1 and 9 ms for a 100-μm droplet and a small droplet, respectively [231]. According to the theoretical calculations regarding solvent evaporation (Figure 4.2) and droplet coalescence (Figure 4.6), these times are too short to observe a remarkable change in drop diameter. Thus, the fractions of liquid evaporated volume could be estimated to be 0.00001, 0.01 and 14% for droplets with 100, 10 and 1 μm, respectively. The expected change in drop diameter caused by coalescence is also negligible. The volume increase fractions would be 0.08 and 0.2% for droplets with 10 and 1 μm, respectively, being negligible for 100-μm droplets. Therefore, the modification in drop diameters after the impact surface would be mainly caused by the action of this surface.

Throughout this discussion about droplet impacts, secondary processes such as droplet fractionation have been neglected. This point together with the assumptions made as the impact theory is developed is mainly responsible for the discrepancies between the expected and the obtained results.

As mentioned at the beginning of this section, there are other processes responsible for droplet impacts. Gas turbulences can induce losses of droplets, the diameters of which are included in the range 1–10 μm. In this case, the rate of droplet deposition is proportional to the square of the droplet diameter and the acceleration suffered by the droplets. The characterization of this mechanism is actually difficult, because turbulences can act in any spatial direction.

Other processes, such as gravitational losses, also induce a removal of coarse aerosol droplets although they are not the dominant processes [84, 129, 231].

When particular spray chamber designs are used (i.e. double-pass or cyclonic) without impact beads, this study is somewhat complex. In those cases, only a fraction of the aerosol coarse droplets is directed towards the impact surface, whereas the remaining ones have a chance of suffering additional processes such as evaporation or fractionation and to be finally transported towards the plasma. Furthermore, aerosol characteristics are continuously changing because of the solvent evaporation and coalescence.

Droplet impact may be favoured by the use of impact devices such as impact bead or baffles. Droplet impact remains the major source of coarse droplet removal.

4.3 DIFFERENT SPRAY CHAMBERS DESIGNS

4.3.1 Double-pass spray chamber

This design has been the most widely used in ICP techniques for many years and is still commonly employed in a lot of instruments [21, 807].

Double-pass spray chambers have two concentric tubes (Figure 4.1). The central tube promotes the droplet elimination through impacts and reduces the turbulences and fluctuations in the aerosol generation caused by pump pulsing or renebulization. The inner tube separates the two aerosol fluxes, thus isolating the aerosol passing through the volume left between the two tubes from the turbulences associated to the nebulization process.

The double-pass spray chamber is especially efficient in transforming the turbulent aerosol stream generated by the nebulizer into a laminar flow aerosol that will be introduced into the plasma.

In addition, double-pass spray chambers are excellent in eliminating coarse droplets. Figure 4.7 compares the drop size distribution curves of the aerosol produced by a pneumatic concentric nebulizer (Meinhard® A-type), primary aerosol and that obtained at the exit of a $100 \, cm^3$ inner volume double-pass spray chamber. The main change induced by the use of a double-pass chamber with respect to the primary aerosol consists of a shift of the curve towards low aerosol diameter. Figure 4.7 confirms that this design is conceived to remove those droplets coarser than $10 \, \mu m$. With regard to the proportion of small droplets, it is observed that the AL volume contained in droplets with diameters lower than $1.2 \, \mu m$ (not shown in Figure 4.7) is much higher for tertiary (48%) than for primary (8%) aerosols.

The use of a double-pass spray chamber also promotes the appearance of droplets with diameters not present in primary aerosol. Thus, in the particular case considered in

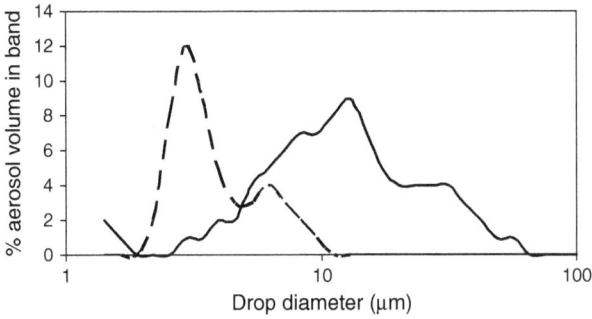

Figure 4.7 Comparison of the drop size distribution of the aerosol generated by a pneumatic concentric nebulizer (primary aerosol, continuous line) and that leaving a double-pass spray chamber (tertiary aerosol, dashed line). The nebulizer is operated at $1\,mL\,min^{-1}$ and $0.7\,L\,min^{-1}$ liquid and gas flow rates, respectively.

Figure 4.7, primary aerosols do not contain droplets with diameters ranging from 1.9 to 2.55 µm. Nonetheless, the maximum for the drop size distribution curve of tertiary aerosol is located near to this particle diameter.

The double-pass spray chamber also dampens the changes in the primary aerosol drop diameter as a parameter (e.g. the gas flow rate) is varied. This role is illustrated in Figure 4.8 in which the Sauter mean diameter ($D_{3,2}$) is plotted versus the nebulizer gas flow rate for primary and tertiary aerosols. Therefore, it can be concluded that an increase in the gas flow rate leads to a change in the net amount of solution leaving the spray chamber rather than in the tertiary aerosol characteristics. The same can be stated concerning the liquid flow rate.

The simulation method known as computational fluid dynamics (CFD) has been applied to the characterization and optimization of the double-pass spray chamber [699, 736]. CFD has proved to be very useful to understand the working principles of this device as well as to

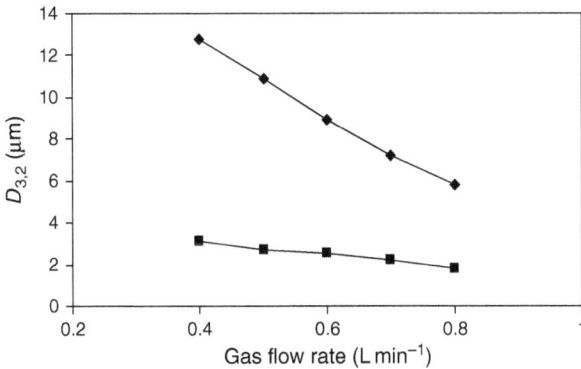

Figure 4.8 Variation of the surface mean diameter (Sauter diameter, $D_{3,2}$) with the nebulizer gas flow rate for the primary aerosol generated by a pneumatic concentric nebulizer (upper line) and the tertiary aerosol at the exit of a $100\,cm^3$ double-pass spray chamber (lower line). Liquid flow rate: $0.6\,mL\,min^{-1}$.

improve the analytical results. This procedure allows achieving conclusions in situations in which no experimental data can be obtained, such as determining the main function of every zone of the spray chamber in terms of droplet selection. The whole CFD method is composed by three main stages: firstly, the chamber is divided into a high number of small control volumes with different density according to the expected velocity or pressure gradient at particular chamber locations. For example, density in control volumes close to the nebulizer tip must be higher than at the exit of the spray chamber. Secondly, the trajectories of the aerosol droplets and the gas flow behaviour are calculated from a suitable physical model. There is a database containing different physical models. Those required for modelling spray chamber are related to compressible, turbulent two-phase flow. However, the program permits introducing new non-considered models. Finally, the results obtained are plotted in a three- or two-dimensional fashion. With the CFD, all the processes taking place inside the spray chamber (evaporation, coalescence, impacts, etc.) can be simultaneously considered. This method has been initially developed for the analysis of fluid flows, heat transfer and related phenomena by numerical techniques, and also allows taking into account the interactions among the different aerosol transport phenomena. A thorough description of the CFD procedure can be found in ref. [699].

Figure 4.9 plots the vector field of the gas velocity inside the spray chamber. Following this simulation, it has been observed that turbulences inside the double-spray chamber are very intense when a pneumatic nebulizer is used [699]. Strong eddies are found near the nebulizer and at the exit of the chamber central tube whose position is continuously changing [736]. In agreement with previous observations [231], a recirculating flow is observed behind the nebulizer tip. The eddies found at locations close to the end of the inner tube demonstrate the existence of small droplet inertial impacts in this chamber zone. Finally, it is also possible to detect the changes in the flow dynamics observed when the gas trajectory is reversed.

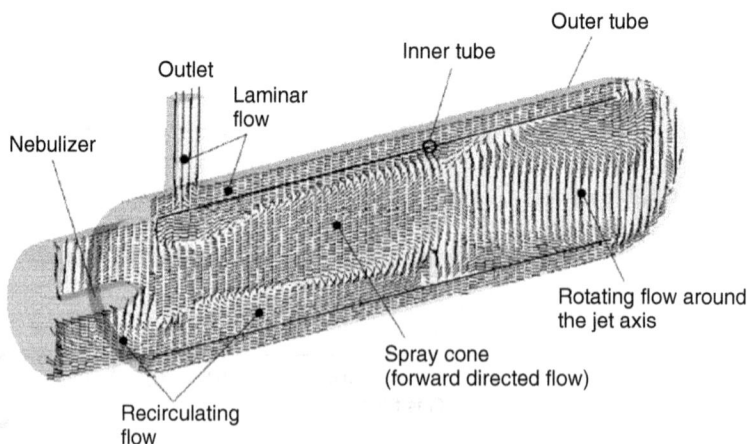

Figure 4.9 Normalized vector field of the velocity in the spray chamber central plane. (Taken from ref. [699] with permission.)

The CFD allows determining the exact volume of solution being deposited on each spray chamber segment. As the chamber was separated into 25 000 different elements, Schaldach et al. [699] divided the device in six different regions to evaluate the point responsible for the collection of the majority of the aerosol. As expected, it was found that the critical part of the double-pass spray chamber was the inner tube. Almost 90% of the solution deposited on the inner walls of this tube. The outer tube was responsible for the removal of less than 6% of the aerosol. In agreement with the problem of solution renebulization, it was found that 2% of the aerosol deposited on the nebulizer.

There is experimental evidence of the role of the inner and outer chamber tubes [623]. Thus, for a pneumatic concentric nebulizer operated at 0.5 mL min^{-1} and 0.7 L min^{-1} liquid and gas flow rates, respectively, $D_{3,2}$ of the primary aerosol was 10.9 μm. At the exit of a double-pass spray chamber inner tube, this parameter took values of 2.0 μm, with the $D_{3,2}$ value of the tertiary aerosol being 1.7 μm. These results demonstrate that the highest change in primary aerosol characteristics occurred inside the inner chamber tube. The corresponding drop size distribution curves are included in Figure 4.10. According to this figure, droplets with diameters higher than 20 μm are completely removed in the central tube. The remaining chamber volume only serves to yield to a decrease in the proportion of aerosol volume contained in coarse droplets with a concomitant increase in the proportion of small droplets.

Figure 4.11 shows a picture of a double-pass spray to which a dye solution has been introduced to evidence the chamber zones of maximum droplets impact. In agreement with the previously discussed simulations, the central chamber tube is the main one responsible for droplets removal. The wall in front of the inner chamber tube is also a source of droplet losses, although as it is observed the colour in this zone is much less intense, thus indicating that lower aerosol mass impacts at this chamber point. Note that, in concordance with the CFD predictions, only 2% of the aerosol mass is lost at this chamber location.

Modelling with the CFD is also useful to determine the most selective spray chamber zone in terms of aerosol drop size. The conclusions drawn on this subject are quite

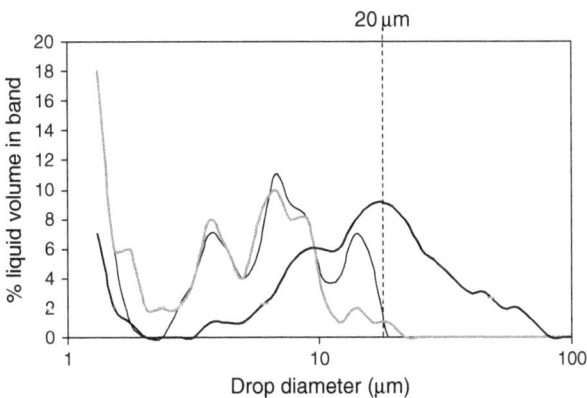

Figure 4.10 Volume drop size distributions in band for the aerosols produced by a pneumatic concentric nebulizer (black coarse line), those at the exit of a typical double-pass chamber inner tube (black fine line) and tertiary aerosols (grey coarse line). Liquid flow rate: 0.5 mL min^{-1}; gas flow rate: 0.7 L min^{-1}.

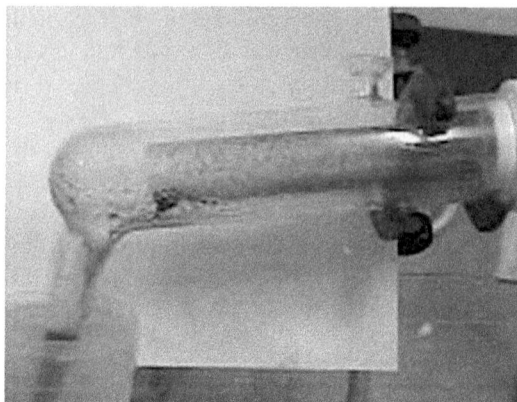

Figure 4.11 Picture of a double-pass glass spray chamber highlighting the zones of droplet impact. A methylene blue solution has been introduced.

interesting [699]. Firstly, it has been observed that the inner tube is not selective in removing aerosol droplets according to their diameters. In other words, coarse as well as fine droplets are lost in this part of the chamber. This has been observed despite the fact that, for concentric nebulizers, coarsest droplets travel through the outermost aerosol cone region (see Chapter 3). The nebulizer tip and the chamber wall located in front of the inner tube are the most selective devices. In the former case, small droplets can follow the trajectory of the recirculating flow, whereas in the latter one, coarse droplets are not able to follow changes in the gas trajectory and are lost because of their inertia.

Double-pass spray chambers suffer from several drawbacks: (i) due to the high dead volume and the long aerosol path inside it, memory effects are severe thus giving rise to long wash-in and wash-out times and (ii) analyte transport efficiencies are very low. About 98–99% of the solution is lost through drain.

The wash-out times for this chamber are long because it has some dead volumes difficult to wash. Droplets can be trapped there for long periods of time, thus increasing the extent of memory effects. Figure 4.12a illustrates the dead volumes of a double-pass spray chamber [104]. To reduce the extent of memory effects, special double-spray chambers have been developed in which these volumes are removed. Figure 4.12b depicts one example of low dead volume double-pass spray chamber.

The spray chamber dimensions have been modified to increase the analyte transport efficiency. Thus, the combination of an iterative evolutionary strategy with the CDF procedure allowed the performance of this chamber to be improved [736]. The simultaneously studied chamber parameters were the inner tube length and diameter, and the chamber total length. With regard to the effect of the inner and outer tube lengths, the calculations indicated that short tubes (with lengths ranging from 60 to 100 mm) might induce the transport of a higher mass of small droplets than long ones. Thus, for a chamber having a 60-mm inner tube length, the relative mass transport efficiency for droplets with diameters below 10 μm was four to five times higher than in the case of chambers with 120- to 160-mm length tubes. As a result, measured limits of detection were lower for

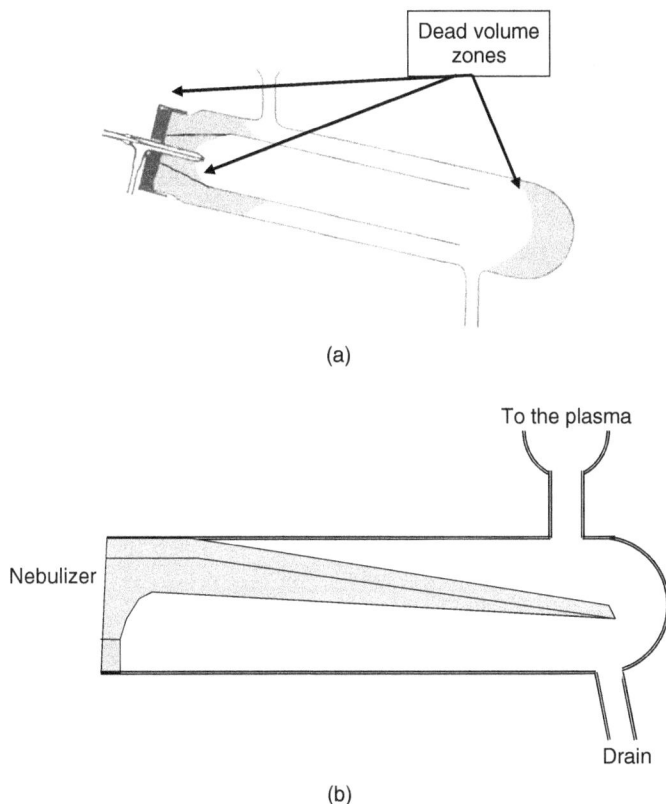

Figure caption text within figure (a): Dead volume zones

(a)

To the plasma

Nebulizer

Drain

(b)

Figure 4.12 Dead volume zones of a double-pass spray chamber, adapted from ref. [104] (a) and reduced memory effects double-pass spray chamber (b).

small than for big chambers. As expected, the calculated wash-out times were longer for long than for short inner tubes.

The effect of the inner tube diameter was also studied and, according to the simulations, a reduction in this dimension from 32 to 28 mm would lead to an improvement in the analytical performance, whereas a further decrease would produce a remarkable change in the transport efficiency for small droplets. The explanation to this trend is not clear because of the existence of two opposite effects. First, decreasing the tube diameter means that there is a higher volume left between both tubes. Second, the drop impacts are more frequent for narrow than for wide tubes. The calculated wash-out time for chambers with similar inner volumes was longer for narrow tubes than for others having higher diameters.

By applying the evolutionary procedure, a short double-pass spray chamber was designed. Figure 4.13 compares the dimensions of the optimized and a conventional spray chamber. Unlike the original chamber, the simulation indicated that there were no large eddies [736].

It is interesting to observe that there is a much higher amount of solution deposited on the nebulizer walls and cell for the short spray chamber than for the conventional one

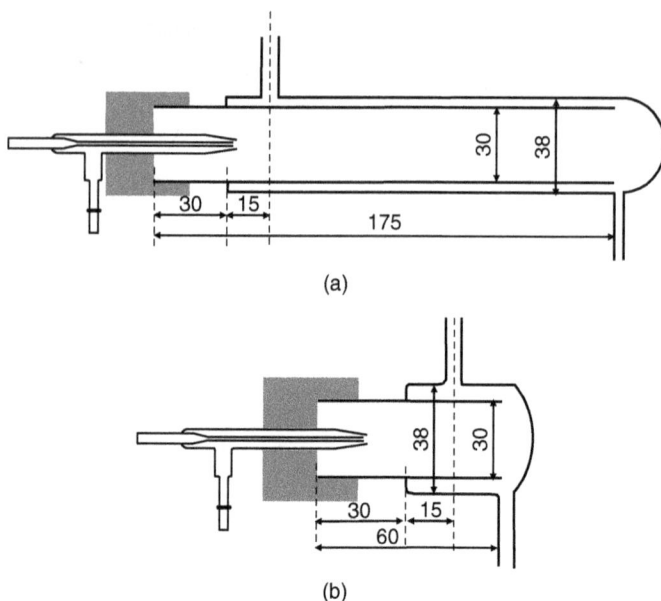

Figure 4.13 Comparison of the geometry and dimensions of a conventional double-pass spray chamber (Taken from ref. [699] with permission) (a) and the optimized spray chamber following an evolutionary CFD simulation procedure (b). (Taken from ref. [736] with permission.) Dimensions are given in millimetres.

Table 4.1

Comparison of relative amounts of aerosol (%) deposited on the inner walls of a double-pass spray chamber at different locations for a conventional double-pass spray chamber and the optimized design according to the procedure followed by Schaldach et al.

System zone	Conventional double-pass chamber [699]	Optimized double-pass chamber [736]
Nebulizer surface	1.9	8.2
Nebulizer cell	0.1	7.4
Chamber inner tube	89.4	64.2
Chamber dome	2.0	17.4
Outer tube	5.6	4.8
Outlet	1.0	4.6

(Table 4.1). This appears to suggest that renebulization and drop deposition near to the nebulizer are to be more significant for the optimized chamber. Also it is worth to mention that the droplet impacts against the front chamber wall (dome) are expected to be more pronounced for the short chamber, what seems obvious provided that the dome is closer to the nebulizer in the second case. Finally, it can be indicated that the role of the inner tube is less important in removing droplets for the optimized chamber.

With the developed chamber, better analytical results were reached [736]. According to the computer simulation, the mass fraction of aerosol leaving the chamber was about two times higher for the optimized chamber as compared to a conventional double-pass design. As a result, the measured ICP–AES limits of detection were from 1.4 to 6 times lower for the optimized device.

One of the merits of the work developed by Schaldach et al. was that the optimization of the chamber design could be done without the need for experimental data. It was only necessary to verify that the optimum chamber design predicted by the simulation provided satisfactory data.

The material of the chamber affects its performance through chemical processes [730]. Thus, it has been observed that for the analysis of Ir, Ag and Pd volatile species through ICP–MS, both the shape and the intensity of the peak depend on the material. Volatile species of these elements are generated through the participation of a reducing agent. Particularly, for Ryton double-pass spray chambers, no noticeable memory effects were found. Meanwhile, in the case of glass chambers, 10-fold signal depression together with an increase in the wash-out times were found. The position of the chamber also appears to have a remarkable effect. Vertically placed double-pass chambers provide better results than almost horizontally placed ones. With regard to the chamber geometry, it emerges that double-pass devices are preferred over single-pass ones in terms of sensitivities [730]. All these effects can be explained by taking into account the atom generation mechanism. Thus, a two-step process is believed to occur in which the analyte ion is first reduced to the metal atom. Then it is quickly released in vapour form as an intermediate unstable in solution, which is adsorbed onto the inner chamber walls. The hydrogen in excess generated from the reducing agent subsequently reacts, thus promoting the intermediate desorption and its further transport towards the plasma [707].

The original design of the double-pass spray chamber could easily be modified through simulation to improve sensitivities and to reduce wash-out time. It is then surprising that the old design is still used on commercially available ICP systems.

4.3.2 Cyclonic type spray chamber

In cyclonic spray chambers, the aerosol is tangentially introduced and first impacts against the front walls to be subsequently transported towards the injector.

Cyclonic spray chambers used in ICP techniques have several configurations such as the so-called Sturman-Masters (baffled), rotary and vortex [553]. The cyclone configuration is widely employed in engineering applications to separate particles suspended into a gas stream, to recover products in synthesis processes and as sampling systems for particles of different sizes [482, 365].

The nebulizer is tangentially adapted to the cylindrical part of this chamber (Figure 4.14). Tertiary aerosol emerges through the exit at the top of the chamber whereas drain is located at the chamber bottom. According to the cyclones normally used in engineering applications, the aerosol is tangentially introduced following a double con-centric spiral trajectory. According to this flow model, the aerosol moves downwards in an external spiral movement, close to the spray chamber walls. Then a second inner spiral carries the droplets towards the top of the chamber. This model is supported by the measurements of drop velocity profiles inside an industrial cyclone in which it is observed that the radial velocity sign of the droplets in the innermost chamber locations is opposite to that found in the outermost zone of the cyclone [445].

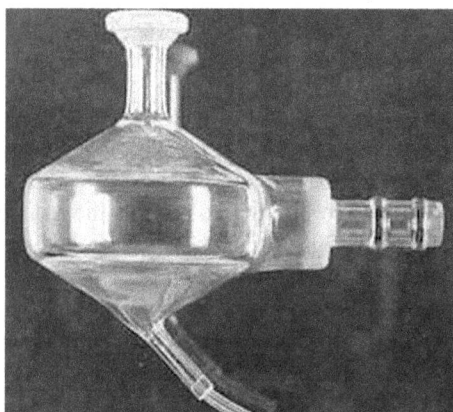

(a)

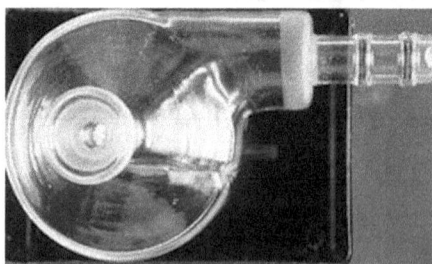

(b)

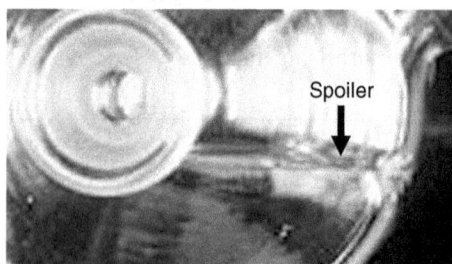

(c)

Figure 4.14 Pictures of a conventional cyclonic spray chamber. (a) Side view; (b) top view; (c) detail of the flow spoiler of the spray chamber.

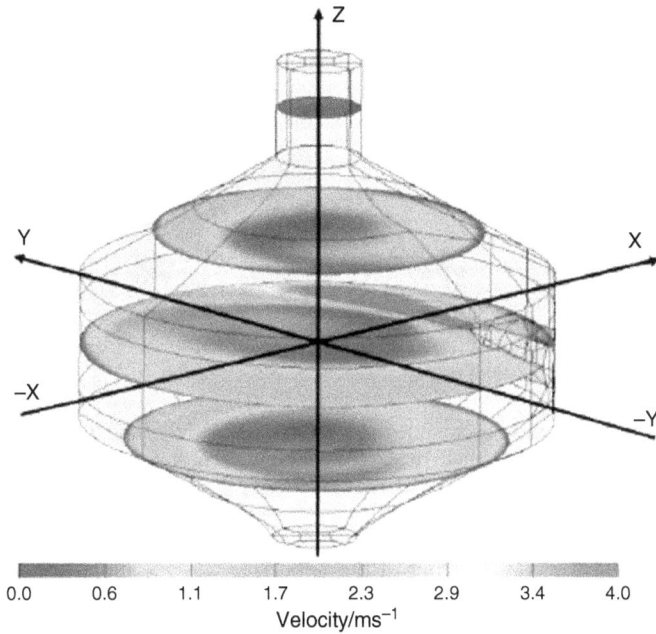

Figure 4.15 Argon velocity distribution inside a cyclonic spray chamber for four different horizontal planes. (Taken from ref. [685] with permission.)

The typical swirling flow dynamics inside a cyclonic spray chamber, similar to those used in industrial applications, has also been described following the CFD method [685]. Figure 4.15 shows the argon velocity profiles obtained from the numerical solution of the argon flow pattern for four different planes inside a cyclonic spray chamber. Because of the symmetrical character of the chamber, the profiles in the first and third plane from the bottom are equal. It is also noteworthy to indicate that for all the planes, except that at the exit of the chamber, an eddy is observed, which is not centered with respect to the spray chamber. Finally, as can be seen from Figure 4.15, the velocity profile at the exit of the spray chamber is homogeneous. Therefore, the turbulences associated to the nebulization process are virtually eliminated inside this spray chamber. The velocity profiles in the second plane from the bottom are asymmetric, mainly if considering them with respect to the x-axis. As mentioned before, the zone of the minimum velocity does not correspond to the centre of the cyclonic spray chamber, which is likely because of the fact that the aerosol is tangentially introduced into the chamber. The two eddies found at the top and bottom of the spray chamber preclude the drop size selection inside cyclonic spray chambers. As a result of the appearance of eddies, the gas velocity along the z-axis shows several fluctuations. Furthermore, droplets are accelerated in the x/z and y/z planes towards the top and the bottom of the spray chamber.

As for double-pass spray chambers, Schaldach et al. [685] calculated the mass fraction of aerosol deposited on the different parts of a cyclonic spray chamber.

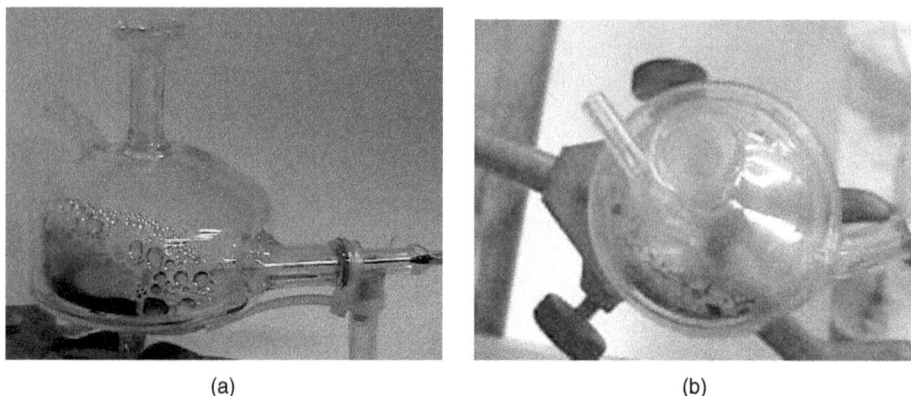

(a)   (b)

Figure 4.16 Picture of a glass cyclonic spray chamber highlighting the zones of droplet impact. (a) Lateral view; (b) top view. A methylene blue solution has been introduced.

They observed that the wall in front of the nebulizer was the main one responsible for coarse as well as fine droplet losses (a 36% of the mass of the aerosol was deposited in this part of the chamber). This is in agreement with the visual observations made with a methylene blue solution (see Figure 4.16). In this case, it can be seen that most of the solution impacts against the curved surface between the top and the bottom of the spray chamber. The calculated fraction mass of aerosol trapped at this location is about 54% [685]. Finally, the amount of aerosol lost as droplets impact at the bottom of the spray chamber is less significant (29.5%), whereas at the top the aerosol mass fraction lost is 14.9%.

Cyclonic spray chambers employed in ICP techniques do not act as a engineering cyclones in which coarse particles are deposited on the inner walls of the cylindrical portion of the chamber (ring) [231]. This is likely because of the lower gas velocity inside the ICP chambers. Thus, it has been observed that the wall in front of the nebulizer is not selective in terms of separation of coarse droplets. Conversely, the rest of the chamber ring is especially effective in removing fine droplets. The most selective area of the chamber appears to be its top, because mainly coarse droplets are removed in there.

Globally speaking, cyclonic spray chambers are excellent to remove droplets with diameters higher than 10 μm, although they are not suitable devices to transport small droplets.

These devices act primarily as an impactor and then the remaining aerosol is subject to the action of the eddies appeared inside the chamber [685].

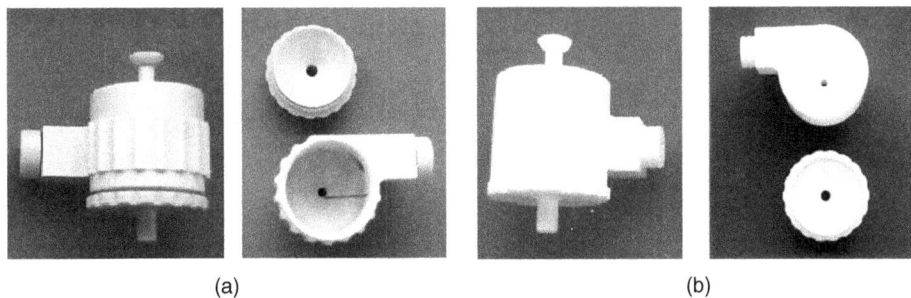

(a)   (b)

Figure 4.17 Cyclonic spray chambers made of polypropylene (a) and PTFE (b).

Cyclonic spray chambers of several materials are commercially available. Figure 4.17 shows the pictures of two of these spray chambers made of polypropylene (Figure 4.17a) and PTFE (Figure 4.17b). The material of the spray chamber has an effect on the wash-out times for some elements. For example, experiments in ICP–AES have demonstrated that wash-out times for a 10 μg/mL palladium solution have been 26 s for a glass and polypropylene chamber and 38 s for a PTFE cyclonic device [703]. Apparently, palladium is retained on the surface of PTFE and hence it is more difficult to wash. Consequently, it is not recommended to use chambers made of this material for palladium analyses. Moreover, some plastic materials (PEEK, PFA, etc.) also reduce the background levels as well as the spikes observed when rinsing the system (i.e. memory effects) [766].

In the case of cyclonic spray chambers, the nebulizer position must be optimized to achieve the optimum signal. If the nebulizer is too close to the walls in front of it, the inertial losses will be too intense. As it has been previously mentioned, this part of the chamber is not selective in removing coarse droplets. If, on the contrary, the nebulizer is too far from the walls, the flow regime established inside the chamber can be disturbed by the action of the nebulizer gas. The different nebulizer positions can explain why for two chambers with similar critical dimensions, emission signals can differ by a factor of up to 2 [570].

In a further modification of the cyclonic design, a flow spoiler has been introduced (Figure 4.14c). As it can be observed, the spoiler is located in the third quadrant of the aerosol introduction zone. This modification has been presented together with a reduction in the inner volume and a vertical operation of the spray chamber [536]. The spoiler deviates the flow inside the chamber, bringing the aerosol stream towards the chamber central axis. This change in the flow dynamics becomes more pronounced as the spoiler approaches to the nebulizer tip. Opposite to this for a chamber with no spoiler, the computed aerosol mass distribution inside the main recirculatory stream is quite symmetrical. In the spoiler, a given fraction of aerosol can be lost through impacts. The position of the spoiler has also an effect resulting in shorter wash-out times if the spoiler is placed in the second chamber quadrant [724]. However, at conventional liquid flow rates, the lowest limits of detection are obtained for a cyclonic chamber with no spoiler.

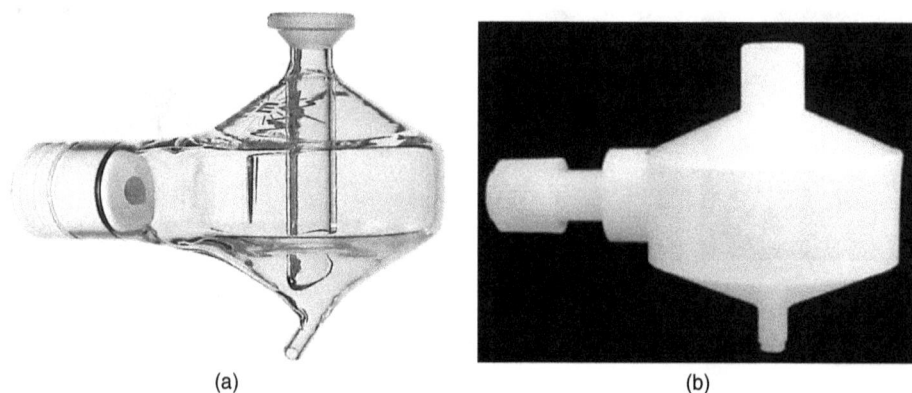

(a)   (b)

Figure 4.18 Picture of the glass (a) (Taken from http://www.glassexpansion.com.) and PFA (b) baffled cyclonic spray chambers (taken from http://www.elementalscientific.com)

The particles trajectory inside a cyclonic spray chamber has also been modelled and it has been verified that there exists a huge number of different trajectories. In refs [724, 745], several examples are given that illustrate the enormous difference in droplets residence times inside the chamber: droplets having a 2-µm diameter are lost through impacts only 70 ms after their generation, fine droplets that require up to 50 s to finally emerge through the chamber top. In the opposite situation, there are some droplets, having diameters of 11 µm, which follow a very simple path inside the chamber and exit it only 2 s after their production.

A way of lowering the mean drop diameter of the aerosols leaving a cyclonic spray chamber is the introduction of a vertical tube at the chamber axis (Figure 4.18) [478, 804]. This is achieved at the obvious cost of a 20% drop in the mass of solution transported towards the plasma with respect to a conventional cyclonic spray chamber. When comparing with a double-pass spray chamber (http://www.meihard.com), the baffled spray chamber provides an enhancement in the ICP signal by a factor in the range 1.2–1.5 and also shortens the wash-out and equilibration times. However, the presence of the tube generates a solution built-up at its bottom, which can affect the signal stability.

A recent variation of cyclonic spray chambers is shown in Figure 4.19. In this case, two inlets are adapted at a conventional glass cyclonic spray chamber with 50 cm³ inner volume to fit two different nebulizers [819]. This solution is especially interesting to compensate for interferences or to introduce analytes as hydrides into the plasma. With this kind of devices, tandem calibration techniques can be applied. These techniques have been applied by means of the use of two different sample introduction systems [527, 715]. In these instances, the aerosol transport efficiency for the two systems must be the same. Otherwise, the so-called relative transport efficiency should be calculated from the signal data. In the system based on the use of the modified cyclonic spray chamber presented in Figure 4.19, the two nebulizers used (i.e. HEN)

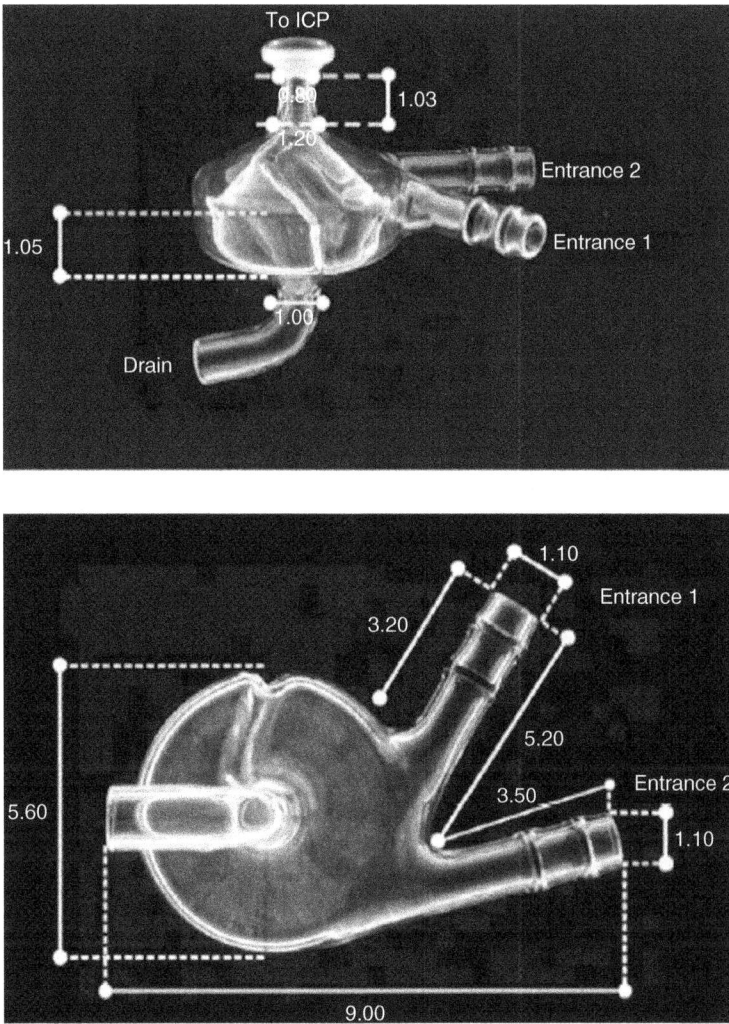

Figure 4.19 Pictures and main dimensions of the cyclonic spray chamber used for dual microebulization. (Taken from ref. [819] with permission.)

produce primary aerosols and they are merged inside the chamber. Therefore, one nebulizer is used for the introduction of the sample and the other one serves for the standard introduction. The dual nebulizer setup is indicated for the easy application of the standard additions method in a quick fashion [819].

The use of cyclonic spray chambers is increasing because of a low dead volume leading to fast wash-out and equilibration times while keeping good sensitivities.

4.3.3 Single-pass spray chambers

The single-pass spray chambers are generally very simple devices in which the aerosol is introduced into a tube or conical conduction and describes a direct trajectory towards the torch injector [234, 253].

These devices are also known as cylindrical type, or direct spray chamber. As a result, coarser droplets than for double-pass or cyclonic spray chambers can be transported away from the chamber. The single-pass spray chambers give rise to higher analyte transport efficiencies than double-pass or cyclonic chambers. However, the signal can be noisier because of the coarser tertiary aerosol.

A scheme of this design is shown in Figure 4.20. In this case, the aerosol selection is produced through impacts against the chamber walls. This design is used with systems that do not require a strong filtering action of the aerosol. It must be borne in mind that with this chamber an important mass of aerosol reaches the plasma, which may cause plasma deterioration.

It has been reported that the most critical variables for the performance of this design are the chamber length and diameter [255, 284]. The chamber inner volume is a function of these two variables and hence they preclude the significance of droplet losses. For a given chamber inner diameter, the signal peaks for a chamber length of 5 cm (Table 4.2). These results are explained by taking into account that for short chambers the inertial impact losses against the wall in front of the nebulizer are intensified. Nonetheless, if the chamber wall is too far from the nebulizer, a noticeable fraction of small droplets can impact against the lateral walls because of other processes such as turbulences. Table 4.2 also reveals that for a given length, the lower the inner diameter, the weaker the emission intensity, because the aerosol droplets impact more frequently against the inner chamber walls.

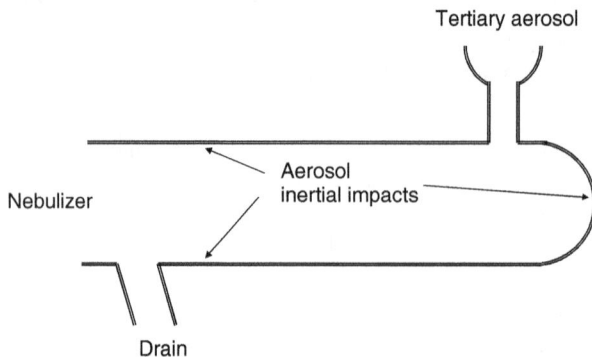

Figure 4.20 Layout of a single-pass spray chamber.

Table 4.2

Relative sensitivites for single-pass spray chambers with different critical
dimensions (taken from [255, 284])

Inner diameter (cm)	Length (cm)			
	3	5	10	15
1.3	88.6	90.6	86.7	77.1
1.8	90.3	96.4	88.3	81.1
2.2	94.1	100.6	89.5	87.6
3.1	—	111.2	100.3	100.0

The inner volume of the spray chamber affects the wash-out times. This parameter
increases as both the chamber length and inner diameter rise (Table 4.3).

An alternative to the single-pass chambers are the so-called direct injection spray
chambers. This design consists of a curved tube connecting the nebulizer to the injector
tube in the plasma torch. Figure 4.21 shows a drawing of this type of spray chamber.
In this case, an impact surface is placed in front of the nebulizer to remove the coarsest
aerosol droplets [67]. The function of an impact bead is illustrated by the measurements of the aerosols before and after this surface. For a pneumatic concentric
nebulizer operated under typical flow rates, the aerosols after a spherical impact
surface have $D_{3,2}$ values close to 3 µm, whereas under the same nebulization
conditions, the reported value of surface mean diameter for primary aerosols is about
11 µm [623].

Single-pass spray chambers are normally used with efficient nebulizers such as the
MAK and the USN (not considered in the present book). A single-pass spray chamber

Table 4.3

Wash-out times (in min) for single-pass spray chambers with different dimensions
(taken from [255, 284])

Inner diameter (cm)	Length (cm)			
	3	5	10	15
1.3	0.68	0.70	1.10	1.26
1.8	0.86	0.98	1.52	1.99
2.2	1.05	1.16	1.77	2.62
3.1	—	1.58	1.94	2.71

Wash-out time is defined as the time required for the signal to drop to 20% of its steady value.

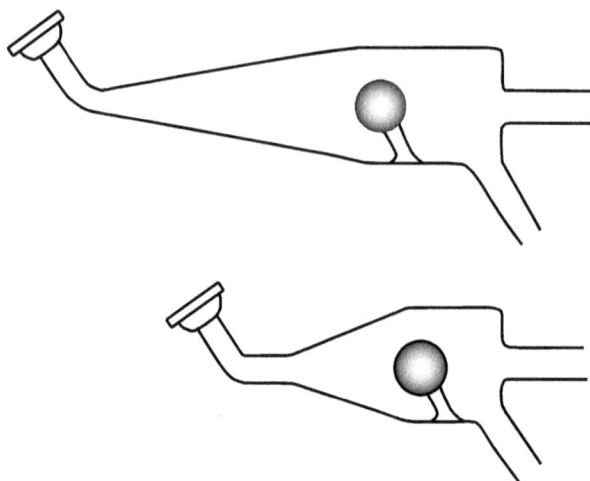

Figure 4.21 Single pass–direct injection spray chamber equipped with an impact surface.

with impact surfaces has been used to characterize peak broadening in flow injection analysis applications [472]. It has been found that the droplets diffusion is not the mechanism responsible for peak dispersion, instead aerosol recirculation phenomena appear to contribute to the increase in the peak width or, in other words, the memory effects extent.

Although strictly speaking the donut-shaped spray chamber cannot be considered as a single-pass type one, the aerosol trajectory inside it is similar to that for one of these chambers equipped with an impact surface [549]. Figure 4.22 shows a scheme of this chamber. The volume of this accessory is about $100 \, cm^3$, and the nebulizer is radially adapted to the donut chamber. The primary aerosol impacts against the central hollow cylinder. It is then separated into two different streams that merge before leaving the chamber.

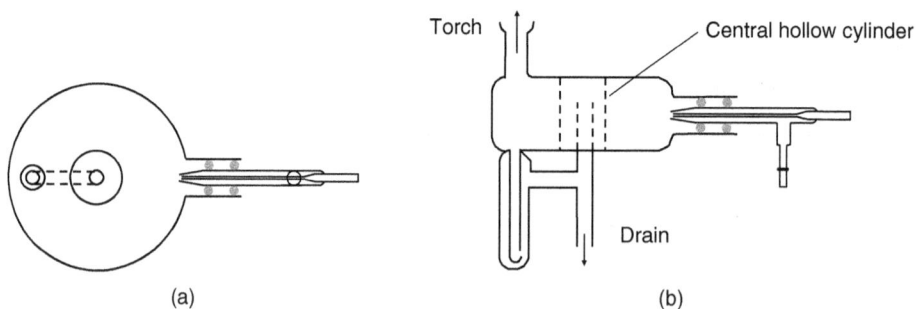

(a)   (b)

Figure 4.22 Layout of the donut-shaped spray chamber. (a) Top view; (b) side view. (Taken from ref. [549] with permission.)

The donut-shaped spray chamber provides higher sensitivities, lower limits of detection, better precision and lower background levels than a double-pass spray chamber.

4.4 COMPARISON OF CONVENTIONAL SPRAY CHAMBERS

Generally speaking and taking into account the drop size selection mechanisms mentioned before, the aerosol leaving a cyclonic spray chamber is expected to be coarser than that at the exit of a double-pass spray chamber. This is illustrated in Figure 4.23, where it is demonstrated that for a set of nebulization conditions, tertiary aerosols have higher D_{50} values for cyclonic devices. The double-pass spray chamber is an excellent design in removing coarse droplets. The higher cutoff diameter found for tertiary aerosols in the case of cyclonic spray chambers can be explained by taking into account that when the aerosol is generated, a portion of it directly impacts against the walls of the chamber. However, there is a fraction of the aerosol cone that is introduced in a more centered chamber location. The coarse droplets located in these positions have a chance to escape the spray chamber and hence to lead to increases in the tertiary aerosol mean drop size.

At their optimum nebulizer position, many cyclonic spray chambers provide higher analyte transport rates (W_{tot}) values than the double-pass design. Reported transport efficiencies close to 8% can be found in the literature [464]. Thus, for a given set of nebulization conditions, W_{tot} took values about two times higher when a glass cyclonic spray chamber was used as compared to a double-pass spray chamber [570]. This fact leads to sensitivities more than two times higher for the former case. These results have

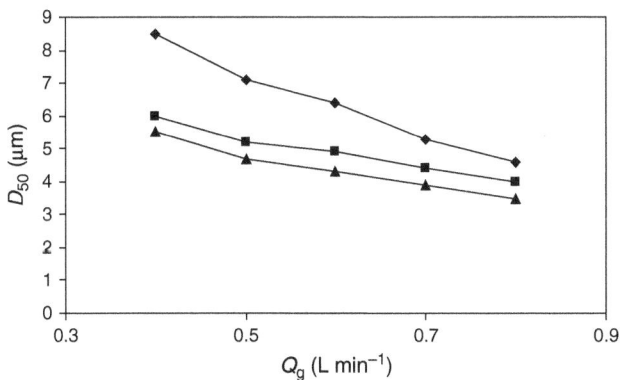

Figure 4.23 Variation of the median of the volume drop size distribution curve (D_{50}) with the nebulizer gas flow rate for the aerosols leaving a double-pass spray chamber (▲), a glass (◆) and a polypropylene (■) cyclonic spray chamber. Liquid flow rate: $0.6\,mL\,min^{-1}$.

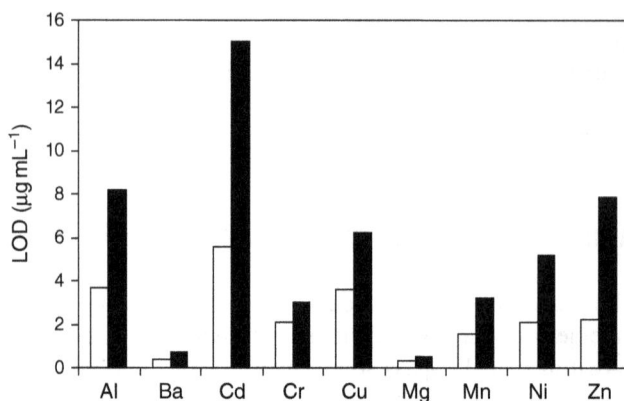

Figure 4.24 Comparison of the limits of detection found for a double-pass spray chamber (black bars) and a glass cyclonic spray chamber (white bars). (Adapted from ref. [570].)

also been corroborated in CE−ICP−MS coupling [602]. However, cyclonic spray chambers have two negative effects on the signal stability: on the one hand, the mass of solvent reaching the plasma is higher than for a double-pass spray chamber, and on the other hand, tertiary aerosols are coarser. Despite this, lower limits of detection are obtained with cyclonic spray chamber. These latter results are represented in Figure 4.24. It can be observed that, under a given set of nebulization conditions, the cyclonic design leads to ICP–AES limits of detection from 1.4 (Cr) to 3.6 (Zn) times lower than the double-pass device.

Because of the increase in the mass of solvent reaching the plasma, the use of a cyclonic spray chamber in ICP–MS could induce an increase in the oxide production. A final advantage of cyclonic over double-pass designs is the shortening in the wash-out times. For example, under the same conditions, reported wash-out times are 20 and 30 s for a glass cyclonic and double-pass spray chambers, respectively. This derives from about two times lower inner volume and different aerosol trajectory in the case of the former device [570].

In Table 4.4, an estimative comparison is established between the three main chamber designs. Taking into account the data appeared in the literature, cyclonic and single-pass spray chamber are preferred if sensitivity and memory effects are the key points.

The suitability of the spray chamber design to the work with slurries has been recently addressed for the analysis of antibiotics [842]. In their work, it has been observed that the combination of a V-groove nebulizer with a cyclonic spray chamber compares favourably against a cross-flow nebulizer adapted to a double-pass spray chamber. Two main reasons can be argued to try to explain this behaviour: (i) the V-groove nebulizer can work with more concentrated slurry than the cross-flow nebulizer and (ii) the cyclonic chamber affords higher analyte transport efficiencies than the double-pass design. As a result, the ICP–AES sensitivity taken as the slope of the calibration lines was up to 3.5 times higher for the former design. Other reports demonstrate the suitability of the cyclonic devices for the analysis of coffee and cocoa slurries [799].

Table 4.4

Comparison of the spray chambers most commonly used in ICP–AES and ICP–MS

Chamber	Inner volume (cm^3)	Analyte transport efficiency (%)	ICP relative wash-out times
Double pass	100–120	0.5–2	1
Glass cyclonic	~50	1–4	0.8
Plastic cyclonic	~60	0.5–4	0.8–2
Baffled cyclonic	~50	0.6–3	
Single pass	4–100	<4	0.3

Cyclonic and double-pass spray chambers have also been compared as interfaces for HPLC–ICP–AES with either concentric [533] or cross-flow nebulizers [803]. Cyclonic chambers appear to be more suitable than double-pass ones, because they provide higher sensitivities. Interestingly, it has been observed that the peak dispersion is not significantly modified irrespectively of the spray chamber used, in spite of the fact that the inner volume of double-pass spray chamber is much higher than that for the cyclonic type one. These results support the fact that the peak broadening and distortion occur mainly in liquid rather in aerosol phase [80]. In a wider comparison, it has been found that an MCN coupled to a cyclonic spray chamber is more appropriate than a DIN and an ultrasonic nebulizer to carry out studies of selenium speciation [798].

4.5 LOW INNER VOLUME SPRAY CHAMBERS

Simplified, low inner volume spray chamber can be designed to be used in conjunction with micronebulizers. Solvent evaporation becomes a major phenomenon, and drain may be reduced, or even suppressed.

As mentioned in Chapter 3, the analysis of microsamples has been an appealing subject, mainly over the past 15 years. Although recycling systems in which the drains were renebulized were previously suggested to reduce the sample consumption volume [162, 356], a solution widely accepted to cope with micronebulizers has been to reduce the inner volume of the spray chamber to shorten wash-out times and to increase the sample throughput when very low liquid sample flow rates are used. This fact is likely due to the problems found with recycling systems such as the drain enrichment in both analyte and matrix with the concomitant signal drift [180]. To solve this problem, several solutions have been proposed, among them is the argon stream saturation with water [178] or the use of cooled spray chambers so as to prevent the solvent evaporation from the aerosol [280].

4.5.1 Aerosol transport and signal production processes at low liquid flow rates

At low liquid flow rates (i.e. below $100 \, \mu L \, min^{-1}$), the extent of production of the aerosol transport phenomena is modified. Both the solvent evaporation [763] and the droplet coagulation [828] have been modelled for these conditions. The main conclusions are that the solvent evaporation inside the chamber increases, whereas the drop coalescence is produced in a less intense fashion as compared with the situation observed at conventional liquid flow rates (i.e. $1 \, mL \, min^{-1}$).

Figure 4.25 shows the complete absolute drop size distribution curves at three different times for water at a $30 \, \mu L \, min^{-1}$ liquid flow rate. At 25°C, the theoretical data indicated that the maximum amount of water that could evaporate to saturate the argon stream was $20-30 \, mg \, L^{-1}$ of argon [103, 242]. It can be verified that, at $30 \, \mu L \, min^{-1}$, 6 s after the aerosol generation, the argon stream becomes saturated with water. This period of time is much longer than that calculated for conventional liquid flow rates (ca. in the order of 30 ms). The simulated drop size distributions are, thus, considered at three different times: 0 s, which corresponds to the primary aerosols; 1 s, which is roughly the aerosol residence time in a low inner volume spray chamber (i.e. ranging from 10 to 20 cm^3); and 6 s, which corresponds to the time required to saturate the argon stream.

The data presented in Figure 4.25 are quite different to those previously discussed for conventional liquid flow rates in Figure 4.2. Thus, the theoretical calculations indicate that, at low liquid flow rates, 1 s after the aerosol generation, all droplets with diameters lower than about 2 μm are completely evaporated. Droplets with diameters higher than about 9 μm, in turn, do not modify their diameters in an appreciable way. By considering

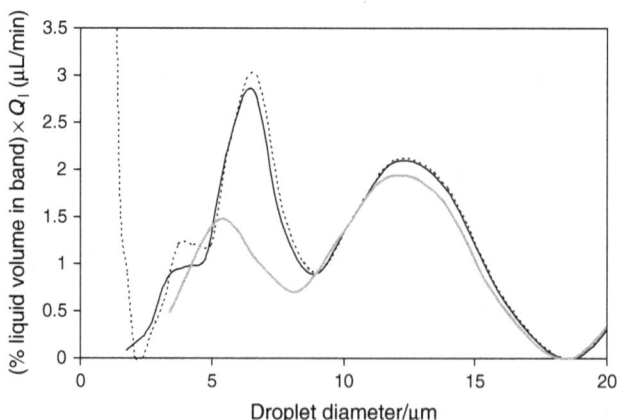

Figure 4.25 Simulation of the evolution of the drop size distribution with time caused by solvent evaporation for distilled water at $Q_1 = 30 \, \mu L \, min^{-1}$ and $Q_g = 0.7 \, L \, min^{-1}$. Dotted line: primary aerosol distribution; continuous line: simulated aerosol distribution 1 s after the aerosol generation; grey line: aerosol distribution 6 s after the aerosol production. Nebulizer: PFAMN. Calculations have been performed at 25°C.

the time required to saturate the argon stream (grey line in Figure 4.25), it could be concluded that the change in the drop size distribution produced by solvent evaporation becomes remarkable. Thus, for example, the AL volume contained in droplets with diameters of about 6.7 µm is approximately two times lower 6 s after the aerosol generation than 1 s after the nebulization is produced. Furthermore, according to the calculations, droplets with diameters lower than 3.4 µm are completely evaporated. In contrast, at conventional liquid flow rates, the change in drop size distribution caused by solvent evaporation is much less significant (Figure 4.2). As a conclusion, it can be indicated that at low liquid flow rates, it is very important to use nebulizers able to generate droplets with diameters lower than 2 µm to achieve high solution transport efficiencies. It has been indicated that, for a given liquid and gas flow rates, the percentages of AL volumes contained in droplets having diameters below 2 µm are 10.4, 21.6 and 31.6% for the MMN, PFAN and HEN, respectively. Therefore, the most appropriate micronebulizer design would be the HEN.

The immediate consequence of solvent evaporation is the disappearance of fine droplets, and globally speaking, aerosols become coarser [231]. The median of the simulated volume drop size distribution (D_{50}) can be calculated from the data generated through the procedure mentioned above. For a 10 s residence time, the calculated D_{50} value is about 6 µm. With regard to the experimental data, measured D_{50} values for the aerosols leaving a double-pass spray chamber are 3.5 µm. Therefore, it can be concluded that evaporation is not the most important process leading to a change in the characteristics of the aerosols. It is interesting to notice that a significant fraction of the solvent could evaporate from the solution deposited on the inner walls of the chamber [532]. This solution mass contributes to a reduction in the amount of solvent evaporated from the aerosol. Furthermore, the simulation performed in this study had some inherent approximations.

From all the above considerations, it can be concluded that to promote complete aerosol solvent evaporation inside the spray chamber, thus favouring the transport of solution towards the plasma, two factors should be born in mind: firstly, the mass rate of water nebulized must be lower than 20–30 mg for a 1 L min^{-1} argon stream, and secondly, the aerosols produced by the nebulizer must be sufficiently fine. There are several evidences indicating that this goal is achieved when very low liquid sample flow rates are used [728].

Thus, for a 2000 mg L^{-1} yttrium solution introduced into the plasma at 20 µL min^{-1}, the initial radiation zone (IRZ) at the plasma base was located at positions far upstream with respect to the observation made at 100 µL min^{-1}. Furthermore, high-speed videos demonstrated that, at 100 µL min^{-1}, there were occasional distortions of the IRZ tip as large red droplet clouds were formed. A second piece of evidence derives from the fact that there is no solution drain needed when working under these conditions.

In general terms, at low liquid flow rates, the transport of solution is favoured with respect to that obtained at conventional values of this parameter, also because coalescence is less frequently produced than if the liquid flow rate rises [809, 544]. This is a clear result of the different drop number density. This parameter is about 10^{13} and 3×10^{14} particles m^{-3} for a 0.05 and 1.5 mL min^{-1} liquid flow rate, respectively [828]. To illustrate the importance of coalescence, it may be noted that a conventional pneumatic concentric nebulizer operated at 1 mL min^{-1} generates five times as much primary aerosol volume contained in droplets with diameters lower than 5 μm than a micronebulizer operated at 50 μL min^{-1}. However, the analyte transport rate is just two times higher for the former device than for the latter one. Because of the much higher aerosol droplet density at 1 mL min^{-1}, fine droplets grow through coalescence and they are removed from the aerosol stream inside the spray chamber.

All these calculations help to understand why decreasing the liquid flow rate by a given factor does not result in a corresponding proportional decrease in the emission signal. For example, one of the earlier studies carried out with the HEN [388] claimed that the HEN operated at a 50 μL min^{-1} liquid flow rate afforded limits of detection similar to those measured for a conventional pneumatic concentric nebulizer operated at a 1 mL min^{-1} sample flow rate. Further measurements of the efficiency of the analyte transport towards the plasma (ε_n) indicated that for the HEN, this parameter took a value close to 60% at 10 μL min^{-1}, whereas it dropped down to 8% at 120 μL min^{-1} [465, 576]. At low liquid flow rates (ca. 2 μL min^{-1}), aerosol transport efficiencies close to 90% were reported. Discrepancies can be found in the literature. The solvent transport efficiency is also higher at low liquid flow rates. For example, for an MCN coupled to a cyclonic spray chamber, the solvent transport efficiency ε_s at a 100 μL min^{-1} delivery liquid flow rate and a 1 L min^{-1} gas flow rate generated values close to 85% [526].

Working at liquid flow rates below 100 μL min^{-1} has several consequences in terms of signal production. As the absolute solvent plasma load is generally low, the thermal conditions of the plasma are not deteriorated when plain water solutions are analysed [388]. In fact, the magnesium net ionic emission intensity to net atomic emission intensity ratio (MgII/MgI) [268] is lower at high (ca. 1 mL min^{-1}) than at low (ca. 50 μL min^{-1}) liquid flow rates. This result is in full agreement with the fact that the argon emission intensity is also higher at low rather than at high liquid flow rates [388]. However, if the solvent mass reaching the plasma is too low, its thermal conductivity decreases, thus leading to a poor energy transfer to the sample and therefore to a deterioration of the plasma thermal characteristics [470].

The trends in signal can be explained according to the considerations of solvent evaporation, droplet coagulation and thermal effects of the plasma made before. Thus, it has been found that, for an MCN, the ICP–MS signal increased with liquid flow rate up to 150 μL min^{-1} and then decreased significantly beyond this liquid flow rate [628]. These results can be partially explained on the basis that, because the drop number density is higher at high liquid flows, drop coalescence is more noticeable. As a result, coarse droplets are produced and the solvent evaporation is dampened. Therefore, the signal increases linearly with the liquid flow rate only at very low liquid flow rates.

> There is no sensitivity penalty to work at very low solution delivery rates, that is below $100 \, \mu L \, min^{-1}$.

4.5.2 Low inner volume spray chamber designs

The design of the spray chamber has a crucial effect on the analytical performance of ICP spectrometers mainly at low liquid flow rates [724]. To carry out the analysis of liquid microsamples through ICP techniques, the spray chamber design has evolved towards small and simple devices [156, 164, 203]. The efforts have been mainly devoted to the shortening of the wash-out times. There are two main designs of low inner volume spray chambers: the cyclonic and single pass. Figure 4.26 shows the pictures of two of these chambers. In the case of cyclonic chambers, also known as Cinnabar (Figure 4.26a), the inner volume is on the order of 20 mL. Single-pass spray chambers, in turn, can be smaller with volumes down to 5 mL (Figure 4.26b).

One of the precursors of the Cinnabar spray chamber is the rotary design [350, 391]. The Cinnabar can be made of either glass or polyethylene. This device is especially efficient in the removal of small droplets. Thus, tertiary aerosols have similar

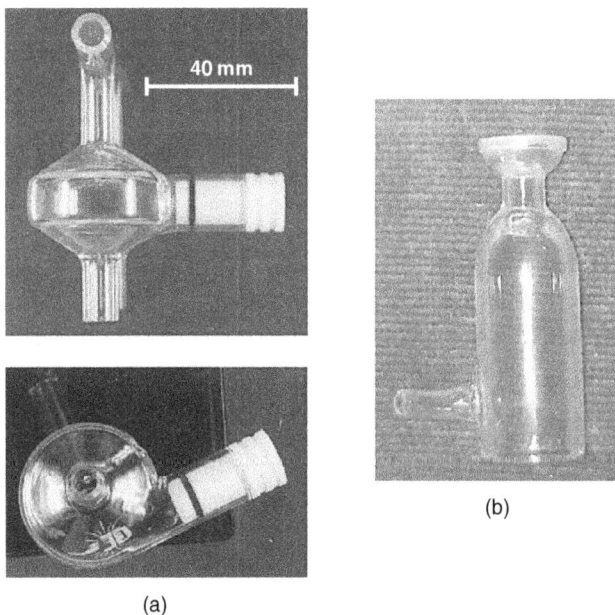

Figure 4.26 Pictures of the most commonly used low inner volume spray chambers. (a) cyclonic, cinnabar design; (b) single-pass design.

characteristics as those found at the exit of a double-pass spray chamber [613]. Furthermore, the mass of analyte delivered to the plasma and hence the ICP–AES sensitivities and limits of detection are also similar for both devices. As expected, the wash-out times at low liquid flow rates are about two times shorter for the Cinnabar than for the double-pass spray chamber. This is especially important for elements difficult to rinse such as iodine [487, 603, 716]. As a result of its features, the Cinnabar is a good CE–ICP–MS interface [631, 709].

Single-pass spray chambers are becoming popular designs for the analysis of liquid microsamples through ICP techniques. There is an interface for CE–ICP–MS coupling, which is based on the use of a single-pass spray chamber [567, 577]. The nebulizer used in this case is an MCN with a narrower capillary to avoid any laminar flow inside the CE capillary. The inner volume of the single-pass spray chamber used in this case is about 4–8 cm^3 [676, 710, 787]. The aerosol is generated at a very low liquid flow rate (i.e. 2–12 µL min^{-1}), and hence, the solvent totally evaporates inside the spray chamber. As a result, this interface is considered as a total sample consumption system and there is no need for an exit drain. In a recent modification, the inner volume of the spray chamber has been decreased to 4 cm^3 and the nebulizer capillary inner diameter has been reduced [777]. The interface based on the use of a single-pass spray chamber is perfectly indicated to the work under gradient conditions, because it is not sensitive to changes in matrix composition. This interface has been efficiently applied to CZE [713] and to HPLC–ICP coupling [710]. An octopole reaction cell has been used to permit the determination of sulphur [706] and phosphorous [723] in biological samples through a CE–ICP–MS system equipped with the above-mentioned interface.

In a single-pass spray chamber, the aerosol follows a direct path until it reaches the chamber exit. If the primary aerosol is fine enough and the liquid-to-mass volume ratio is low enough, the totality of the solvent contained in the aerosol will evaporate before the droplets reach the walls of the chamber. The chamber acts, therefore, as an interface to promote solvent evaporation before the aerosol enters the plasma, instead of being used as a drop size selection system. Single-pass spray chambers have been used in ICP–AES at low liquid flow rates [680]. Compared to double-pass and conventional cyclonic spray chambers, single-pass designs with inner volumes included within the 8–20 cm^3 range provide about two times higher emission intensities. In contrast, single-pass spray chambers can degrade the signal stability, as coarse droplets can be introduced into the plasma. The simplicity of the aerosol trajectory inside the chamber also leads to a shortening of the wash-out times [339]. If sample throughput must be increased, the inner volume of the spray chamber can be further lowered. Nevertheless, the use of smaller spray chamber with an inner volume of ca. 4 cm^3 can lead to a severe drop in the analytical signal [255, 288]. This fact is mainly due to the aerosol losses produced by impacts against the front chamber inner walls. Wash-out times can be further shortened with this kind of chambers by flushing them with a washing solution [632] or a gas stream [393].

The simplicity of the spray chamber and the torch configuration has made it possible to integrate both components into the so-called torch-integrated sample introduction system (TISIS) [686]. As can be seen in Figure 4.27a, two Teflon pieces are used: one to adapt the nebulizer to the base of the torch and another employed to fit the injector

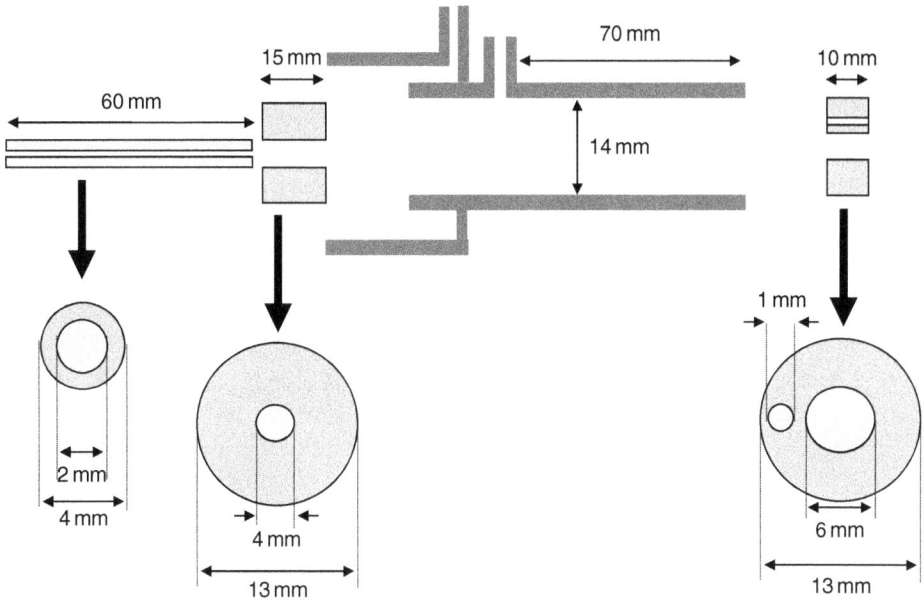

(a)

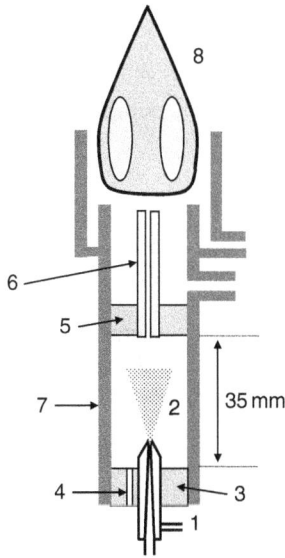

(b)

Figure 4.27 Scheme of the disassembled (a) and assembled (b) torch-integrated sample introduction system (TISIS). (1) nebulizer; (2) evaporation cavity and aerosol; (3) PTFE adapter for the nebulizer; (4) drain exit; (5) PTFE adapter for the injector; (6) plasma injector; (7) torch main body; (8) plasma. (Taken from ref. [828].)

torch. The TISIS can be operated either in a vertical or in a horizontal position. The nebulizer produces the mist at low liquid flow rates inside the cavity left between the two Teflon adapters.

Because the liquid-to-gas volume ratio is very low, the chamber acts as an evaporation cavity instead of working as a drop size selection device. In fact, it has been visually observed that at liquid flow rates below 20–30 μL min^{-1}, the spray chamber walls remain dry. Under these circumstances, there is no need for a drain.

At liquid flow rates above this value, the spray chamber walls become wet, and it is necessary to draw out the accumulated waste solution. Satisfactory results have also been obtained when coupling microfluidic chips with ICP–MS [847] or when using it as a μFI–ICP–MS [832] and nano-FI–ICP–MS [787] interface.

In a recent modification of the TISIS, a sheathing gas stream has been tangentially added at the chamber [846]. This has been done to lower the nebulizer gas flow rate, thus minimizing the droplet inertial losses. The sheathing gas has been used to efficiently draw the aerosol into the plasma. By using this gas stream, it has been possible to lower the nebulizer gas flow rate with a concomitant increase in the ICP–AES sensitivity.

The versatility of the system allows easy modification of the inner volume of the spray chamber to optimize the analytical figures of merit [729]. This is achieved by adapting an easy-to-adjust polyethylene cylindrical body at the torch base. Large cavities (ca. 20 cm^3 inner volume) are preferred to obtain high sensitivities, whereas small ones (ca. 5 cm^3) provide short wash-out times. Compared to a double-pass spray chamber and a Cinnabar, the TISIS ICP–AES emission intensities are enhanced up to five times [690]. With a low inner volume cavity, memory effects for the TISIS are less severe than those for a Cinnabar spray chamber.

The results obtained with the TISIS or systems based on a single-pass spray chamber in ICP–MS have also been very promising [760]. In summary, the new system affords the following advantages over a conventional sample introduction system: (i) a 30-fold shortening of wash-out time, (ii) a threefold lower signal depression caused by seawater matrices and (iii) higher absolute sensitivities. A further advantage of the TISIS over other systems, such as direct injection nebulizers or direct injection high-efficiency nebulizers, is the lower oxide ratios.

Evaporation cavities are very promising devices as they can act as total-consumption sample introduction systems, with enhanced sensitivities, low-memory effects, along with a very low cost when compared to direct injection nebulizers.

4.5.3 Tandem systems

The combination of two series spray chambers has been evaluated for use in ICP–MS. In this tandem arrangement, a double-pass spray chamber is usually coupled to a cyclonic type one (Figure 4.28) [568, 669]. The aerosol leaving the cyclonic spray chamber goes through the double-pass one, also known as homogeneization chamber, before being introduced into the plasma. The second spray chamber is used to dampen the instability of tertiary aerosols and to favour the production of fine droplets, thus the signal short-term standard deviation increases. This is very important mainly for precise isotope ratio determinations. With these tandem systems, the optimum operating conditions are different to those commonly used with conventional systems. Thus, higher nebulizer gas flow rates should be used for the former system. Because of the more homogeneous drop size distribution of the aerosols reaching the plasma, higher ICP–MS sensitivities are reported with the tandem system than with a setup in which only a chamber is used. In an approach, a PEEK cyclonic spray chamber has been adapted to a PTFE double-pass design [806]. The reported enhancement signal factors range from 2.5 to 3. The short-term ICP–MS signal RSD is enhanced by a factor of 3 when the tandem system is used. Another benefit of the tandem system arises in the fact that the oxide formation is lower because of the drier aerosols introduced into the plasma.

However, it has been observed that the long-term stability can be degraded because of the solution build up in the connexion between the two spray chambers. The proposed solution has been to heat the former spray chamber and to cool the second one. With this assembly, the aerosol is desolvated and hence a gain in terms of sensitivity by a factor ranging from about 4 for light elements to 2.5 for heavy ones is obtained [622].

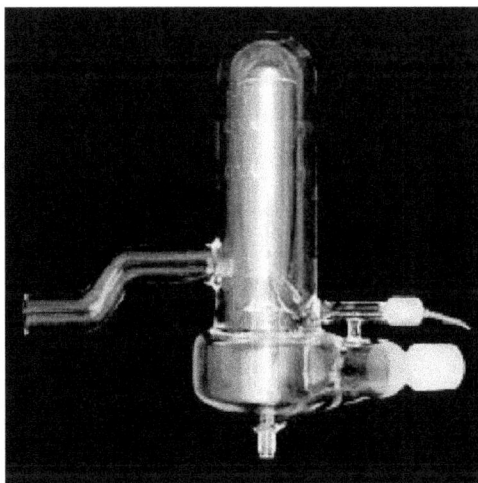

Figure 4.28 Scheme of the tandem system consisting of a cyclonic spray chamber coupled to a double-pass type one. (Taken from http://www.elementalscientific.com.)

4.6 CONCLUSIONS ON SPRAY CHAMBERS

Advantages and limitations of double-pass, cyclonic and single-pass spray chambers are summarized in Table 4.5. If the trend is to increase the use of cyclonic spray chambers, the donut spray chamber and the evaporation cavities seem very promising in terms of analytical performance and low-memory effects and wash-out times.

Table 4.5

Advantages and limitations of the spray chambers currently used with commercially available ICP instruments

Advantages	Limitations
Double-pass spray chamber	
Efficient removal of coarse droplets	Old-fashioned design with a large dead volume
Design can be easily improved	Only the inner tube is concerned with droplet deposition. External control temperature is not efficient
	Most of the aerosol is lost
	High-memory effects
	Long rinsing times
	Position-dependent (horizontal or vertical)
Cyclonic spray chamber	
Less aerosol lost leading to higher analyte signals and better limits of detection	Critical position of the nebulizer tip
Efficient cutoff of large droplets ($> 20\ \mu m$)	Signal can be noisier
Efficient temperature control	
Several designs and volumes	
Low dead volumes leading to low memory effects and short rinsing times	
Single-pass spray chamber	
Less memory effects than a double-pass chamber	Poor aerosol selection leading to coarse droplets
High sensitivity	Need a robust plasma (high hf power)
May be combined with an impact bead	

– 5 –

Desolvation Systems

5.1 INTRODUCTION

Because of the low tolerance of the plasma to solvent load, on the one hand, and the possible production of interferences caused by the solvent, on the other hand, it may be appropriate to lower the mass of solvent reaching the plasma. It can also be interesting to remove water to avoid the formation of oxides in ICP–MS resulting from water dissociation. This lowering can be obtained by (i) reducing the amount of aerosol, and therefore the amount of solvent, and (ii) removing a significant part of the solvent after aerosol formation through a desolvation system.

To enhance the analytical performance of ICP spectrometers, efficient nebulizers or spray chambers could be used. These devices are able to provide higher W_{tot} values than the conventional ones. However, the solvent plasma load may be too high (see Chapter 2), and hence, plasma degradation can be produced. As a result, improvements in the signal are lower than expected by only considering the analyte transport rate values. However, reduction in the amount of aerosol may be necessary with volatile organic solvents, for which nebulization efficiency is significantly improved. This can be obtained by reducing the solution delivery rate or by cooling the spray chamber.

For aqueous solutions, an appropriate alternative is the use of a desolvation system, which consists of aerosol heating followed by solvent trapping either in a condenser or via a membrane. Ideally, a dry aerosol constituted only by solid particles would be obtained. Because less solvent is introduced into the plasma, analyte mass transport rate is increased, leading to an improvement in sensitivity.

After going into more detail about solvent effect in the plasma, processes occurring through the aerosol heating and solvent removal, in particular solvent evaporation kinetics and nucleation phenomenon, are described in the following section. Then, several designs of desolvation devices will be described and commented.

5.2   OVERVIEW OF THE EFFECT OF THE SOLVENT IN ICP–AES AND ICP–MS

In general terms, the presence of solvent into the plasma can lead to different phenomena that can alter the analyte excitation process:

1. When tertiary aerosol is injected into the plasma, the solvent suffers from the same processes as the analyte. This fact means that the solvent will require a given amount of energy for its vaporization, dissociation, excitation and ionization. The energy required for the solvent dissociation is included within the 10–100 W range. This amount of energy can induce a decrease in the analyte excitation efficiency. Note that just a 10% of the total plasma energy is available for the analyte excitation [408].
2. In some instances, it has been verified that the solvent can induce a decrease in the total plasma volume. As the power supplied to the plasma by the generator remains constant, the plasma energy density increases. This phenomenon is called thermal pinch [67, 408]. The contraction is because of the increase in the plasma thermal conductivity observed as the solvent molecules dissociate, thus leading to increased heat conduction from the plasma to the environment. As a result, the outermost plasma zone cools.
3. The solvent modifies some fundamental plasma properties such as the plasma excitation temperature, electronic number density and ion temperature.
4. The solvent dissociation can lead to molecular as well as atomic interfering species. They can be excited and/or ionized, hence interfering in the analyte signal. The presence of these species can also induce an increase in the background [121], thus degrading the limits of detection.
5. The presence of solvent can also lead to a modification in the signal stability. The solvent dissociation is a source of background and analyte signal noise [461].

The magnitude of the detrimental effects caused by the solvent on the spectrometer performance depends on many factors such as the mass of solvent load, the aerosol drop size distribution and the physical form of the solvent delivered to the plasma. With regard to the last factor, it has been observed that the transport of water in vapour form can induce an increase in the plasma thermal conductivity similar to that caused by the addition of hydrogen [233] or helium. As a result, the plasma electron number density and ionic temperature will increase [168]. However, if the solvent is delivered to the plasma in liquid form, the opposite is observed [242].

By a combination of both W_{tot} and S_{tot}, it is possible to determine the mass of solvent delivered to the plasma in liquid form. This can be done by applying the following mathematical relationship [359]:

$$S_{liq} = \frac{L - S_{tot}}{(LC_0/W_{tot}) - 1},$$ (5.1)

where L is the solution mass flow rate and C_0 is the analyte concentration in the solution used to determine W_{tot}. The fraction of solvent transported under vapour form is determined by subtracting S_{liq} from S_{tot}. The proportion of solvent liquid and vapour reaching the plasma depends on the nebulization properties as well as on the solvent nature [460]. Figure 5.1 plots the variation of S_{tot}, S_{liq} and S_{vap} versus the nebulizer gas (Figure 5.1a) and liquid (Figure 5.1b) flow rate.

Because, for a given gas flow rate, the mass of solvent transported towards the plasma in vapour form remains more or less unchanged (Figure 5.1b), an increase in the liquid flow rate rises the mass of solvent transported towards the plasma in liquid form (Figure. 5.1b). The data in Figure 5.1, obtained from eqn. 5.1, suggest that, under the conditions studied, the argon stream is not saturated with water. It should be reminded that thermodynamic calculations indicate that the amount of water required to saturate a $1\,L\,min^{-1}$ argon stream lies in between 20 and $30\,mg\,min^{-1}$ at 20°C.

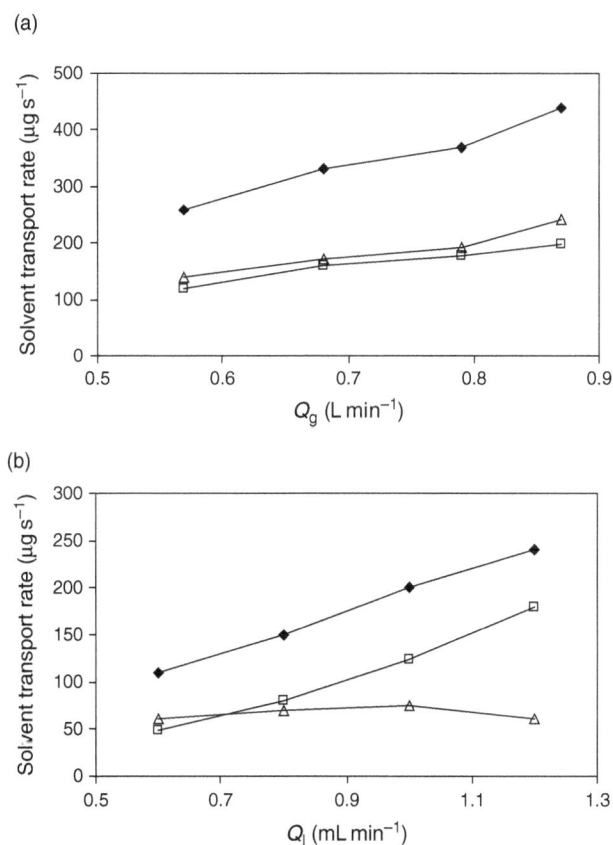

(a)

(b)

Figure 5.1 Mass of solvent reaching the plasma per unit of time versus the nebulizer gas, Q_g (a) and sample delivery, Q_l (b) flow rates. Sample introduction system: pneumatic concentric nebulizer coupled to a double-pass spray chamber operated at room temperature. $Q_l = 1.2\,mL\,min^{-1}$ (a); $Q_g = 0.57\,L\,min^{-1}$ (b). (△) S_{vap}, (□) S_{liq}, (◆) S_{tot}.

5.3 PROCESSES OCCURRING INSIDE A DESOLVATION SYSTEM

Figure 5.2 depicts a basic two-stage desolvation system as well as the processes responsible for the final solvent removal. Besides the aerosol transport phenomena described in
Section 4.2, evaporation and solvent nucleation take place inside a desolvation system.

5.3.1 Solvent evaporation

The principles of solvent evaporation from droplets have been previously mentioned
(see Section 4.2.1). When a desolvation system is used, the aerosol is heated and solvent
evaporation is produced in a more significant way. This fact makes it possible to transport,
through the aerosol heating step, the solution contained in coarse droplets that, due to the
droplet diameters, would not be transported otherwise. Figure 5.3 illustrates this fact with
three representative examples: a droplet of 1 μm that will be transported towards the
plasma at room temperature, another one with a 10 μm diameter that would correspond to
the coarsest droplet transportable through a double pass spray chamber and finally a
droplet that would be lost through impacts because of its high diameter (i.e. 20 μm). For

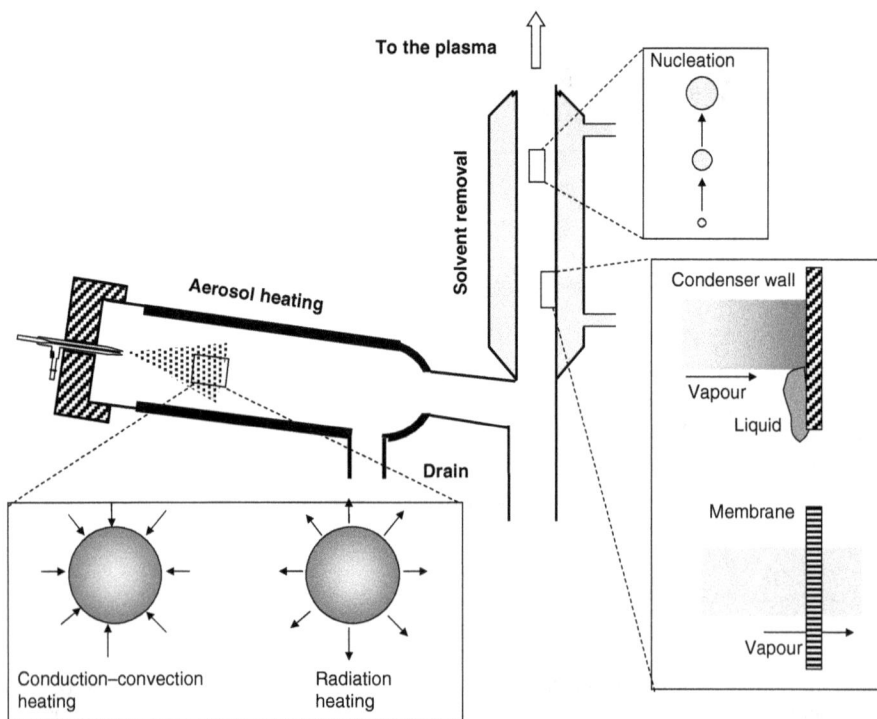

Figure 5.2 Scheme of a conventional desolvation system and the processes involved.

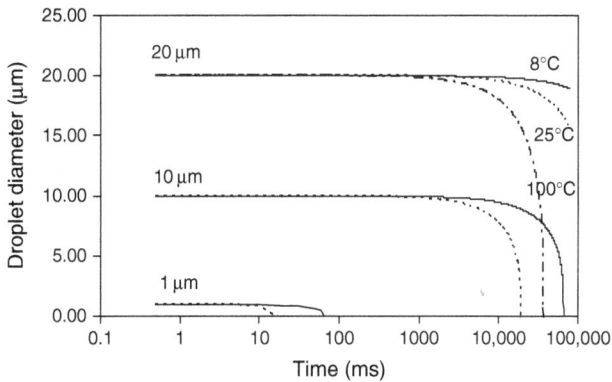

Figure 5.3 Variation of the drop diameter with time caused by solvent evaporation for three representative initial diameters at several temperatures.

the sake of clarity, a logarithmic x-axis scale is considered. As expected, for a given droplet, an increase in the temperature causes rapid complete droplet evaporation. Thus, a 1 μm diameter droplet would evaporate completely after 70 and 10 ms at 8°C and 25°C, respectively. A 10 μm diameter droplet would require 66 and 18 s for the complete evaporation, respectively. In the case of a 20-μm droplet, the complete evaporation at 25°C would require a time longer than 2 min for complete evaporation. In Figure 5.3, a rough estimation of the evolution of the drop diameter with time is presented for a 20-μm diameter droplet at 100°C. This temperature is close to those typically used in some desolvation systems. In this case, the time required for the complete droplet evaporation would be 36 s. By assuming that the residence time of an aerosol inside a conventional heating unit of a desolvation system is close to 10 s, it is concluded that it would be necessary to heat the aerosol at temperatures above 150°C to reduce this droplet diameter down to 10 μm.

Aerosol heating is a difficult task because, as mentioned in Chapter 4, solvent evaporation causes a decrease in the temperature at the droplet surface. Furthermore, pneumatic nebulization is an endothermic process that adds more difficulty to the aerosol convective heating.

The aerosol evaporation efficiency depends on several factors:

(i) The nebulizer gas flow rate that has opposite effects. An increase in this variable leads to a lowering in the aerosol mean drop size, thus increasing the evaporation efficiency. Furthermore, the mass of solvent that can evaporate to saturate the argon stream also increases. In contrast, the higher the nebulizer gas flow rate, the shorter the aerosol residence time inside the heating step, thus decreasing the aerosol evaporation efficiency.

(ii) *The sample delivery flow rate.* The amount of energy required to evaporate the aerosol droplets increases as this variable goes up. However, the solvent can evaporate until the gas becomes saturated. At this point, the evaporated solvent mass reaches its maximum value.

(iii) *Heating temperature*. The evaporation factor (Chapter 4) depends directly on solvent physical properties such as the vapour pressure, surface tension and solvent diffusion coefficient that, in turn, are a function of the environmental temperature. All these parameters rise as temperature does. As a result, the evaporation factor increases, thus accelerating the droplet evaporation.

The combined result of all these factors is that the mass of analyte transported towards the plasma increases when increasing both the nebulizer gas flow rate and the heating temperature [502].

The effect of the liquid flow rate is less regular, because W_{tot} follows a maximum pattern as this variable is increased.

5.3.2 Nucleation or recondensation

When the medium is saturated with solvent vapour, this vapour can condense on the aerosol particles, thus re-increasing the mass of solvent reaching the plasma.

Solvent nucleation or recondensation is a process that leads to the increase in the mass of solvent delivered to the plasma. This phenomenon is produced when the hot aerosol reaches a low-temperature medium. Under these circumstances, the solvent vapour initially generated can condense on the aerosol particles [101, 412]. A condition that must be fulfilled is that the medium must be oversaturated in solvent vapour.

The extent of the solvent nucleation depends on the so-called saturation ratio (r) that is defined as

$$r = \frac{p_v}{p_s}, \qquad (5.2)$$

where p_v is the solvent vapour pressure at the aerosol temperature and p_s is the solvent equilibrium vapour pressure at the same temperature. The solvent nucleation is produced if $r > 1$, whereas evaporation occurs when the saturation ratio takes values lower than unity.

Note that homogeneous nucleation is related to vapour condensation on droplet, whereas heterogeneous nucleation is related to vapour condensation on dry particles. Another important factor influencing nucleation is the particle size. For pure water, the higher the droplet diameter, the lower the saturation ratio required to cause nucleation. This fact means that nucleation is more conspicuous for coarse than for small droplets.

This situation can be very different if a salt is dissolved in the nebulized sample. In this instance, it is observed that the saturation ratio initially increases with droplet diameter (i.e. solvent nucleation is less likely for fine than for coarser droplets) and then it decreases with this variable. Finally, if the salt concentration is high, the saturation ratio grows as the drop diameter increases. In other words, nucleation becomes favoured for small particles.

5.4 AEROSOL HEATING

In many cases, the first step of aerosol desolvation starts with solvent evaporation by heating the aerosol. Two mechanisms have been used: aerosol indirect heating and radiative heating. Both are considered in the present section.

5.4.1 Indirect aerosol heating

This step can take place in the spray chamber or in a glass extension tube placed just at the exit of the spray chamber [292, 301]. Heating the spray chamber is preferred in terms of analyte transport rate, as most aerosol losses take place there. Usually, the walls of the spray chamber or the extension tube are heated by means of a heating tape wounded around it. Energy is transferred to the aerosol droplets by a conduction/convection mechanism (Figure 5.4).

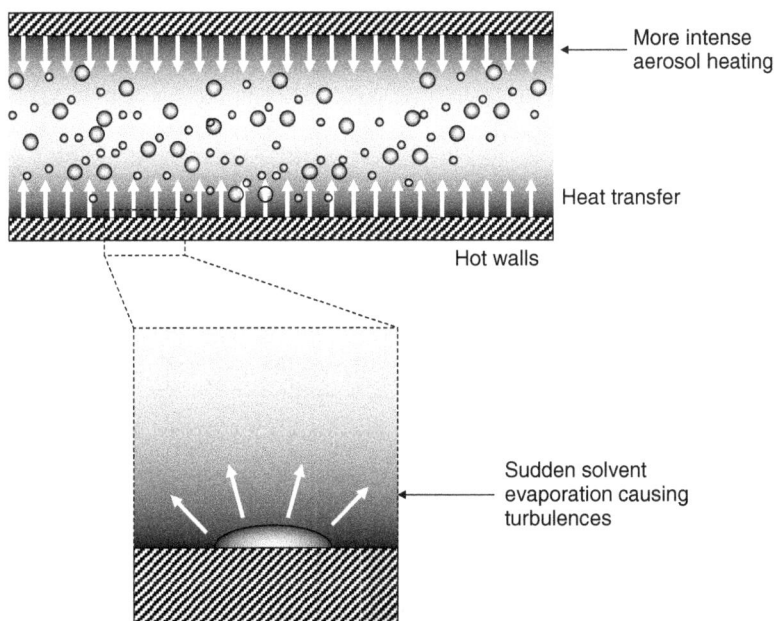

Figure 5.4 Indirect aerosol heating mechanism.

Aerosol heating efficiency in indirect systems is poor. Note that argon thermal conductivity is very low. Therefore, it is necessary to heat either the spray chamber or the conduction at a temperature much higher than that actually reached by the aerosol stream. As a result of the heating mechanism, a radial temperature gradient is established so as the temperature decreases towards the centre of the conduction [401]. When droplets impact against the hot walls of the system, the solvent evaporates suddenly. This tends to increase the background and signal noise owing to pressure fluctuations caused on the aerosol stream [355]. Furthermore, the analyte contained inside the droplets that impact against the inner walls of the system cannot be removed by draining. As a consequence, memory effects become more significant [324]. All these factors deteriorate the analytical response. For these reasons, the aerosol desolvation is sometimes applied at the exit of the spray chamber [292]. The heating efficiency of the tertiary aerosol increases as compared to primary one, because of two main reasons: (i) the liquid mass to evaporate is lower and (ii) the aerosol is finer. Both factors favour the solvent evaporation process.

Indirect aerosol heating has the following effects (Figure 5.5) [502]:

1. As it has been mentioned before, the pneumatic nebulization is an endothermic process. Therefore, it involves a drop in the environmental temperature, thus leading to a decrease in the efficiency of convective aerosol heating. The ability of the aerosol to be heated through this mechanism depends on factors such as the solvent nature, the amount of aerosol and the drop size. Some experimental data have demonstrated that the temperature inside a single-pass spray chamber decreases in the presence of aerosol. Thus at $0.3\,\mathrm{mL\,min^{-1}}$ liquid flow rate, the temperature in the centreline of the chamber is up to 40°C lower than at its walls. Obviously, this temperature drop is more conspicuous as the liquid flow rate increases.

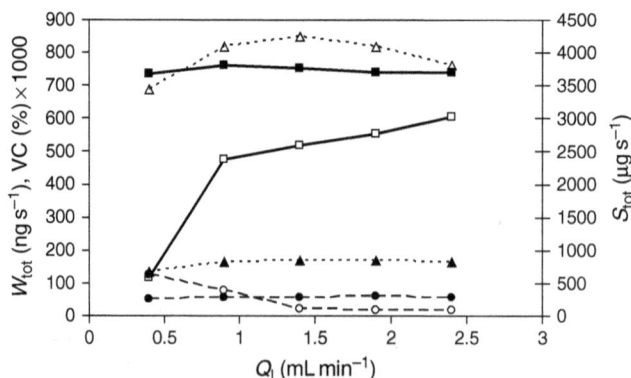

Figure 5.5 Effect of the liquid flow rate on several parameters obtained with a single-pass spray chamber. Aerosol heating temperature: 130°C. Squares: volume concentration (×1000); triangles: S_{tot}; circles: W_{tot}. Full symbols: parameters reported at room temperature; empty symbols: parameters reported after heating the chamber. (Data taken from ref. [502]).

2. The total aerosol volume measured at the exit of the spray chamber decreases with regard to the values reported at room temperature. The parameter considered in Figure 5.5 is the so-called volume concentration (VC), which is defined as the percentage of aerosol that is actually occupied by liquid droplets.
3. The mass of solvent reaching the plasma per unit of time (S_{tot}) sharply increases with the liquid flow rate, the enhancement factor being close to 6.
4. Unfortunately, as it is shown in Figure 5.5, W_{tot} does not increase by a factor as high as S_{tot}. It is interesting to notice that the behaviour of this parameter depends strongly on the liquid flow rate tested. Thus, when working at low values of this variable, W_{tot} at the exit of the heated spray chamber is more than two times higher than when the chamber is operated at room temperature. Meanwhile, at high Q_l values, W_{tot} is about three times lower when heating the spray chamber. These interesting results highlight the negative effects caused by the turbulences created when an excessive mass of water is evaporated from the hot chamber walls. An additional factor is the decrease in the aerosol heating efficiency as the mass of water to evaporate increases.

All these data collectively indicate that most of the solvent leaving the first step of a common desolvation system is present as vapour.

Another point to consider to optimally work with desolvation systems based on a first indirect aerosol heating step is the time required for solvent evaporation stabilization. Thus, it has been anticipated that this parameter depends on the heating temperature and the liquid flow rate and can range from about 30 to 60 min as the heating temperature increases from 60°C to 100°C [704].

5.4.2 Radiative aerosol heating

To solve the problems associated to indirect aerosol heating, radiation has also been used. The spectrum zone covered in these systems goes from the UV-VIS [85] to infrared radiation [375], along with microwaves (MWs) [502]. When aerosol radiative heating is used, the spray chamber walls and the conduction existing at its exit are also warm. Therefore, the aerosol is heated by a conduction–convection mechanism. Furthermore, a fraction of the radiative energy is directly absorbed by droplets. Overall, the aerosol radiative heating is faster and more efficient than resistive heating and reduces the drawbacks of the latter.

UV-VIS desolvation systems [331] use a halogen lamp placed inside an aluminium housing. With this setup, the aerosol is heated at the exit of the spray chamber.

In the case of IR aerosol heating [11, 473, 483], a globar source (a silicon carbure rod) is placed close to the spray chamber. High voltages are applied to the rod that emits light with a black body spectrum. This system is sensitive to changes in the matrix composition. Thus, it has been observed that S_{tot} and W_{tot} for an IR desolvation system vary with the acid concentration and nature [588]. Thus, S_{tot} increases with the acid concentration, whereas W_{tot} does not change or decreases by raising this variable. Besides, perchloric and sulphuric acids afford lower analyte transport rates than hydrochloric and nitric acids. Even low concentrations of the formers can induce drops in W_{tot} close to 60%.

MWs have also been used as an aerosol heat source for further desolvation [502]. So far, there are two basic designs of MW heating units: (i) those based on the use of a multi-mode cavity [588] and (ii) those in which a monomode cavity has been used [822]. In the first case, a single-pass spray chamber is introduced into a domestic MW oven. This system has several basic problems, among them are the facts that the energy distribution inside the cavity is not homogenous and is badly defined. Therefore, it is difficult to place the chamber in the maximum radiation zone. Another drawback of this kind of setup is that there is a lack of control of the incident and reflected power. Finally, the volume and configuration of the oven makes it necessary to place a noticeable length of the sample capillary inside the oven to deliver the solution to the nebulizer. This fact can be detrimental, because the solution can be partially pre-evaporated, thus leading to an increased signal background. Furthermore, it is necessary to use long tubes to drive the aerosol towards the plasma.

To solve the aforementioned drawbacks, a TM_{010} MW cavity, similar to that used for nebulization [616], can be used (Figure 5.6). A modified desolvation system has been

(a)

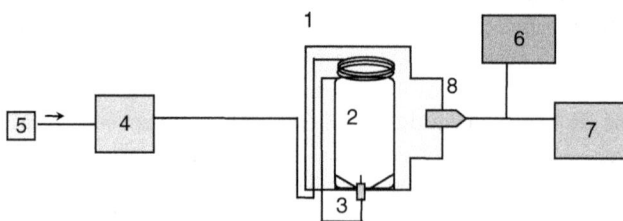

(b)

Figure 5.6 Picture of the TM_{010} microwave cavity in which a single-pass spray chamber has been introduced (a) and scheme of the setup employed to heat the aerosol by means of MW in a TM_{010} cavity (b) 1, microwave cavity; 2, spray chamber; 3, nebulizer; 4, peristaltic pump; 5, sample; 7, MW generator; 8, antenna.

built with one of these cavities together with an adjustable MW source and a waveguide [792]. This configuration shows several advantages against the older one: (i) the spray chamber can be easily placed in the zone of maximum MW energy, because the power distribution pattern is fully known; (ii) the chamber can be placed inside the cavity in a way that the spray chamber walls heating is avoided and thus the aerosol is directly heated; (iii) the interaction between the MW field and the aerosol can be optimized by using tuning devices.

A problem encountered with aerosol MW heating is the fact that the absorption of energy strongly depends on the matrix composition. Thus, for acids such as perchloric and sulphuric, the amount of analyte exiting the heating unit is up to five times higher than for solutions containing nitric or hydrochloric acids [588]. Therefore, matrix effects can be more severe than for a conventional desolvation system. This drawback can be partially solved by using a TM_{010} cavity and lowering the liquid flow rates as it has been recently demonstrated for organic solutions [822].

5.5 SOLVENT REMOVAL

Once the solvent has been partially evaporated, it has to be removed from the aerosol stream. There are two means of eliminating vapour in the desolvation systems used in ICP spectrometry: solvent condensation and solvent elimination through membranes of different kinds.

5.5.1 Solvent condensation

This has been the most widely used method for solvent elimination. Vapour removal is achieved by condensation on cold surfaces [18, 247]. This is the simplest way to reduce the solvent load into the plasma. The mechanism responsible for the solvent elimination is conductive–convective.

5.5.1.1 Nucleation problem in the condenser

It could be thought that an efficient solvent removal would require a low condenser temperature. However, as illustrated in Figure 5.7, it can be observed that the lower the condensation temperature, the lower the mass of the analyte delivered to the plasma. An explanation is related to nucleation [344]. Because of the solvent vapour saturation, the risk of nucleation is high and is increased when the temperature difference between heating and condensation is large. Nucleation leads to droplets that can either move to the plasma or be trapped into the condenser drain after impacts on the condenser walls.

When the droplets move to the plasma, solvent loading is only partially suppressed, which is against what was expected with desolvation. When they are trapped in the drain, analyte is also trapped, leading to a decrease in the analyte delivered to the plasma. This second process is significant and easy to verify by analysing the solution drain.

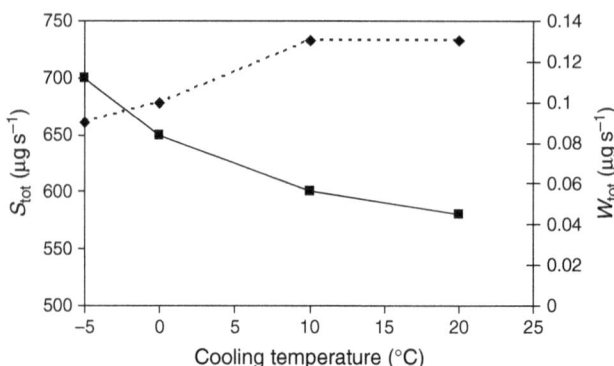

Figure 5.7 Effect of the condensation temperature on S_{tot} (diamond) and W_{tot} (squares) at the exit of a desolvation system. (Data taken from ref. [512]).

> In conclusion, if a single condenser would be used, the condensation temperature should not be too low.

5.5.1.2 *Main condensation systems*

> Three main condensation systems have been employed in ICP spectrometry: (i) Friedrich- [361] or Liebig [358]-type condensers, (ii) multi-step condensers and (iii) devices based on Peltier effect.

In the systems labelled as (i), the condenser is usually placed after the heated spray chamber. Nucleation extent can be reduced by carrying out condensation through two condensers placed in series [system (ii)]. The temperature of the first condenser is kept at higher values than that of the second one. In the first step, the solvent vapour pressure is lowered, thus favouring the condensation. Note that for water, it has been observed that 90% of the evaporated solvent mass can be condensed in a first condenser placed after the heating unit and thermostated at room temperature [412]. This way, oversaturation is always kept at a much lower level than with just one condenser. In these systems, the reduction in the solvent load is accompanied by a more or less noticeable reduction in the analyte transport rate.

In the case of multi-step condensers, spiral-shaped conduction is half placed in a bath at low temperature. In this manner, the aerosol suffers many temperature increase-decrease cycles. As a consequence, a fraction of the vapour solvent first condenses as the aerosol stream goes through low-temperature conductions. Then, a fraction of the solvent that eventually condenses on the particles through nucleation suffers from partial evaporation because the aerosol reaches a high-temperature zone. This cycle is repeated several times, thus giving rise to an extremely dry aerosol at the exit of the system.

Although more costly and more complex to setup and to optimize, multi-step condensers may be preferable when an efficient solvent removal is required.

A main advantage of Peltier-based condensers is related to the temperature that can be feedback-operated and accurately maintained through a large range. The device can then automatically correct for changes in the spray chamber temperature [303, 412]. In the lower temperature level, it has been observed that these condensers cannot be properly operated at values of this variable lower than −40°C [351]. This system has been widely used to undertake the analyses of liquid samples containing volatile solvents such as methanol [412] or carbon tetrachloride [303].

A Peltier-based condensation system has been directly adapted to the spray chamber. The aluminium block that surrounds the spray chamber acts as a heat well and this causes a drop in the temperature. In this way, solvent evaporation is prevented, thus reducing the magnitude of ICP–MS oxide interferences. Figure 5.8 shows the picture of a cyclonic spray chamber used in conjunction with a Peltier-based condensation device. In this case, a baffled cyclonic spray chamber is encapsulated in a heat conductive resin, thus increasing the contact between the chamber and the Peltier heating transfer block [840].

5.5.2 Solvent removal through membranes

Vapour removal from the aerosol stream can also be performed through the use of a membrane, that is a layer between the solvent vapour and a dry argon purge gas.

Figure 5.8 Picture of a cyclonic spray chamber used with a Peltier condensation system. (Taken from http://www.geicp.com.)

Use of membranes have been described for desolvation [237, 347, 404, 424]. Membrane extraction reduces nucleation and avoids contact between the aerosol and the liquid film of condensed solvent that flows down on the inner walls of the condenser. However, it shows a limited capability for vapour removal (i.e. its vapour removal rate is not as high as that of condensation systems).

The term 'membrane' makes reference to a layer placed between two fluids allowing net mass transfer through it. In the case of ICP spectrometry, the aim is to separate the solvent vapour from the aerosol stream with the help of a dry purge gas. Membranes for ICP can be placed either in a tubular [530, 601] or in a flat [514, 558] fashion. The driving force of the solvent removal process through membranes is the difference in concentration existing between the aerosol and the external part of the membrane. Generally speaking, the membranes used in ICP–AES and ICP–MS can be classified as porous and non-porous. In the former (Figure 5.9a), the size of the molecule of the compound to be separated is smaller than the pores. The water transport through the membrane pores is produced because of a change in the pressure, concentration or electrical potential. In the case of non-porous membranes (Figure 5.9b), the removal of water is produced through dissolution–diffusion mechanisms. The latter phenomenon involves the dissolution of the permeable molecules on the membrane surface. Then, the trapped molecules diffuse towards the least solvent concentrated side of the membrane. Afterwards, the solvent diffuses or evaporates from the other membrane side.

Membranes used in ICP can be made of PTFE, silicone or Nafion® (http://www.perma-pure.com). The latter one is widely used for its ability as a cation exchange membrane, but it is also very hygroscopic and therefore can be used to dry gas streams. Nafion® is a PTFE and perfluor-3,6-dioxy-4-methyl-7-octane-sulphonic acid copolymer. As PTFE is resistant to the chemical action, this polymer is able to retain up to 13 water molecules for sulphonic group. Therefore, Nafion® can adsorb a 22% of its weight of water. The vapour removal by means of a Nafion® membrane is produced very quickly. Therefore, it is a very efficient system for eliminating water vapour. One advantage of this material is that as water is removed via an absorption/desorption process and not via transport through pores in the membrane (Figure 5.9), it is possible to use air as the counter-current sweep gas without compromising the plasma performance.

The membrane parameters to optimize are the gas flow rate and the temperature. An increase in the membrane temperature gives rise to a depression in the desolvation efficiency. As the dew point temperature takes values of several degrees below zero, the majority of the solvent leaving the membrane is in liquid form, thus indicating that the vapour retention is poor at high membrane temperatures. To get the best performance, the membrane temperature should be kept at least 5°C above the aerosol heating step temperature. Interestingly, it has been observed that the membrane temperature does not significantly modify the W_{tot} value. However, the higher the liquid flow rate, the higher the S_{tot} value.

When the analyte is present as a volatile species, a fraction of it may not be transported towards the plasma and is lost as it diffuses through the walls of porous membranes. Results supporting this phenomenon have been presented for elements such as copper [768] and selenium [790]. In the case of organo-chlorine, organo-bromine or organo-iodine

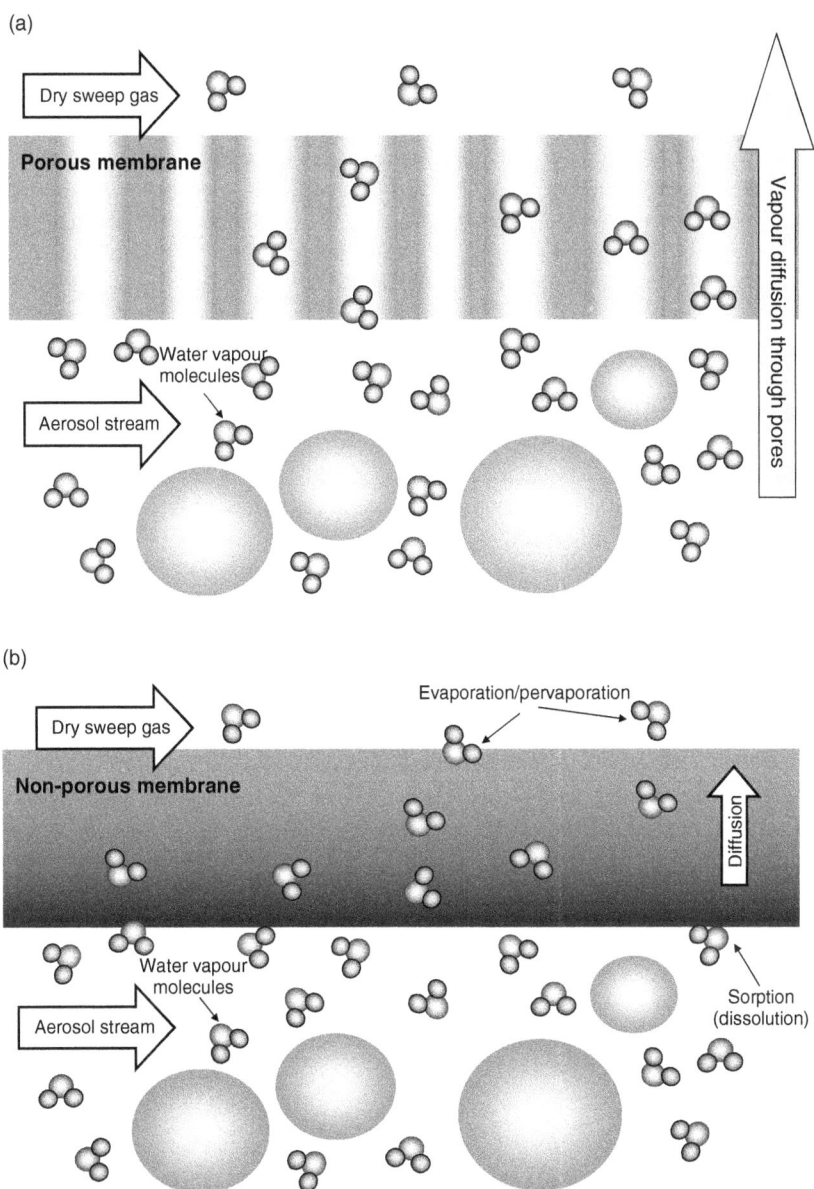

Figure 5.9 Water vapour removal mechanism for a porous (a) and non-porous (b) membrane.

compounds, it has been observed that they are also partially lost in the membrane [726]. A potential problem of these membranes is that if the sample solution concentration is very low, the particle that remains after the solvent evaporates may be small enough to also potentially pass through the pores and be lost.

5.6 DESIGN OF SOLVENT REDUCTION SYSTEMS

5.6.1 Thermostated spray chambers

One of the ways of reducing the solvent plasma load is to decrease the temperature of the spray chamber walls. In this way, the extent of the solvent evaporation from the aerosol is lowered. Vapour partially condenses on the cold spray chamber walls and is removed through the drain. This option is especially interesting when volatile solvents are introduced into the system. Thermostated spray chambers can be double-pass (Figure 5.10), single-pass and cyclonic-type spray chambers. Note that the last two designs are inherently easier to cool.

Thermostating spray chambers have been applied for mitigating oxide interferences in ICP–MS [199, 657]. The population of background species (i.e. ArO^+) is also modified as the spray chamber temperature decreases from 20°C to 5°C [414]. Not only the signal but also the noise of the ArO^+ intensity is lower at temperatures lower than room chamber temperature [794]. In contrast, thermostated spray chambers lead to a decrease in the analyte mass transported towards the plasma as well. By comparing the analyte transport efficiency with a thermostated spray chamber against that found for other desolvation systems, it has been found that the former provides ε_n values about one order of magnitude lower than the latter system [512].

Recently, a cooled cyclonic type spray chamber has been characterized in ICP–MS [794]. In this case, Tygon® tubing has been wrapped to the spray chamber. A cold thermostated distilled water stream circulated continuously through this tube. With this spray chamber, several conclusions were drawn: (i) the ArO^+ signal decreased by 20% as the chamber temperature was lowered from 30°C to 5°C, (ii) for some isotopes, LODs

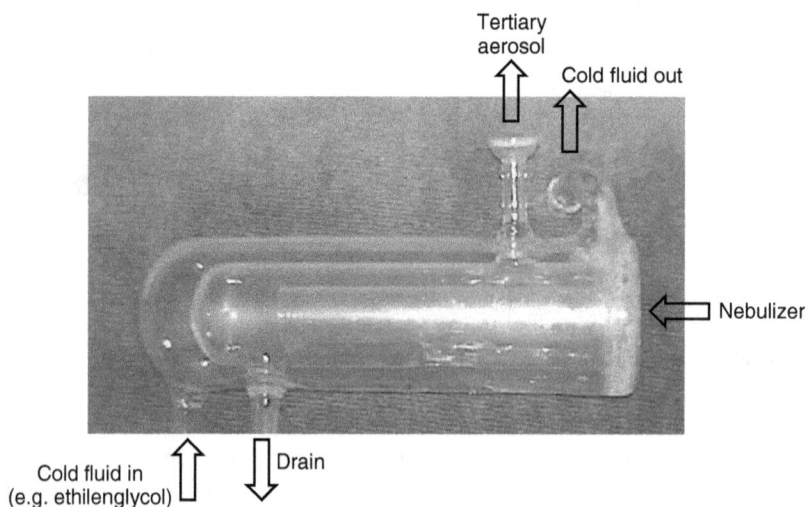

Figure 5.10 Picture of a thermostated double-pass spray chamber.

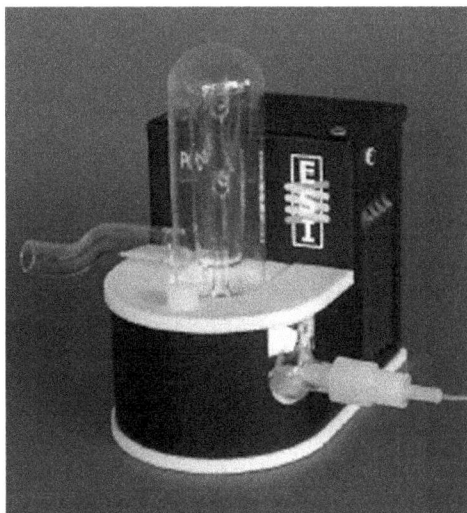

Figure 5.11 Peltier cooled with a double-pass extension to stabilize the aerosol stream. (Taken from http://www.elementalscientific.com.)

were slightly worse at 5°C than at 30°C, (iii) oxides of other elements were lower at the lowest temperature tested and (iv) the background equivalent concentration of the interfering species also decreased with the chamber temperature.

Peltier cooled cyclonic spray chambers have been used to lower the oxide ICP–MS signals (http://www.elementalscientific.com). Another feature of this system is that the wash-out times are shortened with respect to a double-pass spray chamber. The reported respective data on this subject are >3 min (double pass) and <30 s (cooled cyclonic). A claimed advantage of this approach is that the signal long-term stability enhances because the system is less sensitive to changes in the laboratory temperature. RSDs lower than 1% have been found for $1\,\mu g\,L^{-1}$ solutions during measurement periods as long as 1 hour. To further reduce the signal RSD, a second chamber similar to a double-pass one can be adapted at the exit of the cooled cyclonic one (Figure 5.11). With this modification, it is possible to lower the signal RSD from about 0.9% (cyclonic cooled chamber) down to about 0.5%. This feature is highly interesting for isotope ratio determinations.

5.6.2 Two-step desolvation systems

If a cooled spray chamber is useful to overcome the deleterious effect of organic solvents, it is less efficient with aqueous solutions. In this case, a two-step desolvation system that consists of a first heating step in which the solvent is totally or partially evaporated from the aerosol droplets and a second one in which solvent vapour is removed is more appropriate. The desolvation conditions will determine the fraction of solvent and analyte mass transported. An ideal target would be to obtain only dry particles into the carrier gas at the exit of the desolvation system.

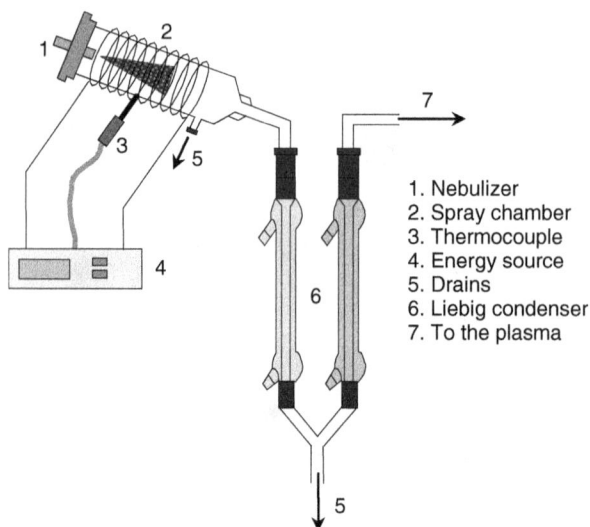

1. Nebulizer
2. Spray chamber
3. Thermocouple
4. Energy source
5. Drains
6. Liebig condenser
7. To the plasma

Figure 5.12 Scheme of a desolvation system based on the use of two series placed condensers.

A common two-step desolvation system is depicted in Figure 5.12. In this case, a single-pass spray chamber is heated by the action of a wounded tape. The aerosol heats, and it is introduced into a condenser adapted at the exit of the chamber. It is certainly the easiest way of setting up a two-step desolvation system. The solvent vapour elimination efficiency of a condenser depends on the liquid flow rate as well as on the heating and cooling temperatures. As seen previously, a more efficient solvent removal is obtained by using two condensers in series (Figure 5.12), the temperature of the condenser being lower than that of the first one. Thus, when heating the aerosol at, for instance, 120°C, if the first condenser temperature is set at 10°C, the solvent elimination efficiency is not 100% but around 90%. When a second condenser is placed in series, it is normally switched at temperatures below 0°C. According to some experiments [704], it has been verified that the solvent removal efficiency of the second condenser operated at −10°C is up to 87%. Thus, S_{tot} takes values of 6000, 375 and 157 µg s^{-1} at the exit of the heating unit (170°C), first condenser (10°C) and second condenser (−10°C), respectively. The overall solvent efficiency removal is typically close to 98%.

However, the condenser at the exit of a heating unit can also be responsible for analyte losses. There are several reasons that can be argued to explain this fact, among them: (i) the solvent rapid condensation on the system cold walls carries some analyte and (ii) nucleation causes drop growing and inertial and gravitation impact losses.

5.6.3 Multiple-step desolvation systems

These setups are also known as cryogenic desolvation systems [274, 494]. In this case, the aerosol leaving a two-step desolvation system is introduced into a spiral tube that is partially immersed into an ethanol bath at −80°C. The part of the tube outside this thermostated bath is heated at a temperature about 40°C above the aerosol solvent boiling point. In this way, the partially desolvated aerosol is introduced into the conduction at -80°C. The residual solvent vapour condenses on the tube walls and the aerosol particles because of the nucleation process. Then, the aerosol is again heated and the nucleated solvent evaporates, thus having another chance to be removed from the aerosol stream through condensation on the conduction inner walls. This process is repeated three or four times [523]. Therefore, cryogenic desolvation minimizes the amount of solvent condensed on the aerosol particles through the nucleation mechanism.

A consequence of desolvating aerosols through cryogenic apparatus is that the spatial distribution of ions in ICP–MS changes. As a result, the optimum operating conditions are modified. In fact, the aerosol gas flow rate yielding the best sensitivity should be increased with respect to that required in the case of other sample introduction systems.

5.6.4 Desolvation systems based on the use of membranes

A membrane can be placed at the exit of a heated spray chamber. Initially, a silicone polymer membrane was used to remove organic solvents such as Freon and chloroform [237, 267]. The solvent removal efficiencies achieved for these solvents ranged from 80 to 100%. To remove water, a membrane chemically modified has been used [305, 306]. Porous membranes have also been used to desolvate aerosols for ICP [450]. Although they are efficient as solvent removal devices, this kind of membranes causes losses of the analyte mass fraction analyte contained in the smallest aerosol particles or as gaseous compounds because they can pass through the membrane pores.

Figure 5.13 plots the variation of S_{tot} versus the liquid flow rate for a system consisting of a heated cyclonic spray chamber combined with a Nafion® membrane [704]. The ability of the membrane for removing water vapour is evidenced from this figure. In the

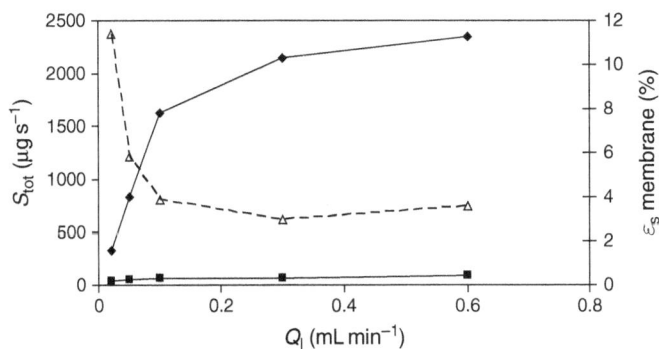

Figure 5.13 S_{tot} values at the exit of a heated cyclonic spray chamber (diamonds) and at the exit of a Nafion® membrane placed after the chamber (triangles) for different liquid flow rates and solvent transport efficiency (squares) through the membrane (right y-axis).

right y-axis, the solvent transport efficiency at the membrane exit (ε_s) is plotted versus the sample delivery rate. In this case, ε_s has been calculated as follows:

$$\varepsilon_s = \frac{(S_{tot})_{\text{membrane exit}}}{(S_{tot})_{\text{chamber exit}}} \times 100. \tag{5.3}$$

From Figure 5.13, it is obvious that the higher the liquid flow rate the lower the ε_s value, thus providing evidence that the membrane eliminates solvent vapour more efficiently at high than at low liquid flow rates. With this system, it is possible to reach a 90% analyte transport efficiency while keeping the solvent transport efficiency at low levels (i.e. in the order of 3%). These figures are in full agreement with previously published work [518] in which it was claimed that the solvent removal efficiency was close to 97%.

A commercially available desolvation device using a membrane is the so-called APEX (Figure 5.14), which is a desolvation system that combines many of the previously mentioned components. In this case, the solution is transformed into aerosol by means of a pneumatic concentric nebulizer (usually made of PFA). A heated (140°C) cyclonic spray chamber is adapted to a first condenser cooled with air at room temperature. Then a second Peltier-based multi-pass condenser is used and its temperature kept at lower values (ca. −5°C). As a result, a dry aerosol is obtained. The second condenser has a N_2 port to lower the oxide interferences observed in ICP–MS [795]. Furthermore, the addition of this gas increases the sensitivity [793], because nitrogen enhances the energy transfer towards the plasma central channel. To remove the residual aerosol vapour, a membrane (e.g. a 3°C cooled Nafion® fluoropolymer membrane [765, 814])

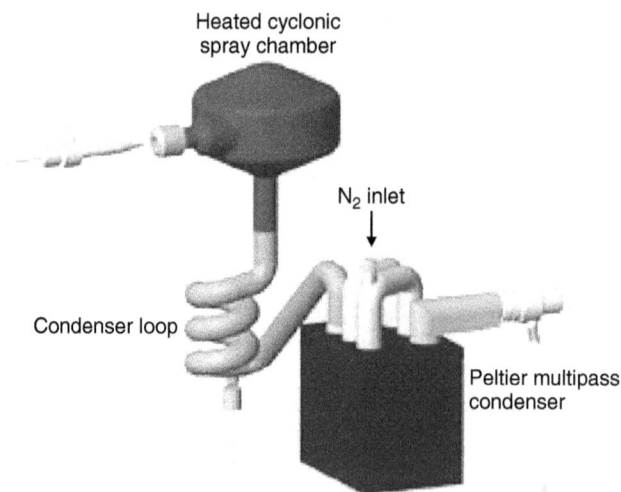

Figure 5.14 Setup of the APEX desolvation system. (Taken from http://www.elementalscientific.com.)

can be placed at the exit of this condenser. The vapour passes through the membrane pores with the help of a counter-current sweep flow of a dry argon or nitrogen stream. Nafion® membranes only remove water and other small polar organic molecules, hence, volatile analytes are not wasted thus being transported towards the plasma. The APEX can also be equipped with Macroporous heated Teflon® membranes. In this case, both water and organic solvents are removed from the aerosol stream.

There exist several versions of the APEX in which different materials for the spray chamber and the condensers are used. In the so-called APEX–IR, all the system flow path is made of quartz and it includes an additional mixing chamber to enhance even more the signal stability. This fact is very interesting when performing isotope ratio analysis. The reported aerosol drying efficiency is close to 95%. This fact means that for a $350 \, \mu L \, min^{-1}$ liquid flow rate, the amount of vapour reaching the plasma is just $17 \, \mu L \, min^{-1}$ (http://www.elementalscientific.com).

When using membrane-based desolvation systems, high salt content or highly viscous solutions can cause membrane blockage [448]. To prevent it, the membrane should be rinsed after every 10 samples. Despite this, there are some successful experiments demonstrating the applicability of this kind of desolvation devices to the analysis of seawater samples [578] and slurries [508].

5.6.5 Radiative desolvation systems

An IR source induces increases in the ICP sensitivity with respect to a conventional sample introduction device consisting of a pneumatic nebulizer coupled to a spray chamber. The signal improvement factor depends on the emission line and can range from 4 to 6 [775]. Concomitantly, the analyte transport efficiency values are 4.1 and 1.5% with and without desolvation, respectively. Again, the advantage of using a condensation unit has been evidenced as the mass of solvent transport rate has decreased from 37 for the conventional system to $8 \, mg \, min^{-1}$ when the desolvation system has been used.

MW-based desolvation systems have been used together with two condensers placed in series to enhance the removal of the solvent vapour. In these works, the condensation temperatures have been kept at 10°C and −1°C for the first and second condenser, respectively. The effect of the nebulization variables depends on the desolvation system used [792]. Thus, for a domestic oven, the emission intensity increases with the liquid flow rate up to a value close to $0.1 \, mL \, min^{-1}$. Above this delivery flow rate, the ICP–AES signal decreases with the sample flow. However, for a TM_{010} cavity, the signal steeply increases with this variable, the variation being much more conspicuous than that observed for a conventional desolvation system based on a conductive–convective heating mechanism. The concentration of the MW energy in a localized zone achieved in TM_{010} cavities yields an enhancement in the aerosol heating efficiency, thus providing higher analytical signals [834]. Comparative studies revealed that ICP–AES and ICP–MS enhancement factors higher than 10 are produced with respect to a conventional desolvation system [792].

5.6.6 Desolvation systems for the analysis of microsamples

Ideally, desolvation systems are a very interesting choice for the analysis of liquid samples at low liquid flow rates. Under these circumstances, the aerosol heating is produced in a very efficient way. At a low enough sample delivery rate, the mass of solvent that can be potentially evaporated is lower than the amount of solvent required to saturate the gas stream at the heating temperature. A confirmation of this expectation can be found in the work by Nam et al. [410], who observed that the ICP–MS limits of detection were generally lower for a desolvation system placed after a double-pass spray chamber than for the spray chamber alone. This analytical parameter was not affected by the liquid flow rate in the $10\text{--}1200\,\mu\text{L min}^{-1}$ range.

As for conventional liquid flow rates, at flow rates below $100\,\mu\text{L min}^{-1}$, porous [316, 456] as well as non-porous membranes provide excellent performance in ICP–AES and ICP–MS [647, 748, 752, 755, 764, 768]. An example of this system is the so-called high-efficiency sample introduction system (HESIS) [591, 666]. In this case, a pneumatic micronebulizer is coupled to a heated single-pass spray chamber. A hot argon stream is introduced longitudinally to the spray chamber. As a consequence, aerosol solvent evaporation is enhanced and analyte transport efficiencies close to 100% are obtained. A membrane dryer is finally used to remove the solvent vapour.

There are two commercially available desolvation systems based on the use of a membrane for the analysis of microsamples (http://www.cetac.com, 2007): the Aridus^{TM} (Figure 5.15) and the MCN6000. Both of them use heated single-pass spray chambers. The chamber heating temperature is kept at a quite low value (e.g. 75°C) likely because of the low liquid flow rate used. This low temperature allows efficient solvent evaporation without compromising signal stability. A fluoropolymer membrane is set up at the chamber exit and its temperature increased up to 160°C. A sweep argon stream is introduced to remove the solvent vapour. As for the APEX, a nitrogen stream can be added, thus increasing the sensitivity [617]. The main difference between the Aridus^{TM} and MCN6000 is that in the latter no drain exit is found in the spray chamber.

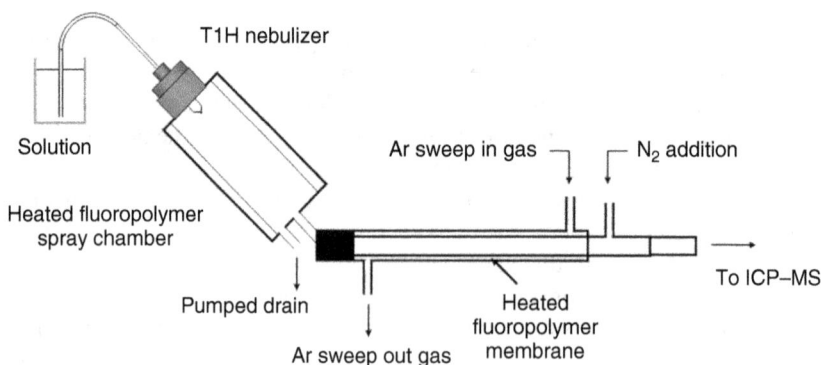

Figure 5.15 Drawing of the Aridus^{TM} desolvation system for the analysis of microsamples. (Taken from ref. [828].)

With these desolvation systems, ICP–MS limits of detection are enhanced by more than one order of magnitude with respect to those indicated in the already existing ICP–MS-based methods [508, 717]. Under certain conditions, the Aridus^{TM} provides limits of detection even lower than an ultrasonic nebulizer equipped with a membrane desolvation system [761]. In a comparative study involving two desolvation systems and a nebulizer coupled to a spray chamber, it has been observed that the APEX provided the best limits of detection [791]. This is likely due to the lower liquid flow rate used for the ARIDUS (0.06 mL min^{-1}) as compared with that employed for the APEX (1 mL min^{-1}). Thus, the LOD for ^{239}Pu were 510, 1000 and 33 Bq L^{-1} for the conventional, ARIDUS and APEX systems. In this case, the conventional system also provided better limits of detection than the ARIDUS, because the former was also operated at higher liquid delivery rates. However, it has been suggested that, because of the large surface areas of the system (note that this system has an about 400 cm^3 inner volume) and the low liquid flow rates used, the memory effects should be more severe for the MCN6000 than for conventional sample introduction systems [508]. Unexpectedly, according to Ma et al. [598], lower wash-in and wash-out times are reported for the former systems. ICP–MS interferences caused by oxide formation are minimized by using these devices. The reported CeO$^+$/Ce$^+$ ratios for a conventional liquid sample introduction system and the Aridus^{TM} were about 3 and 0.05%, respectively (http://www.cetac.com, 2007). In fact, according to the work by Prohaska et al. [617], the only interest in using an MCN6000 was the fact that polyatomic interferences were reduced in ICP–MS with respect to a MCN. This device also provided lower hydride [615] and oxide [508] formation yields than other sample introduction systems. This feature made it possible to use the MCN6000 for such determinations as Pu isotopes in seawater [671] or arsenic in steels [564] with good precisions.

With regard to the long-term stability of the ICP–MS signal, the MCN6000 was able to provide only a 10–15% drift over a 3-day period of operation [719]. The determination of ^{239}Pu and the ^{240}Pu/^{239}Pu isotopic ratio are problematic because of the interference caused by ^{238}UH$^+$ and the tailing of the uranium peak. By using an Aridus^{TM}, this kind of analysis can be efficiently performed without suffering from the problem of hydride generation [693, 725]. In fact, the degree of formation of this hydride was reduced by a factor of 10 with respect to a conventional liquid sample introduction system [761]. Non-spectroscopic interferences caused by the presence of organic compounds (e.g. methanol) are also mitigated when using a membrane-based desolvation system. This point is highly interesting for HPLC ICP–MS coupling, particularly when the mobile phase composition changes with time. This situation is often found when working with gradient elution [790].

Another reported shortcoming of the MCN6000 is the appearance of signal spikes caused by the droplet formation at the nebulizer tip. To overcome this, the original nebulizer and spray chamber have been replaced by PFA nebulizers [719]. In this way, it is possible to increase the aerosol heating temperature from 75°C to 105°C, thus enhancing droplet evaporation. The wash-out times were also shortened when switching from Teflon to PFA. Thus, Regelous et al. [748] found that to achieve an accurate determination of protactinium, the wash solution (i.e. 0.6 mol L^{-1} HCl and 0.02 mol L^{-1} HF) had to be aspirated between samples for 20 and 30 min when PFA and Teflon chambers were used, respectively.

The APEX has been used in conjunction with a micronebulizer for the analysis of low sample volumes. Compared with a conventional sample introduction system with a micronebulizer operated at liquid flow rates from 50 to 70 µL min^{-1}, the APEX generated ICP–MS signals for Fe, which were four to six times higher [786]. In other studies, it was claimed that the ^{226}Ra ionic intensity improvement factor was as high as 12 [781]. Unlike other desolvation systems, no signal spikes were found when the APEX was operated long term. Wash-out times were rather short (i.e. about 80 s were required to reduce the ^{56}Fe signal to background levels) [786]. This desolvation system is recommended for applications in which HF is used, such as geochemical and semiconductor analysis (http://www.elementalscientific.com, 2007).

A double-membrane desolvation system has also been used with ICP–AES and ICP–MS. In this system, a heated (i.e. 80°C) double-pass spray chamber with a MCN is coupled to a two-PTFE concentric membrane tubes having a 3.5 µm maximum pore size (70% porosity). The membranes were placed in such a way that the solvent elimination took place using two countercurrent argon flows. The solvent vapour diffused through the membrane pores and a dry aerosol was introduced into the plasma. A characterization of this assembly was done by Sung and Lim [636] using ICP techniques for the analysis of isopropyl alcohol. An increase in the sweep gas flow gave rise to a drop in the ICP emission intensity for all the elements tested, thus indicating that the analytes were likely retained on the membrane. Simultaneously, it was observed that the carbon emission decreased sharply with this gas flow rate, thus demonstrating the efficiency of the membranes for removal organic solvents.

Finally, it is worth mentioning that the use of membrane-based desolvation systems are beneficial for mitigating non-spectral interferences in ICP–AES caused by inorganic acids [529, 681]. At low liquid flow rates, for matrices containing acids such as hydrochloric and nitric, interferences are eliminated even for concentrated solutions. Sulphuric acid solutions show a more conspicuous matrix effect, although with this kind of desolvation system the extent of the interference is less significant than in the case of a conventional sample introduction system.

5.7 COMPARISON AMONG DIFFERENT DESOLVATION SYSTEMS

In absolute terms, the use of a two-step desolvation system does not improve the efficiency of solvent removal against those based on only a single. However, the analyte transport efficiency is improved by a factor of up to an order of magnitude depending on the solvent tested. This is a clear benefit of the aerosol heating unit. As the solvent evaporation is promoted in this first step, a higher mass of analyte is able to escape to the spray chamber, thus being introduced into the second step.

An APEX–IR system has given rise to an ICP–MS signal enhancement factor of about one order of magnitude with respect to a pneumatic concentric nebulizer adapted to a cyclonic spray chamber [813].

In a comparative study involving two desolvation systems and a nebulizer coupled to a spray chamber, it has been observed that the APEX provides the best performance in terms of limits of detection [791]. This is likely because of the lower liquid flow rate used for the ARIDUS ($0.06\,mL\,min^{-1}$) as compared with that employed for the APEX ($1\,mL\,min^{-1}$).

In a comparison among the different systems, it has been observed that the cerium oxide ratios are 3.5, 0.8, 0.06 and 0.03% for a double pass, APEX, APEX with Nafion® membrane and APEX with Teflon® membrane, respectively. This desolvation system has been successfully used for the analysis of samples containing complex matrices such as serum [813] or road dust [795] with a minimum sample preparation procedure.

Membrane-based desolvation systems have been compared with condenser-based ones [704]. Figure 5.16 summarizes the reported results for two systems equipped with a heated single-pass spray chamber, a condenser and either a second condenser or a membrane. In terms of solvent transport rate, it has been observed that, depending on the liquid flow rate, the use of a membrane lowers S_{tot} by a factor ranging from about 2 to 4 with respect to a two condenser-based system. This fact confirms the benefit of using a Nafion® membrane as a solvent removal unit. The higher solvent removal efficiency is accompanied by a virtually identical W_{tot} value. However, the ICP–AES emission intensity provided by the membrane is lower than that afforded by a condenser. According to the author, this trend is due to the decrease in the plasma thermal characteristics reported when a membrane is used as compared to a condenser. This fact also evidences the beneficial role of water vapour in terms of enhanced plasma thermal conductivity [470].

The results shown in Figure 5.16 suggest that the determining part of a desolvation system is the heated spray chamber. Furthermore, it is also recognized that the analyte losses in the first solvent removal unit are more noticeable than those produced in the second one.

An overall system characterization in terms of solution transport parameters is established in Figure 5.17. Thus, while condensers are responsible for the removal of a fraction of the analyte exiting the heated spray chamber, non-porous membranes allow the pass of virtually the totality of the analyte introduced by simultaneously lowering the mass of solvent reaching the plasma (bold face characters). Another interesting fact is that for single-pass spray

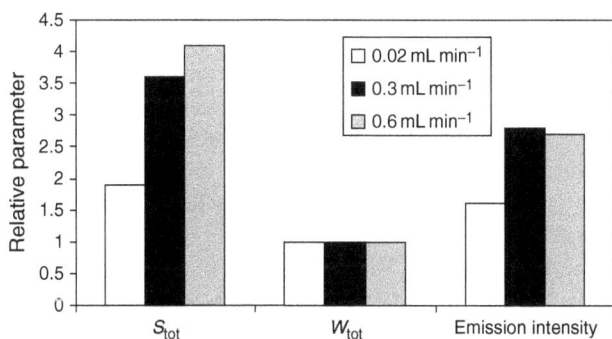

Figure 5.16 Comparison of a system consisting of a single-pass heated spray chamber and two condensers (System 1) with another one based on the use of the same chamber and a condenser followed by a Nafion® membrane (System 2). $S_{tot} = S_{tot}$(System 1)/S_{tot}(System 2); $W_{tot} = W_{tot}$(System 1)/W_{tot}(System 2). Heating chamber temperature: 120°C; Condenser 1 temperature: 40°C; Condenser 2 temperature: 5°C; Nebulizer gas flow rate: $0.6\,L\,min^{-1}$; Membrane gas flow rate: $3\,L\,min^{-1}$.

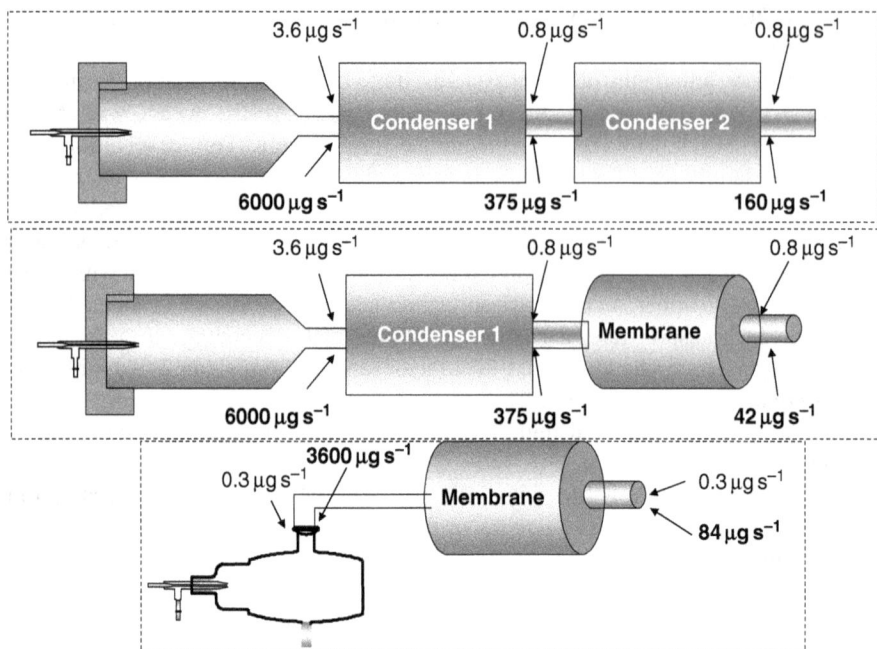

Figure 5.17 Comparison of the W_{tot} and S_{tot} (bold face numbers) for three different desolvation systems at several locations. Sample liquid flow rate $= 0.6$ mL min^{-1}; heating temperature for single-pass spray chamber $= 120°C$; heating temperature for cyclonic spray chamber $= 100°C$; condenser 1 temperature $= 10°C$; condenser 2 temperature $= -10°C$; membrane temperature $= 40°C$; membrane gas flow rate 1–3 L min^{-1}. (Data taken from ref. [704].)

chamber, W_{tot} values are higher than for a cyclonic type one. The use of the former device is beneficial mainly when a desolvation system is used. For cyclonic spray chambers, a fraction of the aerosol droplets impacts a few milliseconds after their generation.

With regard to ICP–AES analytical figures of merit, it has been observed that the systems equipped with a membrane at the exit of either a single-pass or cyclonic spray chamber provide better signal stabilities (i.e. lower RSD) than those in which a condenser is adapted at the heated chamber. This fact would prove that membranes are able to dampen the pulses generated in the aerosol heating step [355, 469]. This has a direct influence on the limits of detection (Figure 5.18). It can be observed that, in general terms, for a single-pass spray chamber, the use of a membrane lowers the LODs with respect to the situation found when two series condensers are placed at the exit of the hot chamber. The data for the cyclonic spray chamber are highly interesting, as it can be observed that it provides similar LODs as the single-pass spray chamber plus two series condensers. Note that, as Figure 5.17 indicates, the mass of analyte delivered to the plasma with this device is almost three times lower for the former than for the latter setup. The effect of the membrane as a pulse dampener is again evidenced. MW radiative desolvation systems provide better limits of detection than conventional devices mainly when a monomodal heating cavity is used (Figure 5.19).

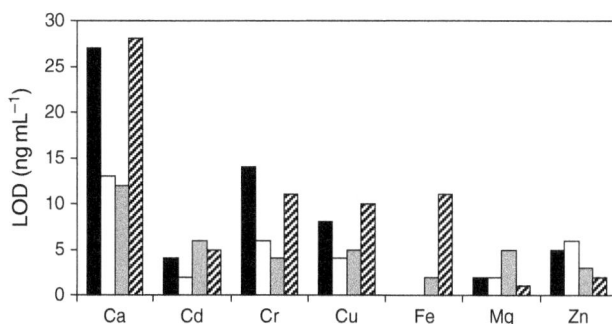

Figure 5.18 ICP–AES limits of detection for several elements and four different desolvation systems. Black bars: Heated single-pass spray chamber coupled to two series condensers; white bars: Heated single-pass spray chamber coupled to a condenser followed by a Nafion® membrane; grey bars: Heated single-pass spray chamber coupled to a Nafion® membrane; dashed bars: Cyclonic spray chamber plus Nafion® membrane. (Data taken from ref. [704].)

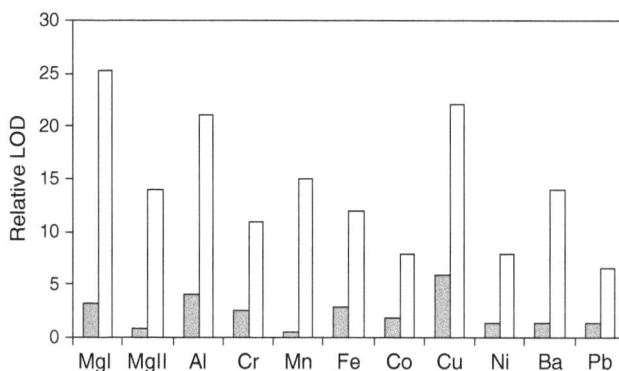

Figure 5.19 ICP–AES LOD enhancement factor incorporated by MW-based desolvation systems with respect to conductive–convective desolvation systems (relative $LOD = LOD_{conductive\ system}/LOD_{MW}$. Grey bars: results obtained with a MW domestic oven; white bars: results obtained with a TM_{010} cavity.

MW and infrared aerosol heating systems have been compared for the analysis of inorganic acids containing samples [588]. It has been observed that both aerosol heating means are sensitive to the presence of acid. For nitric and hydrochloric acids, the infrared-based system affords emission signals from three to five times higher than the MW-based one, whereas for perchloric and sulphuric ones, both systems provide similar sensitivities. However, similar limits of detection have been reported likely because of the higher background stability found in the case of the MW desolvation system.

Figure 5.20 shows the relative intensity (I_{rel}) defined as the ratio between the analyte net emission intensity in the presence of acid divided by that found in the presence of this matrix for these two radiative desolvation systems. Concerning the infrared-based desolvation device, the presence of either nitric or hydrochloric acids has a little effect on the

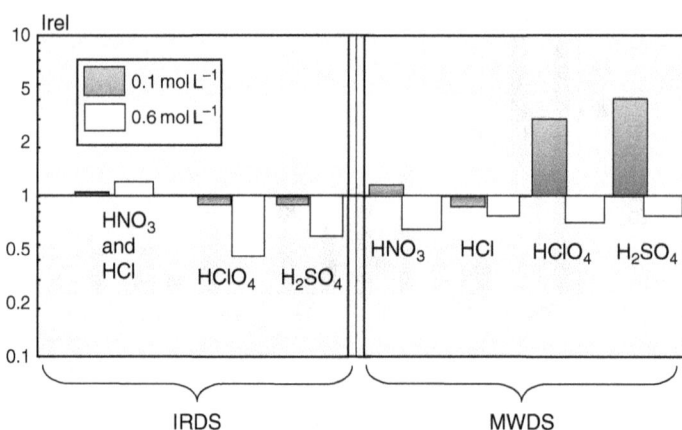

Figure 5.20 Relative intensity obtained for different acid solutions at two different concentrations with two radiative desolvation systems.

emission signal ($I_{rel} = 1$). However, concentrated solutions containing sulphuric and perchloric acids induce a drop in the signal with respect to that found with pure water solutions (i.e. $I_{rel} < 1$). This fact demonstrates that the aerosol heating by means of IR radiation is less efficiently produced when these two acids are present in the solution.

In the case of the MW-based desolvation system, nitric and hydrochloric acids cause a modification in the signal with respect to a plain water solution. Even small amounts of perchloric and sulphuric acids cause a significant change in the signal. In conclusion, it appears that MW desolvation systems are more sensitive to changes in the acid matrix than infrared ones. In terms of limits of detection, both systems provided similar results with plain water solutions [585].

More sophisticated desolvation systems can be built. Thus, for example, a heated single pass spray chamber can be adapted to a condenser. At the exit of the condenser, a membrane can be adapted and its exit fitted to a cryogenic condenser. The use of this complex system does not incorporate any improvement in terms of solvent removal [523]. Instead, it is preferred to adapt a cryogenic condenser at the exit of a conventional desolvation system. The figures indicate that the LaO^+/La^+ ratio is 0.24, 0.08 and 0.02% for a conventional desolvation without any extra device, membrane and cryogenic condenser, respectively. A similar comment can be made in terms of hydride and hydroxide generation. Thus, ThH^+ and TOH^+ signals are attenuated by a factor of 7 and 20 with respect to a conventional desolvation system. Cryogenic condensers compare favourably with respect to membranes in terms of solvent removal [523].

Generally speaking, desolvation systems can be efficient to remove solvent in a significant way. However, they may be (i) complex because of the additions of devices and then costly, (ii) more sensitive to any change in the matrix or in the reagents (e.g. acid concentration) and (iii) subject to significant memory effects.

– 6 –

Matrix Effects

6.1 INTRODUCTION

Matrix effects can result from the sample introduction system, that is aerosol formation, transport and filtration and from mechanisms in the plasma, that is diffusion,
ionization, excitation and so on. Even if the effects arise in the plasma, the sample
introduction system will play a role through the amount of aerosol, the size of the
droplets and the carrier gas flow rate. If some effects can be clearly assigned to
the introduction system, such as solution viscosity and surface tension, matrix effects
are usually a combination of effects due to the introduction system and the plasma.
However, most of the work published in this field deals with mechanisms in the
plasma and neglects what could arise within the introduction system. Alkali elements,
which have been extensively studied as providing significant changes in the plasma,
may also modify sample transport. The subject of this chapter is to deal not only with
matrix effects due to the introduction system, but also with those occurring in the
plasma and related to the parameters used to generate aerosols.

Acids and easily ionized elements (EIE) are the most commonly present matrices in
the elemental analysis by inductively coupled plasma–atomic emission spectrometry
(ICP–AES) and inductively coupled plasma–mass spectrometry (ICP–MS) [679, 828].
High concentrations of these species induce a significant variation in the analytical signal,
thus giving rise to matrix effects. Studies dealing with matrix effects are of crucial
importance, because they give support to the selection of a given analytical technique
for a given elemental determination. To overcome this problem, the sources of the matrix
effects must be perfectly known, so as to find an appropriate method to eliminate or to
compensate for the interferences. This chapter is aimed at describing the effects caused by
acids, concomitants and organic solvents on each step of the analysis by ICP–AES and

ICP–MS. Some sections are also devoted to briefly expose the most important methods used to correct these matrix effects.

6.1.1 Effect of physical properties on the sample introduction system performance

6.1.1.1 Effects on the aerosol generation

For pneumatic nebulization, the solvent's physical properties having more marked influence on the aerosol characteristics are the solution surface tension and the viscosity. There are several studies in which it has been found that the higher the surface tension, the coarser the primary aerosols. The reason for this behaviour is clear: as the surface tension increases, the energy required to break down the liquid bulk goes up. The effect of the viscosity is different. As it has been mentioned (Chapter 3), the pneumatic aerosol production mechanism involves the generation of instabilities on the liquid surface. The growing of this wave yields to the generation of the aerosol. An increase in the solution viscosity gives rise to a dampening of these instabilities, thus favouring the production of coarse aerosols.

The different models developed to predict the aerosol characteristics for pneumatic nebulization take these facts into account. This kind of studies was first undertaken by Nukiyama and Tanasawa [2, 796], who developed an empirical relationship between the droplet diameter and the solution's physical properties and nebulizer characteristics (N–T model).

$$D_{3,2} = \frac{585}{V} \sqrt{\frac{\sigma}{\rho}} + 597 \left[\frac{\eta}{\sqrt{\sigma\rho}} \right]^{0.45} \left(1000 \frac{Q_l}{Q_g} \right)^{1.5}, \qquad (6.1)$$

where $D_{3,2}$ (μm) is the surface mean diameter of the drop size distribution of the aerosol, also known as Sauter mean diameter, σ (dyn cm^{-1}) is the solvent surface tension, ρ (g cm^{-3}) is its density, η (dyn s cm^{-2}) its viscosity, V (m s^{-1}) is the difference between the velocities of gas and liquid streams at the nebulizer outlet and Q_g and Q_l (cm^3 s^{-1}) the volumetric gas and liquid flow rates, respectively.

This widely known relationship has several limitations. For example, under certain conditions, eqn. (6.1) predicted neither the absolute [240] values of the mean drop size nor its trends. This is mainly due to the fact that this model was obtained for a given nebulizer design under special solution physical properties (i.e. $30 < \sigma < 73$; $0.01 < \eta < 0.3$; $0.8 < \rho < 1.2$). Extrapolation outside these ranges was not advisable [30, 84] as it has been experimentally demonstrated [282, 348]. For example, in the case of 70 per cent (v/v) acetic acid solutions, eqn. (6.1) yielded higher $D_{3,2}$ values with respect to water. Note that the results encountered experimentally have indicated the opposite trend [520]. Furthermore, this is not a dimensionally correct model. Despite this, the N–T model has been applied for predicting the characteristics of the aerosols produced by different nebulizers [193].

Some additional empirical models have been developed. The results measured for two different pneumatic concentric nebulizers have been used to predict the $D_{3,2}$ of primary aerosols produced by other nebulizers simply by interpolation of the nebulizer's critical dimensions [240]. This interpolation model has two important limitations: (i) it is useful only for pneumatic concentric nebulizers and (ii) the result gives a rough estimation of the actual $D_{3,2}$ value [436]. In this model, the solution's physical properties have also been considered. Other studies considering the prediction of the drop size distribution characteristic parameters have been adapted from previous considerations made in engineering applications of the so-called atomizers [270]. Two examples of models that have been adapted are the Rizk–Lefebvre [102] and the El Shanawany–Lefebvre [60] approaches. The experimental parameters included in these equations have been modified to apply both models to pneumatic nebulizers commonly used in ICP techniques [687]. Good results have been obtained with modified models for pneumatic concentric nebulizers such as [796]

$$D_{3,2} = 0.48\,d_1 \left(\frac{\sigma_1}{\rho_g V^2 d_1} \right)^{0.5} \left(1 + \frac{Q_1 \rho_1}{Q_g \rho_g} \right)^{0.53} + 0.15\,d_1 \left(\frac{\eta_1^2}{\sigma_1 \rho_1 d_1} \right)^{0.49} \left(1 + \frac{Q_1 \rho_1}{Q_g \rho_g} \right), \quad (6.2)$$

where $D_{3,2}$ (μm) is the Sauter mean diameter, d_l is the liquid outlet diameter, V (m s^{-1}) is the difference between the gas and liquid velocities at the capillary exits, σ_1 (dyn cm^{-1}) is the solvent surface tension, ρ (g cm^{-3}) is its density, η is its viscosity and Q (cm^3 s^{-1}) is the volumetric flow rate. In all cases, the subscript 'l' refers to liquid and 'g' to gas.

The N–T and Rizk–Lefebvre (R–L) models have been modified to apply them to predict the characteristics of the aerosols produced by micro-nebulizers [796]. The results were not successful, because it has been observed that for a direct injection high-efficiency nebulizer (DIHEN), the modified N–T model gives rise to an overestimation of the $D_{3,2}$ measured for several solvents, whereas the predicted $D_{3,2}$ from the R–L model are lower than those measured.

From the experimental results and the models such as those previously discussed, it emerges that the solution surface tension plays a more important role in the effect on aerosol characteristics, that is droplet size distribution, than viscosity [226]. Density and viscosity do not play a significant role.

6.1.1.2 Effects on the aerosol transport

As regards the aerosol mass-transport rate leaving the spray chamber, the physical properties that can modify this parameter are the solution's relative volatility [460] and density [438]. It has been observed that the higher the solution volatility, the larger the mass of both analyte and solvent delivered to the plasma. This is mainly because the solvent evaporation is more intense for volatile than for non-volatile solutions. Furthermore, for the former, as the droplet diameter decreases more efficiently, there is a higher chance for aerosol transport towards the plasma. As regards the solution density, it is recognized that for a droplet with a given diameter, its momentum increases with the solution density. Therefore, the droplet-inertial losses are more pronounced for dense solutions.

Solvent evaporation can be modified in presence of dissolved salts or acids. This is because the vapour pressure of the solution decreases proportionally to the salt concentration. In general terms, the presence of a given dissolved concomitant reduces the solvent evaporation extent from the droplet surface. For example, theoretical calculations [62] demonstrated that for a 10 000 µg mL^{-1} sodium solution, the decrease in droplet diameter would be lower than 5 and 2.5% for droplets with diameters of 0.5 and 1 µm, respectively. This effect becomes less obvious as the salt concentration decreases. Thus, for a 200 µg mL^{-1} sodium solution, a reduction of around 20% in diameter because of solvent evaporation was found for droplets of 10 µm in diameter [57]. The solvent evaporation processes occurring inside the spray chamber are produced in a different way as a function of the drop size. This is because when there is a dissolved salt in the sample at a relatively high concentration, small droplets contribute initially to the evaporation to a greater extent than coarser ones. As a result, the salt concentration increases far above the original value, and then, the evaporation from the finest droplets slows down. At this point, the solvent evaporation is preferentially produced from coarsest droplets and drains on the inner spray chamber walls because the salt concentration is lower than that of the finest droplets [57]. According to this reasoning, finer tertiary aerosols could be expected in the presence of dissolved salts than for a pure solvent [231].

Obviously, the solution transport degree depends on the primary aerosol characteristics. Therefore, when a given matrix causes a modification in the aerosol generation process, it also induces a change in the mass of solution that will escape the spray chamber.

> Besides the primary aerosol droplet size, aerosol transport depends upon the relative volatility and the density.

6.2 INORGANIC AND ORGANIC ACIDS

Acids are often present in solutions, because they are used for solid digestion [124, 506] and liquid pre-treatment [427] and can also be used for sample storage and analyte stabilization. Procedures such as pre-concentration by using ionic-exchange resins [49, 278, 379] involve the elution of the analyte of interest with some acid species (i.e. nitric acid). In other instances, acids (i.e. acetic) are used in sequential extraction of different fractions of a given metal in, for example, soils and sediments [186]. In some cases (i.e. lubricating oils), organic acids (e.g. isobutyric acid) are used to dilute the sample before its analysis by atomic spectrometry [56].

Apart from some important spectral interferences, the effects of the acids can be classified in two general groups [587]. The first one would include all the physical effects arising in the properties that acids confer to the solution, and therefore lead to a modification in the aerosol characteristics as well as in the mass of solution delivered to the plasma. In the second group, all the processes that take place inside the plasma would be considered.

6.2.1 Physical effects caused by inorganic acids

For acid solutions, viscosity, density, volatility and, to a less extent, surface tension are different with respect to water alone. As a result, the processes taking place in the sample introduction system are modified with respect to aqueous solutions. These processes are described in this section.

6.2.1.1 Influence on the sample uptake rate

When an inorganic acid is added to a solution, the density and viscosity generally increase. This gives rise to a decrease in the free liquid uptake rate [246, 297, 396]. Hence, Q_l for 30% (v/v) sulphuric and phosphoric acid solutions were roughly 50 per cent of the value obtained for water when a glass pneumatic concentric nebulizer was used [30]. In the case of acetic acid [297, 396], it has been observed that the Q_l values could be between 38 and 54% lower with respect to that for water, depending on the sample introduction system, gas flow and acetic acid concentration. As a result of the decrease in this parameter, the signal observed for several inorganic acids is lower than for water [30, 42].

6.2.1.2 Influence on the aerosol characteristics

Inorganic acids such as HNO_3, HCl, $HClO_4$, H_2SO_4 and H_3PO_4 do not induce marked changes in surface tension with respect to water. Instead, they lead to an increase in the viscosity and density, mainly when they are present at high concentrations. For this reason, pneumatically generated aerosols are slightly coarser for concentrated solutions of sulphuric and phosphoric acids (the most viscous ones) than for nitric, hydrochloric or perchloric solutions [438].

For organic acids, such as acetic, formic or propionic, different results are obtained. This is mainly due to the fact that they clearly confer to the sample lower surface tension values and higher viscosities than those of water [327, 520].

> In general terms, primary aerosols are finer for organic than for inorganic acids or aqueous solutions.

For example, with a cross-flow nebulizer, the primary Sauter mean diameters were 11.1 and 9.8 μm for water and 10% (v/v) acetic acid, respectively [520]. These results were explained in terms of the reduced surface tension value for the acetic acid solution

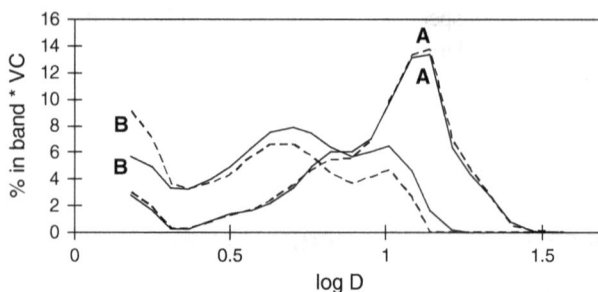

Figure 6.1 Volume drop size distributions in band obtained for distilled water (continuous lines) and 0.9 mol/L nitric acid solution (dashed lines). A, primary aerosols, B, tertiary aerosols measured at the exit of a double-pass spray chamber. (Taken from ref. [187].)

(54.6 and 71.2 dyn cm^{-1} for 10% acetic acid and water, respectively). The effects of other carboxylic acids, traditionally employed as surfactants, on the primary drop size distribution have also been studied [374]. Thus, for pentanoic and hexanoic acids, finer aerosols were generated than for water. Obviously, the reason argued to explain these results was the reduction in the surface tension values.

Acids cause a remarkable modification in the characteristics of the aerosols leaving the spray chamber or tertiary aerosols. It has been generally indicated that even low inorganic acid concentrations lead to an increase in the proportion of fine droplets (Figure 6.1) [307, 829].

Furthermore, the drop in statistical diameters becomes more important as the acid concentration goes up [389]. Interestingly, the amount of aerosol liquid volume is similar for both acids and aqueous solutions [389, 438]. All these effects are dependent on the liquid and gas flow rates as well as on the spray chamber used. Thus, the variation of the VC with the acid concentration has proved to be also more pronounced for double-pass than for cyclonic spray chambers [570]. By taking all these results into account, it has been observed that a possible mechanism explaining these observations can be based on the aerosol droplet fission [700]. According to this mechanism, when acids are dissolved, the aerosol droplets have a given electrical charge at their surface. As the drop diameter decreases through solvent evaporation, the charge density increases up to a value high enough to cause the droplets to be fragmented in other progeny.

6.2.1.3 Effect on the solution transport rate

In general terms, inorganic acids give rise to a decrease in the mass of solvent delivered to the plasma with respect to water. This drop is more pronounced for acids such as sulphuric and phosphoric than for nitric, hydrochloric and perchloric. This is likely because of the lower volatility and the coarser primary aerosols generated for the formers [438, 519]. Meanwhile, the results found for organic acids have indicated that the solvent plasma load is higher than that for water [520, 297].

As regards the mass of analyte transported towards the plasma, the presence of an inorganic acid produces general decreases in W_{tot}. For example, for 30% of both sulphuric and phosphoric acid solutions, W_{tot} values were two times lower than for water [187].

Even diluted sulphuric solutions caused a significant drop in the mass of analyte delivered to the plasma [437]. Hydrochloric and perchloric acids, in turn, produce reductions in W_{tot}, only important at actually high concentrations (i.e. 30% w/w) [438]. The factors affecting the W_{tot} values are the gas flow rate, the drop in this parameter being more pronounced at low than at high gas flow rate values and the spray chamber design [570]. An interesting effect of inorganic acids is related to the change in the analyte concentration depending on the droplet size. Thus, for the finest aerosol droplets, it has been observed that the concentration in acid and analyte are lower than for coarse ones [307, 519]. In contrast, because of the generation of finer aerosols, organic acids lead to higher W_{tot} values when compared to water [327].

Inorganic acids usually lead to a decrease in the mass leaving the spray chamber in contrast to organic acids.

6.2.2 Effects in the excitation/ionization cell

Acids can modify the plasma excitation characteristics. A distinction between robust and non-robust plasma is made. A robust plasma will not be significantly affected by the presence of acids. This assessment is confirmed by the fact that under robust conditions, neither the electronic density number, n_e, nor the excitation temperature, T_{exc}, change when acids are introduced [397, 400]. Under these conditions, any possible change in the signal induced by acids is mainly due to the sample introduction system. However, if the plasma is working under non-robust conditions, the signal variation can also be attributed to changes in the fundamental plasma properties [249, 279, 400]. For these conditions leading to a 'weak' plasma, the acid effect is dependent on the spectral line investigated. For ionic lines, the higher the energy sum value (sum of the ionization and excitation energy), the stronger the acid effect [97, 279, 400]. Besides these spectroscopic properties, there are others that also preclude the extent of the acid effects, such as the strength of the oxide bonds with rare earth elements (REEs) [440].

For robust plasmas, it can be said that the acid effects are mainly due to changes in the amount of solution reaching the plasma, whereas for non-robust plasmas, there are two sources of interferences: the plasma and the introduction system [400].

With the generators currently used, robust conditions are obtained at hf power values higher than around 1.3 kW and Q_g values below 0.5–0.6 L min^{-1} [268]. Non-robust conditions, in turn, correspond to low hf power values (i.e. lower than 1.0 kW) and gas flow rates of around 1.0 L min^{-1} or higher. These values obviously depend on the

Table 6.1

Consequences of the possible variation of the MgII/MgI line intensity ratio and of the
analyte signal with a change in matrix concentration

MgII/MgI	Analyte signal	Plasma conditions	Spray chamber
No change	No change	Robust conditions	No effects
No change	Change	Robust conditions	Effects
Change	Change	Non-robust conditions	?

Taken from ref. [660].

generator efficiency and the selected nebulizer and spray chamber. The MgII/MgI ratio has
been used to monitor the plasma robustness [268]. The higher the hf power and the lower
the Q_g value, the higher the MgII/MgI ratio. Based on the MgII/MgI ratio, Table 6.1 can be
used for guidance on the sources of matrix effects [660].

Organic acids can also modify the plasma thermal characteristics. Thus, there are several
reports indicating that the MgII/MgI ratio is higher for acids such as acetic [396, 520],
formic and propionic [297] than those for water. Besides, organic acids have been reported
to afford higher plasma electron number densities and excitation temperatures. For a given
element, it is apparent that the ionic state is less affected by the presence of organic acids
than the atomic one [297].

From the aspects mentioned in the preceding sections, acids could be classified in
two general groups:

(1) acids that do not modify the physical aerosol generation process in a significant
way. Several acids can be included in this first group (i.e. nitric, hydrochloric and
perchloric);
(2) acids that modify this physical process:
  – Inorganic acids (i.e. sulphuric and phosphoric) that deteriorate the nebulization
  leading to coarser primary aerosols and
  – Organic acids (i.e. acetic, propionic) that make the drop size distributions of the
  primary aerosols to shift to smaller diameters.

6.2.3 Effect of acids on analytical results: Key variables

The effect of inorganic acids on the emission intensity depends on many factors although,
in general terms, lower signals are reported in presence of acid [828]. As mentioned
above, the change in signal is less pronounced for robust plasma conditions than for non-
robust conditions.

In ICP–MS, the presence of acids induces signal variations that are more complex and unpredictable than in ICP–AES [373]. On the one hand, spectral interferences are very important, and on the other hand, the extent of these interferences is dependent on the instrument design, the nature of the sample and the operating conditions. Under certain operating conditions [169], the acid effects are more pronounced in ICP–MS than in ICP–AES. Note that the robustness of an ICP can be evaluated in mass spectrometry by using oxide to ion ratio, such as CeO^+/Ce^+.

It should be noted that the carrier gas flow rate is usually higher than $0.9\,L\,min^{-1}$ in ICP–MS, which is considered as providing a non-robust plasma in ICP–AES.

6.2.3.1 Acid concentration and nature

The acid concentration is a very important variable affecting the magnitude of the interference [400, 439, 478, 529]. In general terms, the matrix effects are more severe at high than at low acid concentrations. The reasons for this trend are (i) the increase in the drop size distribution of primary aerosols observed for some acids (e.g. sulphuric and phosphoric); (ii) the decrease in the mass of analyte delivered to the plasma. Note that the higher the acid concentration, the lower the solution volatility and the higher the sample density; (iii) the deterioration in the plasma thermal characteristics. The relative importance of each one of these factors depends on both the plasma and the nebulization conditions. However, there are some exceptions to the acid concentration effect. Thus, it has been claimed that at low acid concentration solutions, the ICP–AES signal can be higher than for plain water solutions [307]. The so-called aerosol ionic redistribution (AIR) appeared to be responsible for these trends [51]. In other instances, a chemical reaction produced inside the spray chamber yields a volatile oxidation product that induces signal enhancements [108, 246].

Stewart and Olesik [540] published a study about the effect of nitric acid on the ICP–MS ionic intensities. Three acid concentrations were tested: 2, 10 and 25% (v/v). They found that at gas flow rates lower than $0.7\,L\,min^{-1}$, the ionic signals for 2 and 25% nitric acid solutions were virtually the same. Nevertheless, on increasing Q_g above this value, the ionic intensities peaked at low gas flow rates for the 25% nitric acid solution. The results reported by Stewart and Olesik suggested a delay in the analyte ionization process, when increasing the nitric acid concentration. The acid effects were more pronounced for elements with high ionization potential (IP) values (e.g. As: 9.81 eV) than with low-IP values (Y: 6.38 eV). An interesting conclusion of the study performed by Stewart and Olesik [540] was that the acid effect was dependent on the operating conditions employed. Thus, at low gas flow rates, the 25% nitric acid solution gave rise to a signal enhancement with respect to that of 2%. According to the authors, this effect could be because of a reduction in the ion diffusion in the plasma, with the subsequent increase in the number of sampled ions, when a concentrated nitric acid solution was employed. Finally, it has been observed that the use of low Q_g and hf power reduced the acid effects to a great extent [540].

Indeed, to mitigate the acid effects, a gas flow rate below the optimized value in terms of sensitivity (at a given hf power) might be used. Thus, the ICP–MS robustness criterion seemed to be different for ICP–AES. In ICP–MS the ions are extracted from a small volume, which implies that any small changes in any aerosol or plasma property will modify the spatial ionic distribution in that plasma region.

The analytical results vary strongly as a function of the acid type. Inorganic acids such as nitric and hydrochloric induce ICP–AES signal drops that are less significant than sulphuric or phosphoric. The effect of acetic acid on the emission intensity depends on the plasma conditions. Under robust conditions, an increase in the ICP–AES signal has been observed due to the presence of acetic acid. Nevertheless, under non-robust conditions, the opposite is observed [520]. Besides acetic acid, other organic acids also cause a signal enhancement (e.g. formic, propionic) [30, 105, 297, 396] that depends on the plasma conditions [520]. Besides, there are other acids (e.g. citric acid) that lead to a decrease in the emission intensity with respect to plain water solutions [105].

Hettipathirana et al. [297] measured the analyte vertical emission profiles for 70% (v/v) acetic acid solutions and observed an increase in the signal with respect to water. The improvement factors followed three different trends as a function of the plasma observation height: (i) a step decrease in the signal improvement, (ii) a constant value of this variable and (iii) a maximum pattern. The existence of one or another behaviour depended on either the hf power or the element line considered.

The effect of acetic acid has also been studied in ICP–MS [258]. It has been observed that the optimum nebulizer gas flow rate shifts to lower values as the acid concentration increases. As regards the magnitude of the ICP–MS signal, no important differences are found under optimum conditions. The interferences of methacrylic acid (i.e. whose use is extensive in the painting, electronics and polymers industry) in ICP–MS have also been reported [328]. Already at a 5% (w/v) methacrylic acid, a noticeable signal reduction was observed. However, once the conditions are optimized, an increase in the acid concentration produces increases in the analytical signal.

Although the acid effect is stronger when increasing its concentration, it may be beneficial to add a large amount of acid in both standard and unknown solutions to ensure that the acid concentration is similar. Even if the effect is significant, it would be the same for every solution.

6.2.3.2 *Effect of the design of the sample introduction system*

As it has been previously indicated, under robust conditions, the acid interference is mainly related to the sample introduction system. Several studies have indicated that the design of this component of the instrument is a key factor in the field of acid matrix effects [437]. Already, Willis [10], and Dahlquist and Knoll [42] claimed that the magnitude of the acid effect was dependent on the sample introduction system used.

Some reports have evaluated the effect of the pneumatic nebulizer design on the extent of the interferences [307, 437]. For low hydrochloric acid concentrated solutions, the use of a cross-flow nebulizer leads to an increase in the analyte signal. The maximum signal was obtained for a slightly higher acid concentration than that obtained for a concentric nebulizer. When a concentric nebulizer was compared with a conespray nebulizer, it was found that the sulphuric acid matrix effects were more pronounced for the latter design than for the former. Meanwhile, for nitric acid solutions, the interference degree was similar for both nebulizers [519]. Some studies made with micro-nebulizers (MCNs) revealed that the effect depended on the particular design used. Therefore, for two different MCNs (Chapter 3), the presence of nitric acid induced a 20–40% drop in the emission signal as compared to plain water standards [529]. Small differences in nebulizer's critical dimensions could lead to these results. Note that, as it has been pointed out in the literature [457], the relative position of the sample capillary is of great importance, and any modification of this position can modify the MCN's long-term operation. On comparing the magnitude of the acid effect for the MCN and for a conespray nebulizer, it has been indicated that, at very low liquid flow rates (i.e. $30 \, \mu L \, min^{-1}$), the latter induced stronger interferences than the former. This fact was more noticeable for sulphuric than for nitric acid [519].

The spray chamber is a very important device affecting the acid effects [525]. Hence, when a concentric nebulizer was coupled to either a double-pass Scott spray chamber or a single-pass one, the average percentage of MgI signal reduction was 19 and 0%, respectively, for several perchloric acid concentrations. When the chambers were used with a conespray nebulizer, these values were 37 and 16%, respectively. These facts revealed that either the spray chamber or the nebulizer design had a very important effect on the acid interference.

Cyclonic spray chambers have proved to be less prone to acid effects. For two of these spray chambers (i.e. Sturman-Masters and cyclone-type), the nitric acid effect was more or less the same [478]. When cyclonic chambers were compared against double-pass chambers, it was observed that the acid effects were less pronounced for the formers. Thus, the signal obtained for a 3-mol L^{-1} nitric acid solution was around 0.9 and 0.7 lower than that obtained for water when a glass cyclonic and Ryton double-pass spray chamber were used, respectively [570].

Many of the processes responsible for the acid effects take place inside the spray chamber. Therefore, some studies were focused on the use of direct-injection nebulizers (DIN) [529]. Under some conditions, the use of the DIN minimizes the ICP–AES interferences for some acid solutions (i.e. nitric and hydrochloric acids, 0.9 mol L^{-1}). However, for other solutions (i.e. nitric 3.6 mol L^{-1} and sulphuric 0.9 mol L^{-1}), a depression in the ICP–AES emission signal was observed with respect to water, which could be attributed to a decrease in the effective liquid flow rate. Note that with the DIN, a gas displacement pumping is generally used.

Another way of modifying the aerosol transport phenomena is to use a desolvation system (Chapter 5). The presence of acids can cause two effects on the desolvation efficiency: (i) a less efficient aerosol evaporation because of the increase in the boiling point and decrease in the evaporation rate because of the decrease in the so-called evaporation factor [214, 215] and (ii) an increase in the solvent leaving the desolvation system. The latter effect is mainly observed when a condenser is employed. In this case, 'solvent nucleation' (Chapter 5) is produced [101]. The nucleation effects are mainly referred to as the solvent condensation on dry particles and droplets of the aerosol as it passes through the condensation unit (i.e. condensation nuclei). In the case of acid solutions, the acid trace impurities may generate solvent condensation nuclei [355].

Desolvation systems have proved to reduce the extent of the acid interferences as compared to conventional liquid sample introduction systems [434]. To solve the problem labelled in the previous paragraph as (i), the aerosol heating temperature must be increased [110]. Thus, it has been found that the use of a desolvation system (i.e. single-pass heated spray chamber followed by a two-step condenser) operated at both 100 and 160°C aerosol heating temperature lowered the effect of $3.6\,mol\,L^{-1}$ nitric acid to undetectable levels [529]. With regard to the sulphuric acid effect, it is only eliminated by using diluted solutions or/and selecting high heating temperature. The correction of the acid matrix effects was more effective when working at low sample consumption rates (i.e. at $30\,\mu L\,min^{-1}$) [529]. The problem indicated above as (ii) can be mitigated by using desolvation systems equipped with a membrane to remove the solvent vapour. These systems are able to reduce the ICP–MS polyatomic interferences to marginal values when sulphuric, hydrochloric and hydrofluoric acid concentrated solutions are analyzed [853].

Most sample introduction systems have been studied to provide high sensitivity along with good precision. It would be worthwhile to provide a sample introduction system really optimized to minimize or even suppress effects due to the most common acids such as HNO_3 and HCl.

6.2.3.3 Effect of the plasma observation zone and observation mode

Some studies [439] have revealed that for REEs, the inorganic acid effects are more pronounced at the plasma base, then decreasing in its higher region. This may suggest an effect similar to the delay in particle atomization, so that, in presence of nitric acid, the plasma spends more time to excite the analyte reaching it. Therefore, the residence time, and hence the plasma observation height, must be considered as an important variable from the point of view of acid effects.

The two possible plasma viewing modes, axial and lateral [485], have been used to study acid effects. Initially, axial viewing had a poor reputation concerning matrix effects. However, it was later verified that, on working under plasma robust conditions, the matrix effects due to the acids for ionic lines could be minimized to the same extent for both observation methods [520, 485, 545]. For organic acids, such as acetic, it has been observed that under non-robust conditions, the axially viewed plasmas are more sensitive to matrix effects than the radially viewed ones. The difference in the magnitude of the effect for the two viewed methods being within the 15–30% range, depending on the energy sum of the line tested. Nonetheless, when the plasma is operated under robust conditions, the results are similar for both observation modes. Note that the results with axial viewing probably depend on the collection beam aperture and the location of the beam waist. Any change in the carrier gas flow rate will change the optimum zone located at the beam waist level.

6.2.3.4 Effect of additional variables

The liquid flow rate has a clear effect on the extent of the interferences. It has been confirmed that the matrix effects are more severe at low than at high values of the liquid flow rate variable [519, 537]. Therefore, procedures such as matrix matching have to be conducted more carefully when low liquid flows are used.

The characteristics of the emission line are also relevant in terms of the extent of the acid effects. Therefore, it has been reported that by increasing the incident power by an amount of 20–30 W, when using ionic lines with E_{sum} between 12 and 16 eV, the acid effects are partially eliminated. When dealing with lines with lower E_{sum} values [397], the increase in the hf power required to reduce the acid effects was far above 20–30 W.

6.2.3.5 Effect on the equilibration time

Acids also modify the time required for the system stabilization when switching among solutions containing different matrix concentrations. These are the so-called adaptation effects or acid transient effects [97, 532, 537]. Thus, Maessen et al. observed that when the system was conditioned with a solution of 20% HCl, and then, when another solution containing 1% of this acid was introduced, the ICP–AES emission signal took about 5 min to reach a constant value. On switching from 0.05 to 10% HCl, this time was at least 3 min. Therefore, to get better analysis precision, longer pre-nebulization times might be necessary when changing from one acid concentration to another. Botto [145] also noticed the adaptation effects for nitric acid solutions. Stewart and Olesik [532] found that when a nitric acid solution was introduced after running the system with a more diluted nitric acid solution, the analytical signal initially decreased sharply and then slowly increased up to a constant value. Conversely, when the diluted nitric acid solution was re-introduced, the analyte signal suffered from a rapid increase ('overshoot') and further, slowly decreased up to a plateau. Figure 6.2 shows an example of transient matrix effects for two different spray chambers. These transient effects led to a variation in the time required to reach the analytical signal constant value (i.e. to perform the analysis). Several factors influenced the acid transient effects: (i) the difference between the acid concentration of the solutions

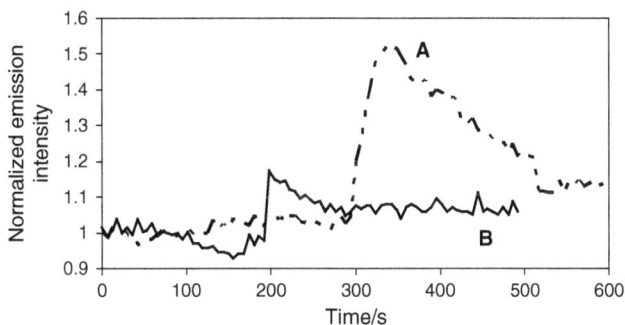

Figure 6.2 Normalized emission intensity versus time when switching from a $2 \, mol \, L^{-1}$ to a $0.1 \, mol \, L^{-1}$ nitric acid solution for a double-pass spray chamber (A) and a cyclonic spray chamber (B). (Taken from ref. [729].)

nebulized (i.e. the higher the difference, the more pronounced the transient effects), (ii) the robustness of the plasma (the higher the MgII/MgI ratio, the lower the transient effects) and (iii) the liquid flow rate (i.e. the higher the liquid flow rate, the less pronounced the transient effects).

The mechanism proposed by Stewart and Olesik [532] was based on considering the preferential evaporation from the chamber walls (and/or aerosol droplets) when different acid concentrations were present. Hence, when the spray chamber was conditioned with 25% (v/v) nitric acid, followed by the introduction of a 2% (v/v) nitric acid solution, the amount of solvent evaporated from the aerosol droplets was higher than that produced from the chamber walls. Note that the vapour pressure for the former solution was lower than for the latter. This fact made the aerosol transport through the spray chamber more likely, thus giving rise to a rapid increase in the analytical signal. After a given period of time, the solution on the spray chamber walls (initially 25% nitric acid) became more diluted because of the coarse droplets of the 2% nitric acid solution. This made the contribution of the aerosol to the solvent evaporation to be less important, whereas the solvent mass evaporated from the chamber walls increased. The process finished when the composition of both primary aerosol and solution on the walls of the spray chamber was the same. At this moment, the system was run under the steady state and the analytical signal obtained corresponded to the 2% nitric acid solution. Further experiments performed by Stewart and Olesik [532] confirmed this interpretation.

Finally, it has been observed that the use of a cyclonic spray chamber reduces, to a great extent, the acid transient effects with respect to a double-pass spray chamber. A possible explanation to this effect could arise due to the fact that the coarsest droplets are quickly eliminated from the cyclonic spray chamber walls, so that the contribution of the solvent evaporation from the chamber walls to the acid effects is reduced with respect to the double-pass one. These results are very interesting when the sample volume available is the limiting factor to perform the analysis. Thus, the volume of sample required to reach a constant signal value for a 3.6 mol L^{-1} nitric acid solution after introducing 0.9 mol L^{-1} nitric acid is reduced from 0.9 (double-pass) to 0.4 mL (cyclonic). Note that a MCN that is operated at 40 μL min^{-1} had been used [828].

The acid effects are mitigated when operating the plasma under robust conditions and with spray chambers with easy aerosol trajectories towards the plasma, such as cyclonic designs. Because of the appearance of transient effect, washing the spray chamber with distilled water between samples is not an appropriate procedure when inorganic acids are present in the solution. In any case, rinsing times are often too short and must be carefully validated.

6.2.4 Methods for overcoming acid effects

According to Botto [145], an ideal method for correcting matrix interferences should meet several conditions: (i) to make it possible to analyse any aqueous sample matrix type using a single set of reference solutions, (ii) to be applicable to both single- and multi-element analysis, (iii) to be applicable to a mixture of matrices, (iv) to be simple and applicable to several compounds and (v) to not require periodic calibration. It should be borne in mind that there is no single method that allows for the complete elimination of the acid effects or, if it does, it is difficult to apply and is time consuming. This highlights the magnitude of the problem associated with the presence of acid.

Matching the acid concentration for both the standards and sample is a very frequently applied procedure [117, 123, 124, 328, 379, 396, 439] and certainly the most efficient. However, the method could be difficult to apply when dealing with acid mixtures or when the matrix was poorly defined.

The standards addition method is widely used to correct the signal variation observed in presence of acids [338, 429]. The main limitation that can be assigned to this method is that it is time consuming.

Internal standardization is also a widely studied method for correcting acid effects in ICP–AES and ICP–MS. This method can be used to improve both repeatability and accuracy [116, 539]. The selection of the element to be used as an internal standard is of crucial importance [42, 68, 112, 116, 326]. The physical properties of the analyte and reference line must be matched in order for the signal ratio to be insensitive to changes in the experimental parameters. Several references emphasized that not a single standard could be used to correct the acid matrix effect. Though in ICP–MS, the selection of the internal standard is partly driven by the mass [328] of the elements and their ionization energies, in ICP–AES the situation is also complex as it is linked to the optical properties of the lines, when the effects are related to change in the plasma.

There are several variations of the IS method that have been applied in ICP–AES as well as in ICP–MS to try to improve the results and to overcome the problems of the traditional IS procedure. The so-called parameter-related internal standard method (PRISM) [138] demonstrated that, by taking only two emission lines with two single-plasma parameters, the responses of 24 elements under 10 operating conditions could be predicted. The use of the PRISM method led to a 70% reduction in the total variance of a routine ICP–AES procedure. As it has been indicated by the Myers–Tracy signal compensation method (MTSC), the selection of the appropriate plasma-operating conditions (especially nebulizer gas flow and observation height) is also of crucial importance. Thus, it was proved that a single element could be used as an efficient internal standard [116]. In other surveys, the background intensity was read [87]. By using the MTSC method in ICP–MS, it was proved that both the precision and the accuracy were improved when nitric, hydrofluoric and sulphuric acids were present in the sample [326].

There are many reports highlighting the importance of the correct selection of the element to be used as IS. Some examples of internal standards are cobalt [42], scandium [439], magnesium [161] and yttrium [68]. In another study, the H_β 486.133-nm line intensity [145], used by Botto, was proposed to correct for the acid effect. He measured the H_β emission intensities for HNO_3, HCl and H_2SO_4 and normalized them with respect to a reference solution. This normalized signal grew as the acid concentration increased. Then, the H_β normalized signal was plotted against the C_i/H_β ratio, C_i being the apparent concentration previously determined for a given element by using a reference solution. Based upon these curves, the matrix effects could be corrected.

As previously mentioned, the mass of analyte transported towards the plasma depends on the presence of acids in the solution as well as on their nature and concentration. Therefore, another method for correcting for acid effects is to normalize the signal with respect to the analyte transport [187] or the laser energy dispersed by tertiary aerosols [220]. Methods based on the factorial design [142] and simplex optimization [172] have been effective for achieving acid effect reduction. The acid elimination is a procedure widely known and applied when dealing with hydrofluoric acid, for example [271]. It consists of the addition of boric acid to the solution and further evaporation until almost dryness. This procedure is, however, not applicable to samples containing *W*, and a precipitate is formed if the amount of HF used for neutralization is high. Volatile oxides production (by the addition of HIO_4 or $HClO_4$) to determine elements such as Os [272] is also a good choice for the elimination of matrix effects in ICP–MS. There are additional methods employed to overcome acid effects, such as the use of a flow injection analysis (FIA)-based system [338] and isotopic dilution (ID) [259, 276], which are not within the scope of this book.

Acid effects in ICP–AES and ICP–MS are very complex and their magnitude depends on many variables. Acids act over each step of the analysis, from the solution aspiration and aerosol generation and transport until the analyte excitation/ionization in the flame or plasma. In addition, when evaluating a new liquid sample introduction system or detection apparatus, its response to the matrix effects must be investigated. The large number of approaches proposed to overcome the acid effects highlights the importance of such effects. In any case, compensating for acid effects is highly challenging

---

6.3   EASILY AND NON-EASILY IONIZED ELEMENTS

The EIEs have traditionally been the most studied in the field of ICP elemental matrix effects [248, 48]. However, not only the low-IP elements do cause a matrix effect but also elements with higher IP have demonstrated to give rise to similar outcomes on the analytical signal. The non-EIEs produce significant changes in the analytical signal as well [72], but these elements must be present at concentrations higher than 0.05 mol L^{-1} for the effect to be significant [137, 139, 165, 197].

The effect of the EIEs in plasma spectrochemical techniques is rather complex [28, 29, 139, 165, 238, 411]. Sentences such as 'Studying the literature on these interferences in the ICP is a *study of confusion*' reveal in an accurate way the actual situation concerning these analytical problems [76, 149]. Blades and Horlick [76] firstly attributed this confusion to the inconsistency of the published data concerning these matrix effects, because some authors found enhancements, depressions or no effects at all on the analytical signal when EIEs were present. Secondly, there is no single mechanism that satisfactorily explains the way in which these interferences are produced. As mentioned by Hieftje et al. [352], the sources of the EIEs and non-EIEs interferences are just partially characterized.

6.3.1 Physical effects caused by easily ionized elements

6.3.1.1 Influence on the aerosol characteristics

In general terms, primary aerosol characteristics are not modified when some elements are present in the sample at high concentrations. For solutions that are enriched with sodium chloride, the primary aerosols generated by a conespray and a cross-flow nebulizer have the same characteristics as for water [545]. These results were later confirmed for concentric and V-groove nebulizers [678].

However, in presence of EIEs, tertiary aerosols are finer than for plain water solutions [654, 678]. These results have been explained in terms of the coulombic droplet fission [605]. Xu et al. [654] stated that there is a critical diameter below which a charged droplet breaks down, thus giving rise to other progeny droplets with lower diameters. Pneumatically generated droplets have net electrical charge on their surface. This electrical charge is higher for ionic solutions and increases with the salt concentration, thus increasing the Coulomb fission frequency and yielding small droplets [654]. The drop size charge distribution in the aerosol depends on the solution composition (i.e. salt nature and concentration). Xu et al. [563] observed that when nebulizing a 0.1 mM $CaCl_2$ solution, the finest droplets did not carry a net charge, whereas the coarsest ones did. However, for a 100 mM NaCl solution, it was found that the finest aerosol droplets had net electrical charge. Experiments performed with a stirred tank to gradually increase the matrix concentration also suggest the fission of droplets through Coulomb repulsion [829]. The change in tertiary aerosol characteristics induced by sodium and calcium depends on the nebulizer conditions and sample introduction system. Thus, for example, for a desolvation system, it has been observed that the presence of sodium causes an increase in the mean size of the tertiary aerosols [386].

6.3.1.2 Effect on the solution transport rate

Traditionally, the changes in the aerosol transport towards the plasma induced by the presence of a given concomitant have been considered to be negligible. It has been observed that similar S_{tot} values have been measured for water and a 10,000 μg mL^{-1} potassium solution [231]. The introduction of a 10,000 μg mL^{-1} sodium solution, in turn, caused a drop in S_{tot} by only around 10% with respect to water [545].

For a cross-flow nebulizer, Lee et al. [170] obtained an analyte transport efficiency of 1.3% for a plain water Mn solution. Neither the addition of Na nor that of K modified this parameter in any significant way. Blades and Horlick [76] concluded that the effect of the presence of sodium on the aerosol transport phenomena corresponded to 23% of the total interference found. Other authors encountered that the presence of EIEs led to a 10% drop in the W_{tot} value as compared to the values obtained in absence of these elements [486].

Few works have tried to elucidate the mechanism responsible for these changes in analyte transport induced by the matrix. One of the possible explanations is based on the so-called Aerosol Ionic Redistribution (AIR) [43, 51, 122, 141, 163]. Borowiec et al. [51] proposed a mechanism to explain the AIR phenomenon, in which the ionic redistribution might occur mainly by an initial stripping of the liquid bulk and a further rupture of the formed ligament. The same effect could take place when an initial large droplet is dispersed into many smaller progeny droplets due to Coulombic repulsions [17]. In fact, Borowiec et al. [51] pointed out that the redistribution of ions at the water–air interface is directly related to the appearance of an electrical double layer. The capability of the cations to migrate towards this double layer depends on their mobility (i.e. hydrated radii). Further discussion and references on aerosol charging can be found in the useful review published 20 years ago by Sharp [231]. There are several evidences of the occurrence of the AIR in ICP. For example, when the spray chamber drains was recirculated and then analysed, it was observed that, in presence of sodium, the analyte concentration in coarse aerosol droplets was higher than in the analysis of a plain water solution [545]. Some experiments performed with a dual nebulizer system also showed the appearance of an AIR in ICP–AES [33].

EIE elements do not modify the primary aerosol but modify the tertiary aerosol during the transport within the spray chamber.

6.3.2 Effects in the excitation/ionization cell

The effect of EIEs on the plasma thermal characteristics can be very different, depending on many factors such as the operating conditions, matrix concentration and even the method employed to determine each of the parameters. In some instances, these elements do not modify in a noticeable way either the excitation temperature or the electron number density [25, 31, 35, 74, 86]. In other instances, it was reported that calcium matrices caused a 100–150 K drop in T_{exc} [165, 534]. To add more complexity to the matrix effects, it was found that cesium solutions induced 300–350 K increases in T_{exc} with respect to plain water solutions [33, 40]. With regard to n_e, some simple calculations demonstrated that the contribution of EIEs to the total electron number plasma density could be considered to be negligible [39, 150]. However, there are some reports indicating that calcium solutions decrease n_e [534], whereas in other cases, lithium induces huge increases in this parameter [152].

The simulations carried out by Holclajtner–Antunovic and Tripkovic [376] help to understand the actual situation. These authors considered that the presence of EIEs only modifies n_e in the low-plasma temperature zone (i.e. from 3000 to 4000 K) where these elements can be ionized, whereas the argon ionization yield is very low. In contrast, in the high-temperature plasma zone, the argon ionization is the main process responsible for the electron production, so that the presence of a concomitant does not modify n_e in a significant way. As for T_{exc}, the effect of the EIEs on n_e depends on the operating conditions. Thus, this parameter is expected to be modified by the presence of an EIE when working at low hf power and high nebulizer gas flow rate (i.e. non-robust plasma conditions) [235]. Under these conditions, the additional electrons supplied by the interfering species could modify the extent of analyte excitation.

Ionic temperature, T_{ion}, can also change when an EIE is present in the solution, but this is produced only when taking the signal from low-temperature plasma locations (e.g. at the plasma top) [376]. Otherwise, no change in T_{ion} is observed, as it has been claimed for sodium chloride solutions in ICP–MS [192]. A similar comment can be made for electron temperature values [300].

The plasma thermal state can also be described by means of the ionic to atomic net emission line intensity ratio. Besides its ability to monitor changes in the plasma conditions, this parameter is easier to obtain than other previously mentioned parameters (i.e. temperatures and n_e). Several elements have been used to obtain this ratio. Among them, magnesium is often chosen [318, 470]. Following Dennaud et al. [660] (Table 6.1), the use of the MgII/MgI ratio allows us to recognize the source of the matrix effects. A general conclusion that can be drawn on this subject is that the magnesium ratio decreases in presence of EIEs [441, 538, 612]. Besides the MgII/MgI ratio, other elements have been selected, such as Cr [572], Sr [98] or Cd, Ni, Pb and Zn [660]. The use of these elements may have some advantages as Mg is present in many samples and in some instances (i.e. environmental samples), the signal for MgII can be out of the instrument linear range. Moreover, small changes in plasma thermal conditions can be more easily monitored. However, some inconsistencies can be found, for example, the presence of sodium and potassium did not modify the CrII/CrI ratio in a significant way, whilst magnesium and calcium lowered it [546]. Dennaud et al. [660] attributed this fact to the different effect of sodium on the chromium atomic line depending on the particular ICP–AES instrument employed.

As a further proof of the perturbation caused by EIEs and the complexity of the effects when they are introduced into plasmas, it is worth mentioning that these concomitants can modify the plasma background. Results suggest that some of these elements can decrease [24, 530], increase [97] or do not affect [42] the plasma background emission intensity.

6.3.3 Effect of elements on ICP–AES analytical results: Key variables

The presence of EIEs in liquid samples can produce either an increase or a decrease of the analytical signal [235] as well as no effect [9, 19, 20]. Fundamental studies carried out by Olesik and Williamsen [248] demonstrated that in the presence of an inorganic matrix, the number of analyte ground state ions decreased with respect to water. Simultaneously, the presence of a concomitant caused an increase in the relative total number of atoms as compared to ions. The relative population of excited atoms also increased in presence of concomitants [98].

There are many variables that modify the extent of the interferences caused by EIEs [23, 31]: (i) the concentration and nature of the interfering element, (ii) the characteristics of the emission line, (iii) the plasma-operating conditions (i.e. hf power and gas flow rate) and (iv) the plasma observation zone.

6.3.3.1 Effect of the interfering element concentration and nature

Globally speaking, the matrix effects are more severe as the concomitant concentration increases [33]. As it may be seen later, these phenomena are spatially dependent. Exceptions to this general trend have also been found.

There are several studies claiming that in ICP–AES, the higher the IP of an alkali, the lower the interference effect [76, 93, 128, 148]. Therefore, the variability in emission signal for a given analyte concentration is highest in presence of cesium and lowest in presence of lithium for the same molar concentration of these interfering elements. In contrast, Thompson and Ramsey [139] reported the extent of the matrix effects in ICP–AES could be arranged as follows: K < Na < Al < Mn < Fe < Mg < Ca. Similar results were obtained by Maessen et al. [97] for Na, K, Ca and Mg. On comparing this order with the IP order for these elements (i.e. K < Na < Ca < Mg), it appears that a non-thermal excitation mechanism was likely to take place. Ramsey and Thompson [165] observed that the variation in the relative Zn signal (i.e. signal in presence of interferent divided by the signal in absence of it) was linearly related to the first IP of the interferent. Nonetheless, the slope for alkali was negative (i.e. the higher the IP the stronger the matrix effect), whereas the slope for alkali earth elements was positive (i.e. the higher the IP the weaker the matrix effect). The same authors defined the so-called matrix energy demand (MED) that was referred to as the amount of energy required to dissociate and ionize the matrices [165]. The relevance of this variable lies in the fact that a linear relationship was found between the signal variation induced by the matrix and the MED.

When several concomitants are present in the same sample, the situation becomes even more complex. In these cases, the matrix effect can be either enhanced or reduced with respect to a single concomitant solution [493, 451, 516]. For example, lithium was found to enhance the Ca signal with respect to a plain water solution. A lithium–aluminum solution also improved it, but the improvement factor was lower than that obtained when only lithium was present. This fact appears to suggest that the use of an ionization buffer does not eliminate the matrix effects produced by concomitant elements [72]. More interestingly, the mixed-matrix effects are not additive [97, 614]. For example, binary combinations of sodium, potassium and magnesium gave rise to stronger matrix effects than the sum of the effect of single-interferent solutions. Meanwhile, binary calcium solutions afforded matrix effects less significant than the sum of the interferences caused by single concomitant solutions. The additive character of binary matrices was studied for calcium–magnesium mixtures. Again, the results demonstrated the complexity of the combined matrix effects. On the one hand, it has been found that for 8 of the 11 analytes studied, the combined effect corresponded to the sum of the isolated calcium and magnesium effects. Nonetheless, for two elements, the matrix effect was lower than the sum of the interferences and, in one case the opposite was found [624]. On the other hand, it has also been indicated that this composite matrix minimized the matrix effects produced by these two elements considered separately for the determination of different arsenic species [661]. The fact that different elements were determined can be argued to try to understand these results.

In any case, matrix effects depend more on the absolute concentration value of the interfering element than on its relative concentration compared to that of the analyte.

6.3.3.2 Effect of the analyte line properties

Although several authors found a direct correlation between the analyte line excitation and/or the IP or the sum of both magnitudes (E_{sum}) and the extent of the effect of calcium [31, 84, 282, 396, 485, 853], lithium [214] and potassium [170], further studies have corroborated that this could not be considered as a rule. For mixed solutions having both Ca and K as interfering elements, an intermediate situation was found. Atomic and ionic lines are differently affected by the presence of a concomitant [106].

6.3.3.3 Effect of the nebulizer gas flow rate and hf power

There is experimental evidence that the EIEs' effects become less severe as the hf power is increased [27, 69, 144]. The relationship between intensity of matrix effects and nebulizer gas flow rate is less regular. This is because, for some values of this variable, the presence of concomitants induces a signal depression, whereas for others the opposite is found [33, 251]. A possible explanation to this fact is that the maximum interference region depends on the gas flow rate used. Thus, for example, for lithium matrices, the region of maximum interference on calcium emission was moved towards plasma upper locations as the gas flow rate was increased [72]. Other studies also point out the importance of these two variables on the matrix effect extent [173].

As mentioned previously, the plasma robustness is a concept that integrates the effect of the hf power and the nebulizer gas flow rate. Under non-robust plasma conditions, the presence of EIEs caused a drop in the value of the MgII/Mg ratio with respect to water [500]. When $10 \, \text{g L}^{-1}$ of sodium or lithium were present in the solution, the use of a non-robust plasma led to 40 and 60% drops in the MgII/MgI ratio with respect to a plain water solution, respectively [660]. Obviously, under robust conditions, these relative drops were lower (i.e. around 10 and 20%, respectively). When the plasma was operated under robust conditions, the effect of sodium was minimized, as it has been found by Dubuisson et al. [485].

The influence of the line characteristics on the matrix effect is prominent when working with a non-robust plasma [87, 326], Stepan et al. [662] confirmed, for a non-robust plasma, the scattering in the values of relative intensity (i.e. intensity with concomitant/ intensity without concomitant) for a set of atomic and ionic lines. The correlation between matrix effect extent and line properties appeared to be rather complicated. Thus, for example, some Ba lines (i.e. 230, 233 and 413 nm) were strongly affected by sodium, whereas others (i.e. 493, 455 nm) were not. Lines with very close excitation energies (i.e. CdII 226 nm, 14.47 eV and CdII 214 nm, 14.77 eV) exhibited a very different response to the presence of sodium. For atomic lines, the effect of sodium was even more complicated than for ionic ones, the trend being completely erratic. When a robust plasma was used, the relative signal dispersion was reduced. Interestingly, a complicated pattern was observed when the percentage of the original signal was plotted against the excitation energy (for atomic lines) or E_{sum} (for ionic lines). The shape of these plots was exactly the same at different interferent concentrations, although it depended on the interfering element.

6.3.3.4 Effect of the plasma-observation zone

The degree of the interferences caused by EIEs depends strongly on the plasma-probed zone [31, 61, 64, 75, 311]. Note that the hf power and the nebulizer gas flow rate will have a crucial effect on the plasma zone, at which the interference will be maximum. Spatially speaking, the plasma central channel can be divided into three different zones [53] a 'Preheating zone (PHZ)'; the so-called initial radiation zone (IRZ) in which the atomic emission begins to be observed and the normal analytical zone (NAZ) that corresponds to the zone at which emission is produced mainly from ions. In general terms, the emission intensities low in the plasma (that is in the IRZ) increase upon the addition of an EIE [144, 109, 149, 235]. A great number of works have focused their attention on the plasma IRZ [40, 151]. This zone should be avoided, as the elemental matrix effect is stronger [25, 40, 52, 59, 72, 75, 149, 311, 534]. According to this plasma division, it can be indicated that for a robust plasma, there is a shift downwards the IRZ [9, 31, 76]. Therefore, the emission intensity is actually taken from the NAZ.

An example of the spatial character of the EIEs' interferences was given by Blades and Horlick [76]. They plotted the variation of the relative intensity (i.e. the emission intensity in presence of sodium with respect to that in absence of sodium) for the CaI and II lines versus the observation height. They found that this parameter was maximum at 7 mm above the load coil (a.l.c.), whereas the matrix effect disappeared for both lines at 17 mm a.l.c. At plasma observation heights above it, the relative intensity dropped. Therefore, a cross-over point was observed in which the matrix effect disappeared and it could be eliminated by a proper selection of the plasma-observation zone [166, 290]. It was indicated that the location of the cross-over point depends on the interferent [235, 413] and the plasma-operating parameters [64], whereas for a given analyte, it is independent of the wavelength [235].

> The cross-over point, that is the location where no matrix effects are observed, will move towards the coil when increasing the power and reducing the carrier gas flow rate. Observing below the cross-over point results in an enhancement of signals, whereas observing above corresponds to a depressive effect. This is the reason why the influence of the operating parameters is so drastic, because of the movement of the cross-over point.

Besides the plasma-observation height, its radial position also modifies the extent of the matrix effects caused by EIEs. Thus, Sesi and Hieftje [471] found that in its centre, the elemental matrix effects led to a signal depression, while there was an off-axis general enhancement in the emission signal. Therefore, studies combining the effect of both axial and radial position on the magnitude of interferences were performed [91]. Hieftje and co-workers [384, 415] showed that for CaI emission, the matrix effects caused by lithium were more pronounced off-axis high in the plasma and both on axis and off axis low in the plasma. For CaII, the interference existed on axis up to 25 mm a.l.c. and off axis from 5 to around 18 mm a.l.c.

The change in the radial matrix effect as a function of the radial plasma position was attributed to the spatial distribution of the aerosol-charged droplets when they entered the plasma [563]. When EIEs were present as concomitants, droplets carrying a net electrical charge suffered from Coulomb fission processes. These small droplets drifted towards the aerosol stream periphery because of the electrical repulsion, whereas large droplets with low net charge remained within the aerosol stream centre.

The modification in the extent of matrix effects as a function of the plasma observation zone is accompanied by changes in plasma fundamental properties, like n_e or temperatures. The reported results indicated that EIEs caused increases in n_e at low plasma locations [353] that are more pronounced in presence of concomitants with low IP. The magnitude of this increase followed the order: potassium (40%) > sodium > lithium > calcium > magnesium.

If the spatial study is done with high resolution, it may be observed that the local plasma conditions are very different in locations close to a droplet or a solid particle [335, 385]. Thus, a fall in plasma temperature by 1000 K was reported at a region from 1 to 2 mm surrounding a non-desolvated droplet. As a result, the electron number density at this location decreased by a factor of 10 [385]. In other words, the matrix effect was more pronounced near the incompletely desolvated droplets or vaporizing particles than away from them.

The plasma-observation position effect also depends on the line tested. In this way, Romero et al. [500] claimed that the effect of the plasma-observation height on the sodium matrix effect for the Ba 233 nm ionic line was different to that described before and no cross-over point was found or it was located higher in the plasma than for the remaining elements. Of course, this happened under non-robust plasma conditions. When a robust plasma was used, the signal changes induced by sodium were the same, irrespective of the ionic line tested [485]. For atomic lines, the changes in the analyte signal profiles induced by sodium were more significant than for ionic ones [235].

> Very robust conditions along with a large injector bore can result in a cross-over position within the coil. This means that with radial viewing, only depressive effects would be observed. In any case, studying the matrix effect at a single observation height may be totally misleading. Note that move of the observation height is not always possible with radial viewing-based instruments, a compromise location being adjusted at the factory.

6.3.3.5 Effect of the plasma observation mode

Interferences caused by EIEs in axially (or end-on) viewed plasmas have been extensively studied [140, 662]. The first studies pointed out that one of the drawbacks of this observation mode was the severity of the matrix effects by comparison to a radially (side-on) viewed plasma [32]. For sodium and potassium solutions in concentrations higher than 1000 and 10 mg L^{-1}, respectively, the signal variation with respect to an aqueous solution was higher in an end-on (i.e. axial) than in side-on (i.e. radial) observation mode [403, 584]. However, Dubuisson et al. [485] observed that, under robust plasma conditions, the matrix effects produced by sodium were similar for both observation modes.

6.3.3.6 Influence of the liquid flow rate

The liquid flow rate (Q_l) modifies the magnitude of the elemental matrix effects. In some cases [586], increases in this variable have led to more severe interferences. However, the opposite trend has been reported for conventional [339] as well as low sample consumption systems. Thus for a high-efficiency nebulizer (HEN) [388], (Chapter 3) the sodium and calcium effect on the ICP-AES emission signal is more pronounced at low (i.e. 5–50 µL min^{-1}) than at high (i.e. 0.8 mL min^{-1}) liquid flow rates [629]. These latter results are in full agreement with those previously discussed for inorganic acids. Therefore, this trend seems to be related with the ionic character of the matrices.

6.3.3.7 Effect of the liquid sample introduction system

As for acids, the design of the nebulizer [33], spray chamber and desolvation system plays a very important role in terms of interferences. For EIEs, the torch design has proven to preclude the final interference degree.

In the case of sodium solutions, the effect for two different pneumatic concentric nebulizer designs was of the same order [244]. When considering different designs, it has been concluded that these effects were more [500] or less [611] pronounced for a cross-flow than for a concentric nebulizer. Other reports indicated that a GMK nebulizer behaves better than both cross-flow and concentric designs [611]. The concentric pneumatic nebulizers specifically developed to handle high salt-content solutions (e.g. seaspray) provided, under some circumstances, less severe sodium effects than the V-groove nebulizer [678].

Some micronebulizers (i.e. MCN) appear to be very sensitive to the presence of saline matrices as compared to a conventional concentric nebulizer [528]. The data concerning the sodium matrix effects in ICP–MS showed that with the MCN, the direction of this effect depended on the nebulizer gas flow rate used [455]. The DIN has been used by LaFreniere et al. [154] to study the effect of aluminum and sodium on the emission intensity of Ca and observed that for a non-robust plasma, the extent of the matrix effects induced by these two concomitants was similar to that found with a pneumatic cross-flow nebulizer. The DIHEN is also sensitive to the presence of sodium in the solution both in ICP–AES [652] and ICP–MS [649].

Concerning the spray-chamber design, it is generally accepted that cyclonic designs compare favourably against double-pass designs in terms of sodium effect. Thus, under plasma robust conditions, Dubuisson et al. [545] reported that a 10 000 µg mL^{-1} sodium concentration led to around 30 and 10% drops in the analytical signal for a double-pass and cyclonic spray chamber, respectively. A low inner volume cyclonic spray chamber leads to a fivefold reduction in these interferences with respect to a double-pass spray chamber [603].

As noted for acid effects, sample introduction systems were not optimized for matrix effect reduction. It could be thought that an efficient combination of nebulizer and spray chamber could minimize EIE effects, at least for a given set of power and carrier gas flow rate.

The use of a desolvation system does not result in a mitigation of the elemental matrix effect [340, 429]. One reason for this observation is that for salty solutions, nucleation becomes important [101, 355]. In presence of a dissolved salt, the number of condensation nuclei and, hence, the solvent transported to the plasma increase. The use of a desolvation system (i.e. a heating and a condensation unit) with a membrane to remove the solvent from the aerosol does not solve the problem [530].

Both the hf power and the outer gas flow rate can be reduced by using a torch of reduced dimensions (i.e. so-called minitorch). With this setup, the spatial interferences appear to be less pronounced in the plasma base than for a conventional torch [52]. Montaser et al. [119] found that the effect of sodium on the CaI emission signal was stronger than for a conventional torch. The low excitation temperature (i.e. about 4 000 K) and electron number density (i.e. about 2×10^{14} electron cm^{-3}) in reference to a normal plasma could make these plasmas more sensitive to the matrix effects. If a water jacket is used to cool the torch's outer tube, the argon consumption is further reduced and a stable plasma can be obtained at $0.9 \, L \, min^{-1}$ argon and 0.7 kW power. In this case, the existing results indicate that matrix effects are extremely severe [45].

All the variables mentioned up to now interact among themselves and lead to unpredictable interferences. This interaction is likely in the origin of the contradictory results reported in the literature. All the operating conditions must be strictly the same to draw conclusions from two independent studies.

6.3.4 Proposed mechanisms explaining the matrix effects in ICP–AES

There are many studies dedicated to possible explanation of the mechanism responsible for the effects caused by EIEs. Apart from nebulization and aerosol transport effects, there are some plasma phenomena that can be a source of interferences [76].

1. Many investigations [33, 98, 136, 149, 155, 175, 179, 209, 263, 311, 321, 322, 352, 384] are focused on the analyte atomization and excitation.
2. Changes in the ionization equilibrium have also been invoked to be in the origin of the interferences [33, 70, 86, 98, 136, 149, 152]. In this case, the rate of electron–ion recombination increases in presence of an interfering element [210].
3. The presence of EIEs may cause an enhancement in the collisional excitation [12, 16, 34, 149]. The increase in the electron number density can also produce a rise in the number of collisions, thus giving support to an enhancement of both atomic and ionic emission intensities.
4. Volatilization effects might be relevant in presence of EIEs [31, 40, 149]. Analyte particles are occluded in solid matrix particles and, hence, there is an additional difficulty for the excitation of analyte atoms. As a result, the analyte emission profiles change.

5. Ambipolar diffusion [149, 155, 175, 179, 209, 322]. Light-charged particles (electrons) tend to diffuse in the plasma more efficiently than heavier ones (analyte, matrix). As a result of these movements, a charge separation can be observed, thus generating an electric field. The potential of this field tends to increase the diffusion rate from lower to higher electron density zones for heavy ions and to decrease this rate for the light particles. Neutral atoms are not affected by this process.

6. The lateral diffusion can also be responsible for these interferences [384]. If the presence of high concentrations of matrix elements causes the analyte particle vaporization rate to increase, the lateral diffusion is enhanced. In this case, the analyte concentration in the NAZ decreases, because the analyte species have had more time to diffuse outwards the plasma.

7. The Penning ionization by metastable argon [26, 38, 100] has also been mentioned as a plausible phenomenon affecting the extent of the EIE interferences. Thus, the high-electron density observed in the plasma (higher than expected from the measured excitation temperatures and assuming LTE) might be related to an overpopulation of argon metastable atoms, which could act as ionization buffer.

8. Other phenomena have been proposed such as changes in the plasma thermal conductivity [318], radiation trapping by argon atoms [96, 99, 131], charge transfer involving argon species [71, 113, 176, 770], a shift from collisional deactivation to radiative decay [735] and loss of thermal capability of the plasma because of energy consumption to vaporize and ionize the matrix [194]. The radiative heating of the sample droplets has been considered as the main process responsible for changes in the spatial distribution of Mg species induced by the presence of high concentrations of zinc and cesium [574].

All these phenomena do occur simultaneously and the relative significance of them depends on variables such as those previously discussed. Furthermore, different processes may take place at different plasma locations [235]. For example, Blades and Horlick [76] concluded that low in the plasma (below 15 mm a.l.c.), a collisional analyte-excitation mechanism could explain why alkalis give rise to increases in the analytical signal, whereas higher in the plasma, the ambipolar diffusion could be the predominating process yielding a suppression of the analytical signal.

A significant amount of work has been devoted to the effects of EIE elements without a clear elucidation of the mechanisms resulting from their presence. Many mechanisms have been suggested, with different locations in the plasma and different kinetics, and unfortunately, not a single mechanism is predominant.

6.3.5 Effect of elements on ICP-MS analytical results: Key variables

Elemental non-spectral matrix effects in ICP–MS are more severe than in ICP–AES [330]. This is likely because of the predominant role of the ion extraction and focusing in the mass spectrometer [373]. Serious elemental interferences are observed in ICP–MS at concentrations at which there is no remarkable effect in ICP–AES [54, 159, 169]. In a classification of the ICP–MS non-spectral interferences, Dams et al. [435] divided them into two categories: (i) reversible, which produce a change in the analytical signal only observed as long as the matrix is introduced into the spectrometer and (ii) irreversible, which are caused by the deposition of matrix salts or oxides at the MS interface. The second kind of interferences cause a steep drop in the analytical signal and are observed even when no matrix is introduced into the instrument.

In general terms, the presence of an element in excess causes a drop in the ionic intensity [107, 177, 191, 204, 205, 206, 218, 219, 224, 229, 236, 257, 262, 332, 333, 363, 380, 402, 405]. These results are not a general rule, because several authors have noticed increases [313, 620] or no modifications [111] in the signal produced in presence of concomitants.

Beauchemin et al. [230] performed a complete study of the matrix effects caused by 0.01 mol L^{-1} lithium, sodium, potassium, cesium, magnesium, calcium, boron, aluminum and uranium solutions on the ICP–MS ionic intensities for several analytes. The trends found were as follows:

1. Na, K, Mg and Ca led to a rise in the signal. An interesting effect was found for sodium. The signal increased by up to a 100% in presence of this interferent. The interferent effects found for potassium were even more pronounced than for sodium. Magnesium and calcium resulted in a less-important matrix effect than alkalis.
2. Li did not modify the analytical signal, though its IP (i.e. 5.39 eV) is similar to that of sodium (i.e. 5.14 eV).
3. B, Al and U caused signal drops. Generally speaking, close to 50%. Note that, interestingly, and despite their different effect on analytical signal, the IP of aluminium is close to that of lithium (i.e. 5.99 and 5.39 eV, respectively).
4. With Na, K, Cs and to a lesser extent Ca and Mg, the extent of the effect was more important as the analyte m/z increased, whereas for the other three concomitants, the extent of the ion-sampling effects was independent of the analyte m/z.

The presence of an element at high concentration also makes the background ionic current to change. Thus, Tanner [432] increased the concentration of bismuth and sodium and registered the variations in Ar$^+$, ArH$^+$, ArO$^+$ and Ar$_2^+$ intensities. For bismuth (and matrix elements with IP above 6 eV) an increase in the concentration caused a drop in the argide ions. Other ionic species as H$_2$O$^+$, H$_3$O$^+$ or NO$^+$ behave in a similar way. In the case of matrix elements with IP below 6 eV (sodium), the argide ion signals were enhanced at high-matrix concentrations, the improvement for the rest of the ions being less pronounced.

6.3.5.1 *Effect of the nebulizer gas flow rate*

The nebulizer gas flow rate has proved to be of crucial importance to these interferences [201]. Some of the contradictory results found in the literature on this subject could be attributed to the measurement of the interference at different values of this variable [377].

Longerich [258] found that sodium induced changes in the plots ionic signal versus gas flow rate. In the same direction, Rodushkin et al. [522] reported that, at a Q_g of 0.9 L min^{-1}, the signal for a given set of elements in presence of a seawater matrix was the same as in absence of it, whereas at $1.2\,L\,min^{-1}$, the signal was about a half of that obtained in its absence. The interferences derived from the so-called space-charge effects are mitigated using conditions leading to a decrease in the relative matrix–argon ion fraction. The conditions suitable for this purpose are low gas flow rate and high hf power [217].

6.3.5.2 Effect of the plasma-sampling position

The spatial profiling of the analyte signal is very useful to get more insight into the matrix effects in ICP–MS [135, 309, 334, 398]. Thus, the signal suppression produced by a given matrix element was more pronounced at the plasma central channel, likely because of the higher matrix concentration encountered in this plasma zone [641].

6.3.5.3 Influence of the interferent and analyte properties and concomitant concentration

The dependence of the matrix effect on the analyte IP is a function of the sampling resolution. It has been reported that the ions going through the sampler cone are those contained in a plasma cylinder of a given size (i.e. 0.3 mm radius × 2 mm length) [228]. The extent of the matrix effects suffered by the analytes present in this plasma region may be the same, irrespective of their IP [219]. In the case of the matrix, it has been found that the lower the first IP, the more pronounced the interferences [260]. To evaluate the effect of the analyte mass, it has been found that the heavier the element, the less severe the matrix effect [189, 522]. Some interesting effects are found as the mass of the interferent increased. Thus, heavy-matrix elements give rise to stronger interferences than light ones [202]. Therefore, the worst case would be a light analyte ion in presence of a heavy-matrix ion [211]. Nonetheless, contradictory results have also been published on this subject. Thus, no correlation between analyte mass [299] or concomitant mass [241] and matrix effect extent have been found. Finally, Fraser and Beauchemin [641] found that, high-mass concomitants contribute more severely to the interferences than low-mass ones due to space-charge effects that will be described later.

The concentration of the matrix has a direct effect on the ionic signal measured for its ionic signal. Tanner employed low hf power values (i.e. cold plasma) and found that an increase in the concomitant concentration led to an initial linear rise in the signals for the interferents [432]. A further increase in the concentration produced a second linear zone. For elements having IP higher than 6 eV, the slope of this second region was close to zero.

Analyte properties such as its chemical form should also be of concern, particularly when comparing a small inorganic molecule to a large organic one, as in speciation.

6.3.5.4 Effect of the spectrometer configuration

The configuration of the mass spectrometer can also have a relevant effect on the extent of the matrix interferences. Thus, Taylor et al. [425] concluded that the use of a sector field MS (SFMS) would reduce the likelihood of interferences with respect to a quadrupole device (QMS). The reason for these assessments could be the acceleration of the ions up to higher energies for the SFMS [406]. The recoveries obtained for several elements in a seawater sample were systematically higher for the SFMS than for the QMS. This fact highlights the importance of the acceleration potential or of differences in the interface design and was in contrast with some published results [448]. In contrast, some other results suggest that the matrix effects are similar for both configurations [428]. Different trends are found for a conventional quadrupole MS and for a twin quadrupole one [476], as the former transmits basically the ions from the beam's outer region, whereas the latter transmits the ions from the beam's central axis.

6.3.5.5 Additional variables

Crain et al. [241] reported interesting variations in the extent of the sodium, potassium and uranium matrix effects as a function of the sampler and skimmer orifice inner diameter. Thus, if the skimmer orifice was larger than that of the sampler's, the higher the analyte IP the stronger the matrix effect. When the skimmer orifice was smaller than the sampler one, the signal suppression factors were higher and the signal depression onset was located at lower matrix concentrations.

The ion lenses settings can also modify the extent of the signal suppression caused by a concomitant [291, 362]. On changing the voltage of the first element of the ion lenses towards positive values, Crain et al. [241] perceived a change in the interferences caused by sodium and uranium. Denoyer et al. [423] mentioned that the matrix effects can be controlled by keeping the extraction lens voltage at low negative or positive values. Other authors have proposed a reduction in the extraction lens voltage to reduce the charge separation (and, thus the matrix effects) caused by a concomitant [372]. Allen et al. [476], on evaluating the effect of the extraction lens potential on the single droplet signal for Li^+ in presence of lead, realized that it changed drastically as this potential decreased from -10 to -130 V. Simultaneously, the peak for Li^+ in absence of lead did not change in a significant way. Chen and Houk [466] mitigated the effect produced by $1\,000\,\mu g\,mL^{-1}$ cesium on Sc signal by applying a slightly positive potential to the first lens. The same authors also noticed that the matrix effects were more pronounced in presence of a photon stop, because it caused a slight drop in the Sc transmission. Therefore, the photon stop voltage should also be modified to alleviate the matrix effects.

The use of a supplementary electron source, consisting of a heated tungsten filament placed between the skimmer and extraction lens (i.e. where the space-charge effects dominate, see later) also has an effect on the interferences extent [594, 595, 597]. Because the electrons are introduced along the beam through the skimmer, the space-charge effects produced between positive ions are virtually eliminated and the dependence of the matrix (sodium) effect on the analyte mass disappears. There are other strategies to minimize the matrix effects in ICP–MS as, for example, the use of a three-aperture interface [394] or keeping the torch shield floating in the case of a sector field instrument [618]. The use of a Pt grounded guard electrode has also been proposed to improve the analytical figures of merit of a SFMS [578, 580].

As in ICP–AES, the use of a minitorch to lower the argon consumption rate has been envisaged for ICP–MS [285, 320]. This choice is interesting because, it has been realized that the sodium effects are similar for the minitorch and for a conventional torch [639].

Rehkämper and Mezger [633] employed a hexapole collision cell to reduce the matrix effects due to K and Nd in Pb isotope ratio measurements. The collision cell technology was mainly developed for the elimination of spectroscopic interferences. Nevertheless, the use of Ar as collision gas effectively blocks the transmission of elements with an atomic mass lower than 40 amu.

6.3.6 Proposed mechanisms explaining the matrix effects in ICP–MS

Several attempts have been made to explain the way in which a given element in excess modifies the analytical behaviour in ICP–MS [211, 217, 299, 377]. Among the possibilities that were tested, it may be found that some of them are already mentioned for ICP–AES. Thus, the ionization suppression could lead to a depression in the ionic current more severe for analytes with low IP. As mentioned by Tanner [432], there is also the possibility of analyte fractionation during the plasma-vaporization process. Hence, some elements would be vaporized at a different rate than others. Clogging of the sampler cone has also been mentioned to explain the interferences [225]. In addition to these processes, other mechanisms invoked as responsible for the EIEs effects in ICP–MS are as follows: (i) Mass-dependent diffusion [211], (ii) changes in the efficiency of the ion transmission from the plasma to the MS detector because of space-charge effects [155, 201, 217, 342] and (iii) analyte deposition on the sampling cones [209, 260].

The phenomena labelled as (ii), which could be partially related to the point (iii), are mass dependent because of space-charge effects [360, 468]. According to some published calculations, it is unlikely that the presence of a concomitant could cause a significant change in the ionization equilibrium position [217, 308]. Nonetheless, the introduction of such a sample causes a mass-dependent rise in the ionic current downstream the skimmer cone. The occurrence of space-charge effects is also supported by additional results showing that the sense of the matrix effects can be modified by changing the DC voltage applied to the ion's optical components [209, 476].

The space-charge effects are considered to be largely responsible for the matrix effects caused by a concomitant in ICP–MS. Interesting experiments have been performed to ascertain the existence of these effects consisting of the placement of a screen in the vacuum chamber behind the skimmer or the ion lenses and the ions collection and observation of their distribution on this screen [433, 466, 474]. A metal collection disc placed at the differential pumping aperture to the third stage has been exposed to the ion beam and further analysed through a scanning electron microscope. The results have shown that a beam broadening is produced in presence of a high concentration of a given element. The spatial distribution for low-mass elements is broader than for heavier ions. Similarly, it has been observed that the beam for Sc is broader than for Ba [547]. However, Chen and Farnsworth [498] concluded that these collection methods exhibited some limitations: not all the elements adhered to the targets with the same efficiency; the analyte stream composition had a direct effect on the rate at which the deposit is formed; non-analyte species (i.e. oxygen) could cause changes in the distribution of the elements on the screen; the deposit could be eroded by the action of buffer gas ions; and, some neutral species could also be deposited onto the screen.

The zone at which the space-charge effects take place was a subject of controversy, because in some cases it has been mentioned that these effects occur in the ion-extraction step [135], whereas in other cases, it has been concluded that they are generated between the sampler and the skimmer cones [332, 333, 334].

Time-resolved ICP–MS signals have been obtained from the ion clouds generated from monodisperse individual droplets through a monodisperse-dried micro-particulate injection (MDMI) system [443, 452]. The results showed that Pb caused a Li^+ cloud broadening, which was more pronounced as the lead concentration was increased. The repulsion of Li ions by Pb ions at the cloud front side made Li^+ to reach the detector earlier than Pb^+. The opposite trend was found at the rear side of the ionic cloud, so that Li^+ arrived after Pb^+. The existence of a strong enough Li repulsion could make Li ionic current to peak off cloud axis. Agreeing with this explanation, the broadening caused by lead on heavier ion as Cu^+ was less severe. Some results also supported that this broadening took place after the ion sampling, for example, the fact that the time-resolved signals in ICP–AES and in ICP–MS were identical in absence of lead, whereas in presence of this concomitant the ICP–MS signals were wider than the ICP–AES ones [452]. Differently, it has been concluded that the effects could take place in the early stages of the ion lens optics [394]. This assessment was confirmed by Allen et al. [476] when studying the effect of the extraction lens voltage on the Li^+ peak shape. These authors also found that the second lens potential had a significant effect on the analyte peak shape, but less pronounced than the extraction lens potential. More exactly, it seems that the space-charge effects are located at a defined area in which the charge separation takes place in the ion beam [583]. According to the space-charge effects, higher signal changes would be expected for the QMS than for the SFMS because of the lower acceleration potential for the former. Several additional strategies have been proposed to overcome the space-charge effects such reducing the inner diameter of the sample orifice [395], so as to reduce the total ion current and, hence, the space-charge effects or removing the ionic lens. [394].

Fraser and Beauchemin [641] performed an evaluation of the different mechanisms responsible for the matrix effects produced by concomitants in ICP–MS. Three possibilities were considered: (i) changes in the ionization equilibrium; (ii) space-charge effects and (iii) ambipolar diffusion. With regard to point (i), a change in the axial distribution of the ions was perceived. Several observations pointed out that ionization suppression occurred, but was not mainly responsible for the observed effects. First, the signal drops caused by potassium or cesium were much more pronounced than those caused by sodium. Note that all three elements are totally ionized in the plasma. Second, the extent of suppression decreased as the sampling height was increased [201, 211, 299]. Third, there was no correlation between the analyte first IP and the signal depression. With regard to point (ii), some results that were inconsistent with a space-charge mechanism were obtained. Finally, two trends gave support to an ambipolar diffusion mechanism (point iii). First, the profiles of several analytes broadened in presence of a matrix element and, second, the broadening degree for doubly charged ions was higher than for singly charged ions.

The operating conditions may also play an important role on the predominant mechanism. Thus, under conditions leading to a rather cold plasma (i.e. low hf power and/or high sampling depth, in other words similar to a non-robust plasma), the ionization suppression could be the main responsible condition of the interference

[260]. The so-called inductively coupled 'cold' plasma mass spectrometry has been attentively studied, since the first approximation to this concept by Jiang et al. [216]. These authors noticed a self-induced matrix effect for potassium at concentrations over $10\,\mu g\,mL^{-1}$. The space-charge effects should play a less important role on explaining the matrix effects under these plasma conditions, because the ion current through the skimmer is lower. As pointed out by Tanner [432], the elucidation of the mechanism responsible for the matrix effects in a cold plasma–MS system would help to know the ion dynamics inside the plasma. This author observed that the interference found for a given analyte was a function of the matrix element. No direct correlation was found between the matrix concentration and the analyte signal drop. Furthermore, although an acceptable correlation was found between the IP and the extent of the matrix effect, there were some concomitants the behaviour of which did not agree with this trend. Therefore, some mass-dependent effects arose. These observations led Tanner not to rule out the possibility of the simultaneous participation of several mechanisms. A likely mechanism that would involve the participation of molecular ions as O_2^+ or NO^+, that would act as reactants transferring their charge to the matrix ions without reaching an equilibrium. Assuming an increase in the plasma–electron density produced by the presence of EIE, the ion signal suppression by an ion–electron combination mechanism would also explain some of the results reported by Tanner. In other cases (i.e. high-matrix concentrations and low-matrix element IP, below 6 eV) an extra enhancement in the ionization could explain the change in the slope of the plots signal versus matrix element concentration for the matrix element as well as for other molecular ions. However, this additional ionization did not seem to modify the response for the analytes.

6.3.7   Methods for overcoming elemental matrix effects

By taking into account the variables affecting the extent of the interferences caused by EIEs, it is possible to advance that these interferences can be mitigated by modifying the operating conditions to achieve plasma robustness. In addition, the plasma observation zone or the liquid sample introduction system can also be modified so as to minimize the impact of the presence of high-elemental concentrations on the analytical results. However, in most cases, there is a residual matrix effect that can be attributed to some of the processes mentioned in the preceding sections.

6.3.7.1   Internal standard and related methods

The efficiency of internal standardization depends strongly on the plasma-operating conditions. For example, it is recognized that for robust plasmas, the extent of the sodium-matrix effect for the several lines can almost be the same [499]. Therefore, a single element can be used as an efficient IS. In contrast, when non-robust conditions are employed, more than one internal standard are required [490]. To select the suitable IS, a method based on principal-component analysis has been proposed [720]. The plasma-observation mode has been evaluated in terms of matrix effects [470, 539]. The best of the situations evaluated in terms of applicability of IS has been the combination of a robust plasma with radial-observation mode.

Several variations of the IS method have been applied in ICP-AES to compensate for changes in the signal induced by concomitants. As for acids, the MTSC method has provided satisfactory results to compensate for the EIEs' effect [116, 223, 298]. Remember that in this case, the IS method was applied under optimized plasma-operating conditions. With the MTSC, the effect of plasma parameters on the noise was studied as well. Under optimum conditions, the noise was lower than 0.1% for several matrices, thus indicating a good time correlation between analyte and internal standard signals. Factors modifying this correlation were the spray-chamber pressure [73], the power control [36] and the amount of sample and the hf power [125].

Other IS modalities are as follows: (i) the so-called generalized internal reference method [118] in which the possible variations in the amount of sample, plasma and carrier flow rates and the hf power are taken into account; (ii) the subtraction of plasma background fluctuations in which a single aqueous standard is used [87] and (iii) analysis of covariance that is very useful to select the appropriate IS [170, 171].

Mathematical IS methodologies are also available. In the parameter-related internal standard method (PRISM) [125, 138], the relative response of the lines to changes induced by several instrumental parameters has been evaluated for 36 elements. Only a single internal standard (ZnII 202.55 nm) is used and a correction equation is obtained. A substantial improvement in accuracy is obtained with the exception of the highest matrix concentrations (calcium 10,000 μg mL^{-1}) for which the correction is incomplete. Another modification of the internal standardization is the so-called proportional correction (PC) method [451, 516]. For a group of selected lines, an approximately linear relationship is obtained between matrix effect extent and the square root of the interferent concentration (C_m). For three different matrices, the interference is reduced by around three times. Al-Ammar and Barnes have developed a method based on the use of two lines of the analyte to compensate for non-spectroscopic interferences [515, 552]. One of them is used as the internal standard line. Once the proper analyte internal reference line pair is selected, a mathematical model is derived. The main advantages of this technique are: (i) it is simple and does not require complicated mathematical calculations and (ii) it is fast and easily adaptable to routine analysis. The major limitation is to find two appropriate analytical spectral lines.

In ICP–MS, the use of an IS has been quite popular for improving accuracy. Thompson and Houk [191] suggested that the IS should have atomic mass and first IP as close as possible to those of the analyte. In other instances, the matrix effects are corrected by using IS with masses close to those for the analytes [491]. However, Vandecasteele et al. [219] have published a study on the effect of increasing amounts of sodium chloride on ^9Be, ^{27}Al, ^{64}Zn, ^{85}Rb and ^{208}Pb. They have called attention to the fact that the reported effects differ greatly. As the regression lines (analyte signal vs. sodium chloride concentration) for all the analytes have not been significantly different as common IS was selected (115). It has been chosen because its mass is in the middle of the mass range normally analyzed by ICP–MS.

Isotope dilution is widely employed in ICP–MS for matrix effects correction. In general, the method is applicable to every element with at least two stable isotopes and to some naturally monoisotopic elements with radioactive isotopes with long lives

[420, 467, 550]. Using this method, the analyte serves as its own internal standard and allows for accurate concentration measurements because matrix effects are properly corrected.

The efficiency and complexity of internal standardization will be related to the major origin of the matrix effect. When the effects result from the sample introduction system, it could be thought that the effect would be rather similar regardless of the element. Any element could be, therefore, used as an internal standard. In marked contrast, when the effects may be assigned to the plasma, change in signals will depend upon line optical characteristics such as excitation energy in AES, and ionization energy and mass in MS. Selection of a single internal standard would be, therefore, far more challenging.

6.3.7.2 Methods based on empirical modeling

In some cases, the signal variation caused by a concomitant can be modelled. A procedure based on mathematical correction by curve fitting to an empirical function has proven to be successful for correcting for matrix effects [139]. The information provided by the obtained equation has given the possibility for a mathematical correction of the interference. The mathematical correction has also been performed by applying experimental design [510]. In this case, the starting point is the assumption that the analyte signal is linearly related to the analyte concentration and also to the matrix concentration. Under these conditions, Kragten and Parczewski [77] have used the factorial analysis for the determination of tantalum in gold. Villanueva et al. [637] applied a two-level two-factor factorial design, Ca and Mg being the main components of the matrix.

6.3.7.3 Methods based on the use of multivariate calibration techniques

The most-used multivariate calibration techniques are the multiple linear regression (MLR), principal-component regression (PCR) and partial least square regression (PLS). All of them establish a linear mathematical relationship between different sets of variables [188, 509]. Villanueva et al. [637] tried to correct for the mixed-matrix effects induced by Ca and Mg in ICP–AES. The MLR method was investigated to overcome variations in the intensity of the analytical signal resulting from the superposition, in different degrees, of spectral and non-spectral interferences. A 2^2 factorial design was used for the calibration set, considering the Ca and Mg concentrations (from 500 to 2 000 mg L^{-1}) as independent variables. The standard error of precision was used to evaluate the performance of the method [370]. The values of the standard error of precision in the validation set illustrated how spectral and non-spectral interferences were combined for the different analytes. Griffiths et al. [626] compared univariate and multivariate techniques. Among the univariate techniques, the univariate calibration with and without matrix matching and the use of inter-element correction factors for the suspected spectral interferences were compared. The multivariate algorithms studied were MLR, PCR, PLS1 and PLS2. The models were built with

248-intensity line data belonging to the most-intense analyte lines and many of the matrix elements. From these techniques, the best predictive accuracy was observed for the PLS1.

6.3.7.4 Sample treatment and other methods

Within this category, it may be found that methods in which the analyte is separated from the matrix sample by, for example, analyte precipitation [286]. Electrodeposition has been used to remove the sample matrix while the analytes remain in the solution [640, 667].

In other instances, the analytes are extracted into an organic phase [55, 160]. Note that the extraction is normally accompanied by an analyte pre-concentration [392]. Cloud-point extraction has become a versatile method for separation and pre-concentration of analytes in different samples such as printed substrate [543], water samples [524, 663] and biological materials [551].

Resins are also efficient in separating the analytes from the sample matrix [559, 581]. Chelating resins of different nature columns were employed by several authors for matrix elimination and/or analytes pre-concentration in different saline samples [345, 368, 378, 381, 569, 638]. Besides, cation- [458] as well as anion- [444, 504, 646] exchange resins have also been used when some elements are present in the sample at high concentrations. This practice is also common in ICP–MS [275, 294, 621].

Additional methods can be used to try to mitigate the matrix effects caused by EIEs. Among them, we can cite the following: (i) sample dilution [277, 295, 317], (ii) flow-injection analysis methods [357], (iii) matrix matching [89, 148, 625], (iv) standard addition [183, 364, 496] and (v) use of an ionization buffer [163].

6.4 ORGANIC SOLVENTS

Most of the samples to be analyzed by ICP techniques are of inorganic nature, that is in aqueous solutions. However, as it has been anticipated in Section 6.1, some samples have a high concentration in organic solvents. Among the different applications in which organic matrices are predominant, we can find the following:

1. Sample treatment consisting of sample dissolution or analyte extraction [573, 821, 684].
2. Hyphenated techniques such as HPLC–ICP–AES or HPLC–ICP–MS [341, 446, 790, 726].
3. In some cases, the sample has an organic nature. For example, in the analysis of petroleum products, most of the compounds present, or used for sample dilution, are organics [571, 682, 845, 852].
4. Some high-viscosity samples must be diluted with an organic solvent [604, 701] or a microemulsion must be prepared [314, 731, 826]. Aqueous emulsions make it possible to mitigate negative effects of organic solvents, although these effects are not completely eliminated [453].

As for the matrices previously discussed, the sources of the interferences induced by organic solvents are related to the sample introduction system as well as to the plasma. Likewise, the interferences can be classified as spectral and non-spectral.

In marked contrast to aqueous solutions, where water properties are only modified by the presence of acids, there is a large variety of organic solvents with significant different characteristics. Results obtained for a given solvent may not be directly extrapolated to another solvent.

6.4.1   Effects on the performance of sample introduction system

Among the different matrices, organic solvents are the compounds that induce the strongest effect on the behaviour of the liquid sample introduction system. The most relevant changes in the solution's physical properties caused by the presence of organic solvents are as follows: (i) a dramatic decrease in the surface tension that, obviously, depends on the concentration and nature of the organic solvent in the solution and (ii) a modification in the solution viscosity, some organic compounds induce a drop in this physical property, whereas for others, higher viscosities are found than for water. The case of organic–aqueous mixtures deserves special attention because the values of viscosity can either increase or decrease as the organic solvent concentration goes up. These two properties severely affect the pneumatic aerosol generation process. In general terms, it could be anticipated that the presence of organic solvents in a significantly high concentration leads to the production of finer aerosols than those found for water [327, 796]. This is clearly demonstrated in Figure 6.3 in which the complete drop size distribution curves are shown for water and two ethanol–water mixtures. It can be clearly observed that finer primary aerosols are generated for the two latter solutions than for distilled water.

There are other properties that are modified in the presence of organic solvents. Solution density uses to decrease when solvents such as methanol, ethanol or acetone are present in the sample. This may favour the transport of solution towards the plasma. Besides, the solution relative volatility also rises, which causes an intensification of the aerosol transport towards the plasma [65]. Once the aerosols are generated, the volatility induces changes in their characteristics. Thus, it has been observed that aerosols for ethanol become coarser because fine droplets disappear as they evaporate almost completely [325]. Finally, it should be considered that both the transport of droplets and the

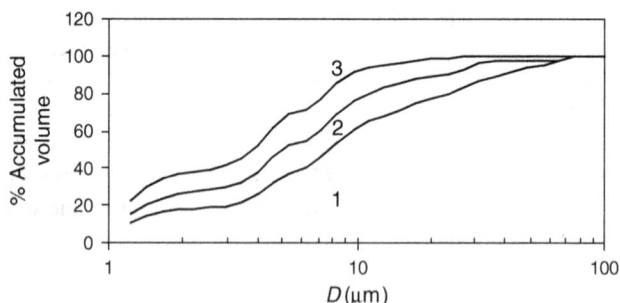

Figure 6.3 Volume drop size distribution curves for primary aerosols generated by a pneumatic concentric nebulizer. 1, water; 2, 20% ethanol solution; 3, 50% ethanol solution.

solvent evaporation are more likely than for water because primary aerosols are finer for organic solvents than for aqueous solutions [460].

> The result of the presence of an organic solvent in the sample is, thus, an increase in the mass of analyte and solvent transported towards the plasma. While the former effect can be considered as positive, then in some cases, too high organic plasma load is detrimental, and it will be discussed later.

6.4.2 Plasma effects

In spite of the fact that under a given set of nebulization conditions, an organic solvent induces increases in W_{tot} as compared to water, many studies have reported lower ICP–AES signals than for water [293]. The high mass of solvent delivered to the plasma when organic solvents are present degrades the plasma thermal characteristics. This is attributed to the fact that the solvent evaporation and dissociation consumes a fraction of the plasma energy [83, 408]. The consequences of this process can be a decrease in the plasma excitation temperature [83, 263, 293] and the electron number density [147]. The solvent dissociation causes an increase in the background noise and level [461]. This may lead to a drop in the plasma excitation conditions, although in some cases, a more energetic plasma has been observed [417] as a result of a mechanism called plasma thermal pinch. In the case of ethanol–water mixtures, it has been observed that the excitation temperature increases with ethanol concentration up to solutions containing 15% in this alcohol and then decreases [426]. These facts can be attributed to the enhancement in the plasma thermal conductivity produced by the pyrolysis products [95]. In some cases, the C_2 dissociation and ionization has been considered as mainly responsible for the enhancement in the plasma thermal conductivity [147].

In ICP–MS, the presence of organic solvents also has an important effect on the ion extraction and focusing. to solve these problems, the plasma should be robust and additional sample introduction systems (e.g. desolvation systems) can be employed [331, 351, 404, 535]. The effect caused by organic solvents on the plasma depends on the solvent nature. Thus, it has been claimed that the H:C ratio precludes the extent of the interference. Solvents with high values of this ratio deteriorate the plasma more severely than others having low H:C ratios [207]. According to Maessen et al. [167], organic solvents can be classified into two different groups: (i) solvents that have low vapour pressure values and, hence, do not affect the plasma stability and (ii) solvents with high vapour pressure values that deteriorate the plasma behaviour. The first group includes water, xylene and MIBK, whereas chloroform, methanol and ethanol would belong to the second group. A different criterion has also been established to classify the solvents based on the carbon content and the molecular weight [263]. In a different study, it has been concluded that the solvent C:O ratio affects the plasma appearance [408].

The ICP–AES spectral interferences caused by organic solvents are more complex than for most of the inorganic matrices because of the complexity of the spectra caused by the pyrolysis products. The most abundant molecules that can be generated in the plasma are C_2, CN and C. Depending on the organic solvent, other species such as CS, CH, NO and CO can appear. The spatial distribution of these products is very important, because it can preclude the zone of analyte emission measurement. It has been found that for chloroform- and toluene-loaded plasmas, C_2 emission appears at low observation heights in the inner zone of the plasma. Meanwhile, CN emits higher in the outer plasma zone [263]. This is because its formation results from the combination of carbon and atmospheric nitrogen. As expected, the distribution of the pyrolysis products depends on the plasma-operating conditions. Therefore, similar to other matrices, the effect of the presence of organic solvents has a spatial character [222]. This spatial character is aggravated by the fact that for organic solvents, a large fraction of the solvent mass is reaching the plasma in vapour form and, hence, the diffusion effects are more conspicuous than for aqueous samples. Therefore, in ICP, it is very important to take into account the physical form in which the solvent is reaching the plasma [293].

The effects caused by organic solvents on aerosol transport and emission intensity are summarized in Figure 6.4. In this figure, the results obtained for water have been taken as reference. The organic solvents cause an increase in the solvent plasma load that depends on the particular solvent. Thus, S_{tot} is increased by a factor as high as six (see black bar for methanol). In the case of butan-1-ol, the increase factor is just 2. Similar results can be observed for W_{tot} (see white bars). The improvement factor in the transport parameters follows the order of solvent relative volatility. Besides, primary aerosols are finer for methanol than for butan-1-ol. The final effect on the emission intensity is also summarized in Figure 6.4 (light grey bars). As it can be observed, the improvement factor in W_{tot} is much higher than the emission intensity (see white bars). This is a clear consequence of the degradation in the plasma thermal characteristics. In fact, under the conditions considered in Figure 6.4, alcohols caused a drop in excitation temperature with respect to water. As a consequence, the background stability is degraded and, consequently, limits of detection obtained for these organic solvents are just slightly lower than for plain water solutions (see dark grey bars in Figure 6.4).

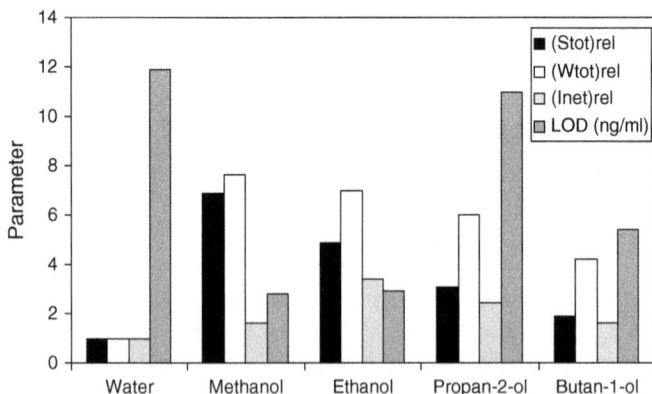

Figure 6.4 Overall effect caused by alcohols on different parameters for a conventional liquid sample introduction system consisting of a pneumatic concentric nebulizer coupled to a double-pass spray chamber.

In the case of ICP–MS, the spectral interferences caused by organic solvents are more complex than in ICP–AES. For this technique, the presence of an organic solvent such as ethanol can either decrease or enhance the ionic signal [258]. Diluted methanol solutions can cause increases in the sensitivity for some isotopes [254, 407]. Other organic compounds that have been proved to increase the analytical signals in ICP–MS are ammonium carbonate [407], ethylene glycol, tartaric acid [503], or triethanolamine and ethylenediamine [575]. The main mechanisms responsible for these enhancements are [369, 431, 557] as follows:

(1) Charge transfer reactions from C^+ species to the analyte ions [589]. In this case, the IP of the ionized carbon species must be higher than that of the analyte. This mechanism appears to be plausible, because all the carbon species present in the plasma central channel (e.g. CO, CO_2, C_2) have high IPs.
(2) Improvement in the aerosol transport efficiency through the sample introduction system. This phenomenon does not fully account for the signal variation observed in presence of organic solvents. Note that for different elements, the presence of organic solvents leads to different signal enhancement or depression factors [776].
(3) Shifts in the plasma zone of maximum ion density [371, 377].

For diluted solutions containing acetone or methanol, it has been observed that the sensitivity is either increased or decreased by a factor that depends on the operating conditions as well as on the mass and IP of the element being monitored. Also, for this technique, the interferences are due to the formation of pyrolysis carbon products [289]. Interestingly, the formation of polyatomic ions resulting from the solvent degradation can compete with the formation of other species that are normally produced from inorganic matrices. Thus, it has been reported that the formation of polyatomic species derived from propan-2-ol can lead to decrease in the levels of interfering species such as $ArCl^+$ [289].

Another effect found when organic solvents are introduced in the plasma is related to the degradation of the system stability caused by the formation of carbon deposits in the injector and/or the cones of the instrument [573, 682]. Organic solvents can also cause problems related to the formation of soot deposits on the walls of the torch. To solve this problem, oxygen can be added to the carrier argon stream. However, if too much oxygen is added in ICP–MS, the cones can quickly deteriorate and the polyatomic interferences can be more severe.

Note that the addition of oxygen is made possible as an accessory from ICP–MS manufacturers. Although it could also be useful in ICP–AES, it is seldom used.

6.4.3 Effect of the operating conditions

The operating conditions have a higher influence for organic than for aqueous solutions [477]. The nebulization conditions determine the extent of the effects caused by organic solvents. This is mainly because, for a given solvent, sample introduction system and plasma conditions, the nebulizer liquid and gas flow rate dictate the solvent plasma load [263, 293, 409, 461].

The operating conditions required to carry out the analysis of organic samples are different from those of aqueous solutions. An increase in the gas flow rate leads to the generation of fine aerosols and the intensification in the solvent evaporation inside the spray chamber, thus leading to higher S_{tot} values. Therefore, a general recommendation is to reduce the nebulizer gas flow rate to decrease the overall solvent plasma load and, concomitantly increase the analyte residence time [359]. However, this variable should be optimized because a decrease in the nebulizer gas flow rate also leads to a drop in the analyte mass transported towards the plasma. In the same way, as the liquid flow rate goes up, the mass of solvent reaching the plasma also rises, simply because the mass of aerosol increases.

As for inorganic matrices, the plasma-operating conditions determine the magnitude and sign of the matrix effects caused by organic solvents. Working at high hf power values, the emission signals obtained when organic solvents are present are higher than for water.

All these variables act simultaneously, and this interaction must be taken into account to optimize the operating conditions. In a factorial design study in ICP–AES, when organic solvents are present [535], it has been verified that the effect of the presence of organic matrices depends on the line nature, that is atomic or ionic. Among the variables tested, the nebulizer pressure had the most significant effect. As the plasma-operating conditions (i.e. hf power and auxiliary gas flow rate) change, the organic effects for ionic lines are more significant than for atomic ones.

6.4.4 Effect of the solvent nature

The solvent nature is a very important affecting factor, from the aerosol generation to the analyte excitation. In terms of aerosol transport, the solvent properties affecting the plasma load are volatility and density. The fractions of aerosol transported, in both liquid (S_{liq}) and vapour (S_{vap}) forms, to the plasma depend on the primary aerosol characteristics and the solvent properties. For two solvents having similar volatilities, S_{liq} will be higher for the solvent producing the finest aerosols. Furthermore, the liquid fraction of S_{tot} will increase as the solvent density decreases. This is because, for two droplets having similar diameters, the droplets having a lower mass (or density) will have less inertia and, hence, will more likely be transported towards the plasma. S_{vap}, in turn, will depend on the relative volatility of the solvent. Thus, solvents such as isopropanol, methanol and acetone evaporate faster than water because their evaporation factors (22.5, 47.2 and 264 $\mu m^3 \ s^{-1}$, respectively) are higher than that for water (13.1 $\mu m^3 \ s^{-1}$). Note that the evaporation factor gives information about the solvent load that can be tolerated by the plasma. In general terms, the plasma does not tolerate S_{tot} values higher than the evaporation factor. This is not the only factor precluding the value of S_{vap}. The aerosol fineness is also important. Note that the solvent vapour pressure is higher for a curved than for a flat surface. This can also be explained by indicating that the higher the curvature radius of a droplet, the higher the vapour pressure [94]. Therefore, a reduction in the aerosol mean diameter favours the solvent evaporation. At this point, it must be indicated that S_{liq} and S_{vap} are related to each other. Thus, an increase in the latter can induce drops in the former.

6.4.5 Effect of the liquid sample introduction system and the related parameters

The correct selection of the liquid sample introduction system and the operating conditions [800] are of capital importance to mitigate the problems associated to the introduction of organic solvents. Furthermore, when the appropriate conditions are selected, a huge increase in the sensitivity is produced as compared to aqueous solutions. For example, a desolvation system can be used to increase the mass of analyte delivered to the plasma and, concomitantly, decrease the solvent plasma load.

6.4.5.1 Conventional liquid sample introduction systems

For a given spray chamber, S_{liq} and S_{vap} depend on the nebulizer used. The effects of organic solvents have been studied with concentric [167, 147, 293], cross-flow [207, 409, 461] and V-groove [263] nebulizers. In a study corresponding to the gradual increase in the ethanol concentration, it has been found that for a pneumatic-concentric nebulizer, the higher the concentration, the stronger the emission signal (Figure 6.5) [829]. This study highlights the fact that, by a correct selection of the plasma-operating conditions, the increase in W_{tot} observed for organic solvents prevails on the detrimental effects caused by the introduction of this organic solvent into the plasma.

6.4.5.2 Low sample consumption systems

In general terms, low sample consumption systems are appropriate devices for the introduction of organic solvents into the plasmas. As the working liquid flow rate is very low, the solvent plasma load is also lower than for conventional designs. This assessment cannot be generalized, because similar S_{tot} values can be found for a non-efficient conventional sample introduction system and an efficient low sample consumption device. For example, for a pneumatic-concentric nebulizer coupled to a double-pass spray chamber operated at

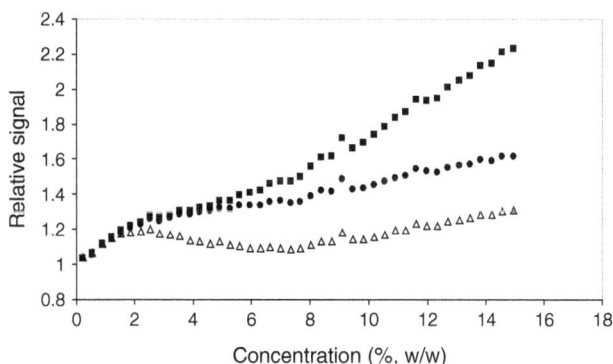

Figure 6.5 Variation of the emission signal versus the ethanol concentration for three different emission lines. (•) Mn 257.610 nm; (■) Cu 324.754 nm; (△) Zn 213.856 nm. (Taken from ref. [829].)

1 mL min^{-1} and 0.7 L min^{-1} liquid and gas flow rates, S_{tot} for methanol takes values close to 170 mg min^{-1}. For direct-injection nebulizers (solvent transport efficiency $= 100\%$), studies have been performed at liquid flow rates close to this solvent load (e.g. 140 μl min^{-1}).

For a concentric micronebulizer coupled to a cyclonic spray chamber, the presence of ethanol in the solution leads to an enhancement in the emission signal with respect to water. This result also indicates that the detrimental effects caused by the increased organic solvent plasma load are mitigated when the system is operated at low liquid flow rates. Note that, under these conditions, the solvent evaporation is favoured as compared to the situation found at conventional liquid flow rates (i.e. in the order of 1 mL min^{-1}). This effect is more noticeable for organic solvents. When comparing the effects of the presence of ethanol in the solution on the emission signal for a cyclonic and a low inner volume single-pass spray chamber, it has been verified that the interferences are less severe for the latter design. In this case, the fact that a single-pass chamber acts as an evaporation cavity for plain water and ethanol solutions, appears to be the reason for this trend [680].

When a DIN is used, it has been found that the atom net emission intensity for copper atomic line is more intense for organic solvents than for water, whereas the intensity for the copper ionic line is lower for the formers, thus proving that the plasma is deteriorated for organic solvents. In fact, it has been observed that the limits of detection for organic solvents are about one order of the magnitude higher than for plain water solutions. This fact has been attributed to the cooling action of the organic solvents. It has been observed that, provided that for the DIN, there is less effect of the solvent nature on the plasma thermal state, this device is more suitable than conventional ones to carry out chromatographic studies in which a gradient elution procedure is used [281]. An advantage of the DIN is the absence of waste.

A solution that has been proposed to solve the problems associated with the introduction of organic solvents with direct-injection nebulizers into the plasma is to work at extremely low liquid flow rates. Thus, by operating the system at 10 μL min^{-1}, it is possible to carry out the analysis of petroleum products with the DIHEN [732]. Another solution to mitigate organic effects with the DIHEN is to add an oxygen stream (10% concentration) to the external argon stream. In this case, it has been found that xylene solutions can be analysed without observing carbon build-up at the DIHEN tip [727].

6.4.5.3 *Desolvation systems*

The main purpose of a desolvation system when organic solvents are present is to lower the mass of the solvent delivered to the plasma. Chilled spray chambers are the simplest desolvation systems. They have proved to be efficient in solvent removal for aqueous solutions [657], although several problems have been found, such as the ice formation or the decrease in the analyte mass transported towards the plasma because of the reduction in the solvent evaporation rate [518]. With these devices, the amount of energy consumed by the solvent decreases as compared to conventional chambers. For a cooled spray chamber, it has been found that for solvents such as toluene, chloroform and 1,2-dichloroethane, the signal-to-noise ratio increases significantly as the spray-chamber wall's temperature is switched at values included within the −5°C to −10°C range [132]. For solvents such as methanol and ethanol, it has been claimed that the chamber thermostat is necessary to obtain a stable plasma.

In the case of two steps desolvation systems, the variables influencing the solvent plasma load are the aerosol heating and cooling temperatures [401]. These variables must be carefully optimized, otherwise the limits of detection for organic solvents can be similar [67] to or higher [133] than that for water. The obvious trend is that the plasma stability improves as the condensation temperature goes down. As regards the aerosol-heating temperature, an increase in this variable leads to growth in the mass of aerosol that leaves the heating unit. If the condenser is switched to an appropriate temperature, the final result will be a decrease in the solvent plasma load together with a rise in the mass of the analyte being injected into the torch. An appropriate condensation temperature is low enough to condense as much solvent mass as possible, but high enough to prevent solvent nucleation on the analyte aerosol particles.

The aerosol heating is very important in modifying the extent of matrix effects caused by organic solvents. For example, in the case of a conventional desolvation system based on a conductive–convective heating mechanism, propan-2-ol solutions cause an enhancement in the ICP–AES signal by a factor close to 3. Meanwhile, if the aerosol heating is produced by means of microwave radiation, this factor falls down to values ranging from roughly 1.8 to 0.7 depending on the liquid flow rate and the design of the aerosol-heating cavity [822].

To lower the solvent plasma load, a condenser or a membrane can be used. For a device equipped with a condenser, it has been found that for chloroform solutions, a decrease in the condensation temperature leads to a rise in the analyte emission intensity and a decrease in the height a.l.c. to which the maximum intensity is obtained [293]. These results are observed because the energy is mainly consumed from the plasma base. When working with highly volatile organic solvents, one-step condensation units are not efficient for solvent removal. In those instances, two-step condensation devices are advisable. The solvent vapour removal is very important to obtain good results. For a Peltier-based desolvation system, it has been observed that the limits of detection can be reduced by up to one order of magnitude [351].

Membranes are as efficient in organic solvent removal as condensers. It has been demonstrated that the solvent-removal efficiency when a silicon membrane is adapted at the exit of a spray chamber is high enough to bring the solvent plasma load below the maximum values tolerable by the plasma. This has been verified for solvents such as methanol, ethanol, chloroform and Freon [267]. In this case, the solvent-removal efficiency (η_{sol}) is given by the following mathematical relationship:

$$\eta_{sol} = 1 - e^{-\frac{ASD}{LQ_g}} \qquad (6.3)$$

where A is the membrane area, S is the solubility, D the solvent diffusivity, L is the membrane thickness and Q_g is the membrane gas flow rate. As it can be seen, to improve the solvent-removal efficiency, it is necessary to use thin membranes and low gas flow rates.

Tubular microporous PTFE membranes have been used for successfully reducing the solvent load for solutions containing hexane, methanol, THF, acetone and dichloromethane [404, 450, 456, 484]. With these kind of membranes, it has been observed that the ICP–AES carbon emission for methanol–water solutions is reduced by a percentage ranging from 80 to 90% [566]. Furthermore, the analytical emission intensity increases as the solvent concentration goes up for methanol, dimethyl sulfoxide and isopropyl alcohol solutions. Other issues of porous membranes are as follows: as the continuous background decreases, memory effects are not very pronounced and wash-out times (0.1% drop in the original emission intensity) are on the order of 75 s [450, 469, 566]. As regards the effect of the operating conditions, it has been claimed that in the presence of organic solvents, the emission intensity decreases as both the sweep gas flow rate and the heating temperature of the membrane are increased.

Therefore, membranes make it possible to decrease the background intensity and simultaneously increase the sensitivity. In addition, the plasma thermal characteristics are improved and, consequently, the limits of detection are significantly reduced [404].

However, two main problems have been described when porous membranes are used: (i) for non-volatile organic matrices (oils, petrol products), the membrane pores can be blocked and (ii) volatile analyte species can go through the membrane pores. In contrast, the use of a desolvation system equipped with a membrane has demonstrated to minimize differences in behaviour as a function of the organic solvent. Thus, a calibration line obtained from hexane standards can be applied for the determination of analyte in pentane, toluene and gasoline. This has been an approach to the so-called universal calibration [456].

As the introduction of volatile solvents such as methanol or ethanol is challenging when using an ICP, classical solvents that are currently used, such as xylene or kerosene, are not highly volatile. However, their solvent capabilities may be less efficient than solvents such as MIBK. The simplest way to reduce the amount of organic solvent reaching the plasma is to reduce the solution delivery rate to the nebulizer and to decrease the temperature of the spray chamber. These devices are now commercially available.

6.5 CONCLUSIONS

Matrix effects are certainly the most-challenging field of research in ICP–AES and ICP–MS. Matrix effects are crucial as they affect accuracy (trueness). A significant number of papers dealt with analytical consequences and possible explanations. Because of the large variety of operating conditions, nebulizer and spray chamber types, and torch design, a fair comparison cannot be obtained between results. After more than 40 years of experimental and theoretical work, it is still difficult to predict and to fully compensate for matrix effects. Although most work dealt with mechanisms occurring within the plasma, it should be kept in mind that the sample introduction system may significantly contribute to matrix effects, either directly through aerosol formation, transportation and filtration, or indirectly through change in the plasma load and residence time.

– 7 –

Selection and Maintenance of Sample Introduction Systems

7.1 SELECTING A LIQUID SAMPLE INTRODUCTION SYSTEM: GENERAL ASPECTS

The sample introduction system must be selected as a function of several requirements: (i) total dissolved solid content; (ii) presence of particles; (iii) chemically inert to reagents (hydrofluoric acid and organic solvents); (iv) limited or absence of drainage (e.g. corrosive or radioactive elements); (v) small available amount of solution or limited delivery rate when coupled with separation method; (vi) reduction in solvent loading (aerosol desolvation or spray-chamber cooling) and (vii) fast response to transient signals. Points (iv) and (v) are obtained with so-called micronebulizers, that is working at very low solution delivery rates. Note that most of the commercially available pneumatic nebulizers have an external diameter of 6 mm, which makes them easily interchangeable. The exception is for cross-flow nebulizer, the body of which usually fits with 35-mm double-pass chamber diameter. It is only necessary to verify whether the ICP instrument can deliver the required backpressure.

Besides, the torch design is a very important issue. This is because the nebulizer gas flow rate and the gas pressure are related to the torch design. Therefore, the selection of

Table 7.1

List of societies involved in sample introduction systems

Burgener	http://www.burgenerresearch.com
CPI International	http://www.cpiinternational.com
Elemental Scientific	http://www.icpms.com
EPOND Switzerland	http://www.spray-chamber.com
Glass Expansion	http://www.geicp.com
Meinhard Glass Products	http://www.meinhard.com
Precision Glassblowing	http://www.precisionglassblowing.com
Savillex	http://www.savillex.com

the nebulizer can be a function of the torch and vice versa. In other words, the quality of the aerosols generated will depend indirectly on the torch type. This will influence the sensitivity and signal stability.

Lists of societies and commercially available nebulizers are given in Tables 7.1 and 7.2, respectively.

7.2   LIQUID SAMPLE INTRODUCTION SYSTEM

Following the previous chapters, general conclusions can be drawn in terms of performance of the conventional liquid sample introduction systems:

1. Aerosols delivered to the plasma are generally too coarse, polydisperse and, in some instances, turbulent.
2. The mass of analyte transport rate leaving the injector is somewhat low and higher values are required in many cases.
3. In some cases, the solvent plasma load is too high.
4. For conventional systems, memory effects for particular samples can be dramatic.
5. Non-spectroscopic interferences arising in the sample introduction system are still present.

To try to solve all the above-mentioned problems is not an easy task. This is because some problems can be overcome at the cost of generating new ones. In the case of point 1, a fine tertiary aerosol can be achieved in several ways:

(i) Using nebulizers that are able to produce very fine primary aerosols. To produce a fine aerosol pneumatically, it is necessary either to increase the available kinetic energy or to enhance the efficiency of the gas–liquid interaction. In both cases, the nebulizer's critical dimensions must be reduced. This has additional problems. On the one hand, the gas backpressure increases and, consequently, an additional gas line is required for the nebulizer, and on the other hand, the aerosol velocity can be too high. Thus causing either an intensification in the aerosol inertial losses (for a system based on the use of a spray chamber) or causing a degradation in the analyte excitation efficiency (for direct-injection nebulizers). Also important is the fact that with nebulizers having reduced critical dimensions, blocking problems are more likely to occur.
(ii) Employing very efficient spray chambers in the selection of aerosol droplets can lead to a low mass of analyte transported to the plasma. Furthermore, the solution proposed to filter large droplets is often to increase the complexity of the aerosol path inside the chamber. This has additional disadvantages such as the intensification of the matrix effects and the increase in the memory effects.
(iii) Using desolvation systems that remove most of the aerosol solvent mass is a good approach to reduce the tertiary aerosol drop size, and simultaneously increase the

Table 7.2

A list of commercially available pneumatic nebulizers – PG: Precision Glassblowing, GE: Glass Expansion, CPI: CPI International

Nebulizer	Material	High solids	Particulates	HF-resistant	Field of applications	References
Concentric	Borosilicate	Tolerates sodium concentration levels <0.5% approx	No	No	General purpose, default device	Meinhard type A and C
	Borosilicate	Moderate	Moderate	No	Developed to work with solutions with a moderate salt content and suspended solid particles	Meinhard type C
					Developed to work with solutions with a moderate salt content and suspended solid particles	PG Aerosalt
	Borosilicate	General purpose		No	Used when low argon flow rates are required and do not block with salty solutions or slurries.	Meinhard type K
	Borosilicate	Moderate		No	Low Ar flow rate	Meinhard type CK
	Borosilicate	<5%	<75 μm	No	Clean solutions and furthermore tolerates salt concentrations up to 5% and particles with diameters close to 75 μm.	GE Conikal
	Borosilicate	<20%	<75 μm	No	Tolerates salt concentrations up to 20% in soluble salts, tolerates slurries with particle diameters of typically 75 μm	GE SeaSpray
	Borosilicate	<1%	Yes, <150 μm	No	Able to nebulize slurries with particles as large as 150 μm size.	GE Slurry
	Quartz					Meinhard TQ-series

(Continued)

Table 7.2

A list of commercially available pneumatic nebulizers – PG: Precision Glassblowing, GE: Glass Expansion, CPI: CPI International—Continued

Nebulizer	Material	High solids	Particulates	HF-resistant	Field of applications	References
	Borosilicate	<15%	No	No	Low delivery rate, analysis of microsamples, HPLC-ICP coupling and Flow Injection Analysis methods	Meinhard HEN, Glass Expansion MicroMist
	Borosilicate PFA	Acceptable		Yes	Low HE flow rate Very good performance for geochemistry and semiconductor industry	Meinhard type HK GE OpalMist
	PFA			Yes	Analysis of microsamples requiring high purity materials or having a strong acid, alkali or organic nature.	CPI MicroFlow
	PFA			Yes	Low delivery rate trough exchangeable uptake capillary	CPI ST
	PFA + quartz				Containing a sample quartz capillary and developed to work with analytes easily adsorbed on plastic materials.	CPI PFA-Q
	Polyimide	<5%	<75 μm	Yes	Able to nebulize slurries with particles having 75 μm in size.	GE PolyCon
	Polymer Polymer	General purpose			Low delivery rate	EPOND Capital EPOND Boreal and Zefyr
Concentric Laval	Polymer Borosilicate		<50 μm		Very low delivery rate	EPOND Lucida EPOND Tyfoon

Cross-flow	Polymer	Acceptable			Used for the analysis of high salt content solutions and viscous solutions	Precision glassblowing
	PFA			Yes	Appropriate for high dissolved solids introduction	Savillex X-Flow
Parallel path	Ruby tips PEEK					PE cross-flow
	PEEK			Yes	Down to 0.2 mL min^{-1}, appropriate for high salt content solutions and slurries but not for acids and some organic solvents	Burgener MiraMist
	PTFE			Yes	Appropriate for high salt content solutions, slurries, acids and organic solvents	Burgener MiraMist
	PEEK	Yes	Yes	Yes		Burgener T2100 (replace T2002)
	PEEK		No	Yes	Developed to handle very low liquid sample flow down to 0.05 mL min^{-1}, but not for working with HF and some organic solvents	Burgener AriMist
V-groove	Borosilicate		Yes	No		Noordemeer, Lichte
	Quartz	Yes, <30%	Yes, <300 µm	No		GE Quartz VeeSpray
	Alumina ceramic	<30%	Yes, <300 µm	Yes	Designed to work with high salt content solutions (30%) and slurries containing particles with diameters up to 300 µm.	GE VeeSpray
	Polymer (PEEK)	Yes			Appropriate for high salt content solutions and slurries	PG Babington V-groove
Direct nebulizer	Borosilicate	No	No		Analysis of microsamples, HPLC-ICP coupling and Flow Injection Analysis methods	Meinhard DIHEN

mass of solvent transported towards the plasma, while decreasing the solvent plasma load. However, for desolvation systems, memory effects are severe and the cost can be high.

Solving the problems indicated above as 2 and 3 is somewhat contradictory when a nebulizer is coupled to a spray chamber. Note that a high analyte transport rate (point 2) accompanies a high solvent plasma load (point 3). High analyte transport rates can be obtained by producing fine primary aerosols, using appropriate spray chambers or employing desolvation systems. From these choices, only the third one is compatible with the need for a low solvent transport rate, although, as indicated before, some additional drawbacks must be considered.

Reducing at the lowest possible levels, the wash-out times (problem 4) is quite difficult, mainly if the problems 1–3 must be overcome. To reduce the wash-out times, a simple solution is to reduce at minimum levels the inner volume of the nebulizer sample capillary. Figure 7.1 (left) shows a picture of the back tube of a glass concentric nebulizer developed for HPLC–ICP coupling. With this tube, the dead volume has been reduced by about an 80% with respect to volume for conventional devices.

Another solution shown in Figure 7.1 corresponds to a concentric nebulizer in which the liquid capillary has been shortened. Finally, micronebulizers have been designed to reduce the dead volume and to be able to work rapidly at low liquid flow rates. If the nebulizer's critical dimensions do not change, these solutions do not modify the characteristics of the aerosol and, consequently, the analytical performance of the system.

In a different approach, a low inner volume spray chamber can be used. In this case, the goal is to reduce the exposure of the chamber surface to the aerosol as much as possible (Chapter 4). Unlike for the nebulizers, the reduction in the spray chamber inner volume has a clear influence on the sensitivity provided by these devices. As a result, aerosol inertial losses can be intensified, which hampers the transport of solution towards the plasma. On the contrary, the aerosol filtration can be inefficient if the chamber volume is reduced without giving consideration to the spray chamber design and flow dynamics.

Figure 7.1 Details of the back concentric nebulizer tube with reduced dead volume (left) and a nebulizer with a shortened liquid input (right). Taken from Meinhard Glass Products (http://www.meinhard.com).

As regards non-spectroscopic matrix effects (problem 5), this is a challenging and a less understood point. The chamber is a source of these interferences. If a simple path is imposed on the aerosol, good results can be achieved. However, tertiary aerosols may be too coarse and the solvent mass transported towards the plasma may be too high. These effects can counterbalance the beneficial role of this kind of spray chamber, thus degrading the accuracy of the results.

7.3 SAMPLE INTRODUCTION SYSTEMS FOR PARTICULAR KINDS OF SAMPLES OR APPLICATIONS

The previously discussed conditions are general and, for a given sample or application, can be accomplished in a different fashion. Therefore, the nature of the sample is a factor that specifies the criterion to select a given device to carry out the analysis.

For aqueous samples containing neither complex matrices nor solid suspended particles, the previously mentioned guidelines are adequate and there are not additional factors that should be considered. Unfortunately, this is not a general case and often the samples are of complex nature that makes it necessary to change the priorities for sample introduction system selection.

When an inorganic matrix is present in the sample at rather high concentrations, some problems emerge. Of course, the system must afford high sensitivities. However, accuracy has to be preserved and, hence, the system should minimize the interferences. For HF, in addition, in order to avoid irreversible chemical attack, the system cannot be made of glass. In the case of acids, the solvent used in cyclonic spray chambers, desolvation systems or direct-injection nebulizers may pose a problem. The final choice depends on whether the analyst is seeking for shortening the wash-out time, lowering the solvent load or lowering the available sample volume. The case of salt solutions deserves special mention, because besides matrix effects, problems due to the nebulizer tip blocking must also be considered. On this subject, several pneumatic nebulizers are conceived to handle these kinds of samples. Similar considerations for acid solutions should be made, although, in general terms, it is worth to mention that a desolvation system does not solve the interferences caused by solutions containing high concentrations in easily ionized elements (EIEs). Another difference between these two matrices is the incidence of the memory effects. Acids cause the so-called transient effects that can be very severe, whereas EIEs do not cause such effects [700]. Therefore, to know what is the best system for the analysis of inorganic acid solutions, one should consider the device affording the shortest wash-out times.

The case of organic samples is different because attention should be paid mainly to the solvent plasma load. It is possible to carry out organic samples analysis mitigating the detrimental effects caused by these compounds under two general circumstances: (i) decreasing the liquid flow rates [455] and (ii) using a desolvation system. Within this second category of devices, we can highlight the use of a cooled spray chamber as one of the most suitable approaches.

Another situation can be found if samples containing solid particles are present. This widely known approach requires the use of nebulizers without narrow conductions.

There are many nebulizers specially developed for analyses of these kinds of samples (Chapter 3).

Finally, coupling of a separation technique with ICP has special requirements in terms of the liquid sample introduction system. In this case, an important factor is the dead volume of the system. To reach a good peak resolution, the nebulizers and chambers must have a low inner volume.

7.4 SELECTING A NEBULIZER

The nebulization step is the most critical, because the results obtained finally depend on the quality of the aerosol produced by the nebulizer. This step has been studied by many authors and, as a result, there are many different designs available commercially. In many cases, the nebulizer design and/or dimensions have been adapted to make it more compatible with the application for which it is used. Table 7.2 gives some examples of available pneumatic nebulizers as well as the specific uses of each.

From the information given in the present as well as in the preceding chapters, a guideline for selection of nebulizer can be given. An initial classification can be made by considering whether the sample volume available to perform the analysis is limited or not. Figures 7.2 and 7.3 help to select the best nebulizer under two different situations, that is when the sample volume is unlimited and when it is low, respectively. In general terms, it can be concluded that to select a pneumatic nebulizer, the sample physical–chemical nature has to be considered. Thus, it is important to know whether the sample contains solid particles and dissolved solids and if it can attack the surface of the nebulizer.

7.5 SELECTING AN AEROSOL TRANSPORT DEVICE

The selection of the aerosol transport device is often dependent upon the nebulizer previously chosen. Thus, for example, if a parallel-path, cross-flow or V-groove nebulizer is our choice, big spray chambers could be beneficial to give a higher chance for droplets to be transported. In the case of concentric nebulizers, smaller chambers could be used while keeping the sensitivity at acceptable levels. This is because, for the former nebulizers, coarser aerosols are generated than for the latter design (see Chapter 3). The decision about whether to use a large or a small spray chamber and whether high sensitivities (or analyte transport efficiencies) are required must be based on aerosol transport considerations. On the one hand, solvent evaporation is more significant if the aerosol residence time inside the spray chamber is long enough. Nevertheless, on the other hand, as illustrated in Chapter 4, coalescence becomes more significant as the aerosol spends more time in the chamber. What the corresponding calculations clearly demonstrate is that working at low liquid flow rates can be considered as a good approach to improve the analyte transport efficiency. These calculations can also help provide an understanding as to why decreasing the liquid flow rate by a given factor does not result in a corresponding proportional decrease in the emission signal. For example, one of the earlier studies carried out with the

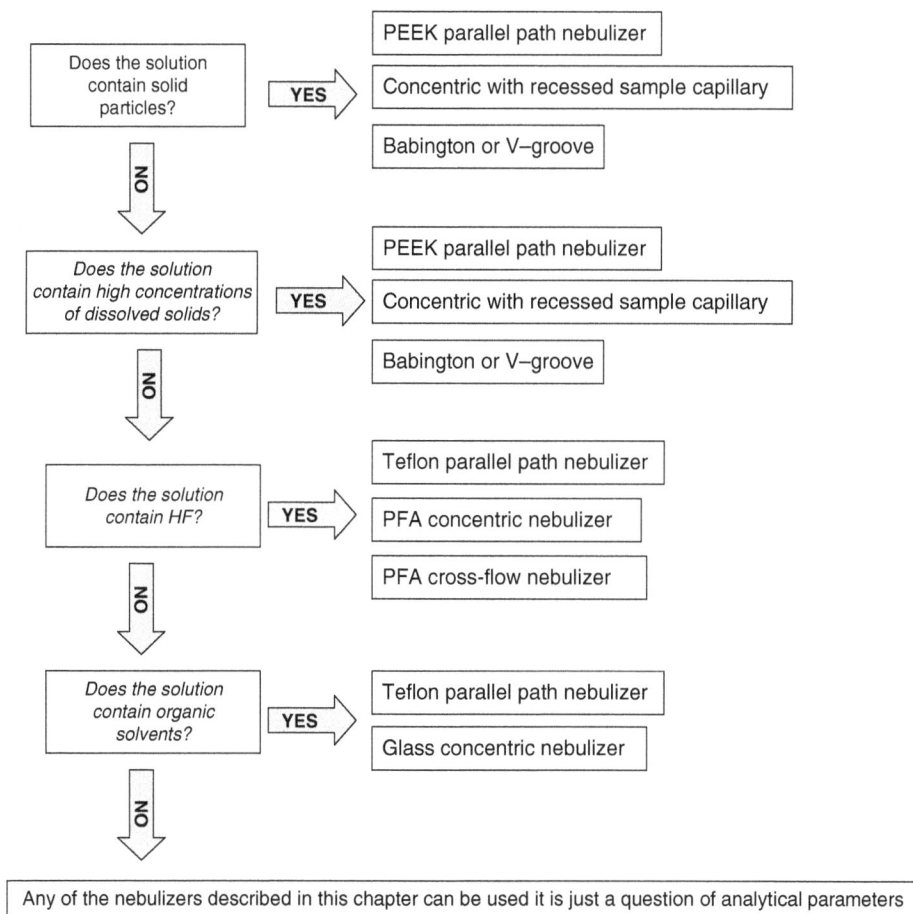

Figure 7.2 Guide for selecting the best nebulizer when the sample volume is unlimited.

high-efficiency nebulizer (HEN) by Olesik et al. [388] claimed that the HEN operated at a $50\,\mu L\,min^{-1}$ liquid flow rate afforded limits of detection similar to those measured for a conventional pneumatic concentric nebulizer operated at a $1\,mL\,min^{-1}$ sample flow rate. Further measurements of the efficiency of the analyte transported towards the plasma (ε_n) indicated that for the HEN, this parameter took a value close to 60% at $10\,\mu L\,min^{-1}$, whereas it dropped down to 8% at $120\,\mu L\,min^{-1}$ [465, 576]. At low liquid flow rates (ca. $2\,\mu L\,min^{-1}$) aerosol transport efficiencies close to 90% were reported. A comparison of the analytical behaviour for different spray chambers is established in Table 7.3 [808].

As regards desolvation systems, it must be indicated that these devices are more efficient in terms of analyte transport for nebulizers generating fine aerosols. This is because the aerosol evaporation yield in the heating unit is higher for fine than for coarse droplets.

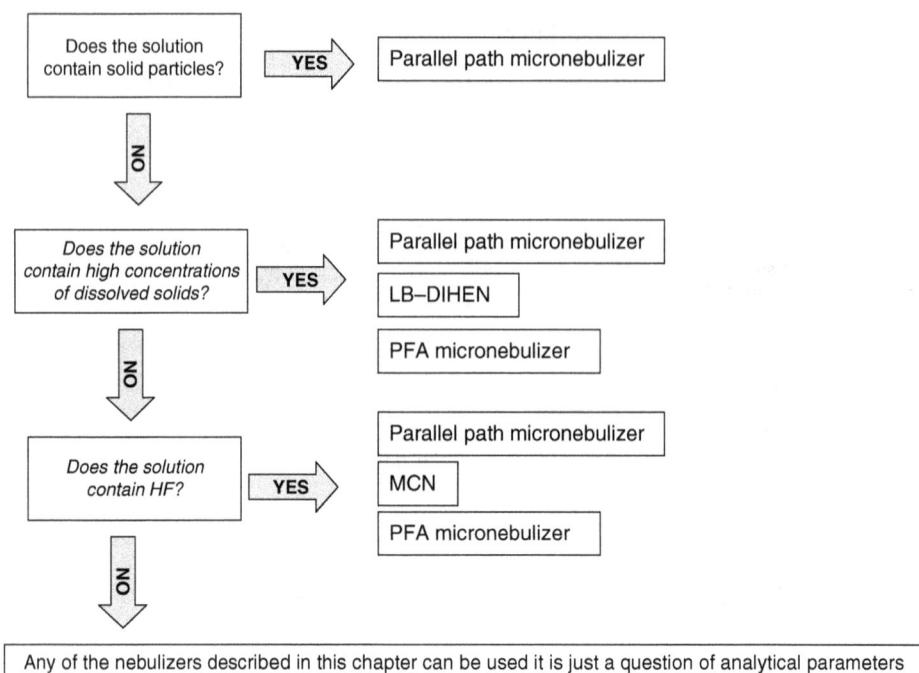

Figure 7.3 Guide for selecting the best micronebulizer for the analysis of microsamples.

Table 7.3

Comparison of analytical behaviour of different spray chambers used at a 50 µl min^{-1} liquid flow rate with a concentric pneumatic micronebulizer.

	Double pass	Cinnabar	Single pass
Maximum drop size transported to the plasma (µm)	~10	~13	~20
Analyte transport efficiency[a]	~20	~20	~30
Relative limits of detection	1	1	0.8
Matrix effect[b]	45	25	10
Wash-out time (s)[c]	120	70	20–40
Transient effects (s)[d]	200	30	10

[a]Transport efficiency $= \dfrac{\text{Analyte mass transported to the plasma}}{\text{Analyte mass nebulized}}$

[b]Defined as the percentage drop in signal induced by a 2 M nitric acid solution:

Matrix effect $= \dfrac{\text{Analytical signal for a plain water standard}}{\text{Analytical signal for a 2 mol L}^{-1}\text{nitric acid concentration standard}}$

[c]Defined as the time required for the signal to drop down to 1% of its steady state value when switching from a standard to a pure water sample.
[d]Time required for the signal stabilization when going from a 0.1 mol L^{-1} to a 2 mol L^{-1} nitric acid solution. Taken from ref. [808]

Table 7.4 shows examples of commercially available aerosol transport devices and some comments indicating the suitability of them for particular situations. Figure 7.4, in turn, represents a scheme of selection of the best spray chamber or desolvation system depending on the characteristics of the application.

Table 7.4

Different commercially available aerosol transport devices and the particular uses of designs

Device design	Device model	Use
Double-pass spray chambers	PFA chambers (http://www.spray-chamber.com) Teflon® chambers (http://www.burgenerresearch.com)	Used for the analysis of corrosive samples and samples containing HF
Cyclonic-spray chambers	PFA chambers (http://www.spray-chamber.com, http://www.geicp.com, http://www.elementalscientific.com) Tracey PTFE (http://www.geicp.com) Twister, Cyclonic with vertical tube (http://www.geicp.com) Low inner volume spray chamber (Cinnabar, http://www.geicp.com)	Useful for the analysis of HF-containing solutions Useful for the analysis of samples containing organic matrices Reduces the solvent plasma load and improves the signal stability Analysis of liquid microsamples. Separation–ICP coupling
Single-pass spray chamber	QuDIN (http://www.spray-chamber.com) MC CE made of Teflon® (http://www.burgenerresearch.com) CEI-100 (http://www.cetac.com)	Reduction of the dead volume for separation–ICP coupling Analysis of microsamples
Tandem-spray chambers	SSI (cyclonic + double pass) (http://www.elementalscientific.com)	Isotope ratio measurements
Solvent-removal devices	Spiro TMD, ACM and AREO membrane modules (http://www.elementalscientific.com, http://www.spray-chamber.com)	Reduction of the solvent load to lower the ICP-MS oxide ratios to values lower than 0.05% Used for the analysis of HF-containing solutions
Cooled spray chamber	PC^3 (http://www.elementalscientific.com, http://www.spray-chamber.com) IsoMist (http://www.geicp.com)	Reduction of the solvent load both for organic and aqueous samples Analysis of Naphta
Desolvation system	APEX-Q, APEX-IR, APEX-E (http://www.elementalscientific.com, http://www.spray-chamber.com) Aridus (http://www.cetac.com)	Removal of solvent load for ICP-MS, Isotopic dilution and ICP-AES, respectively. Analysis of organic solvents Analysis of microsamples with reduced solvent plasma load

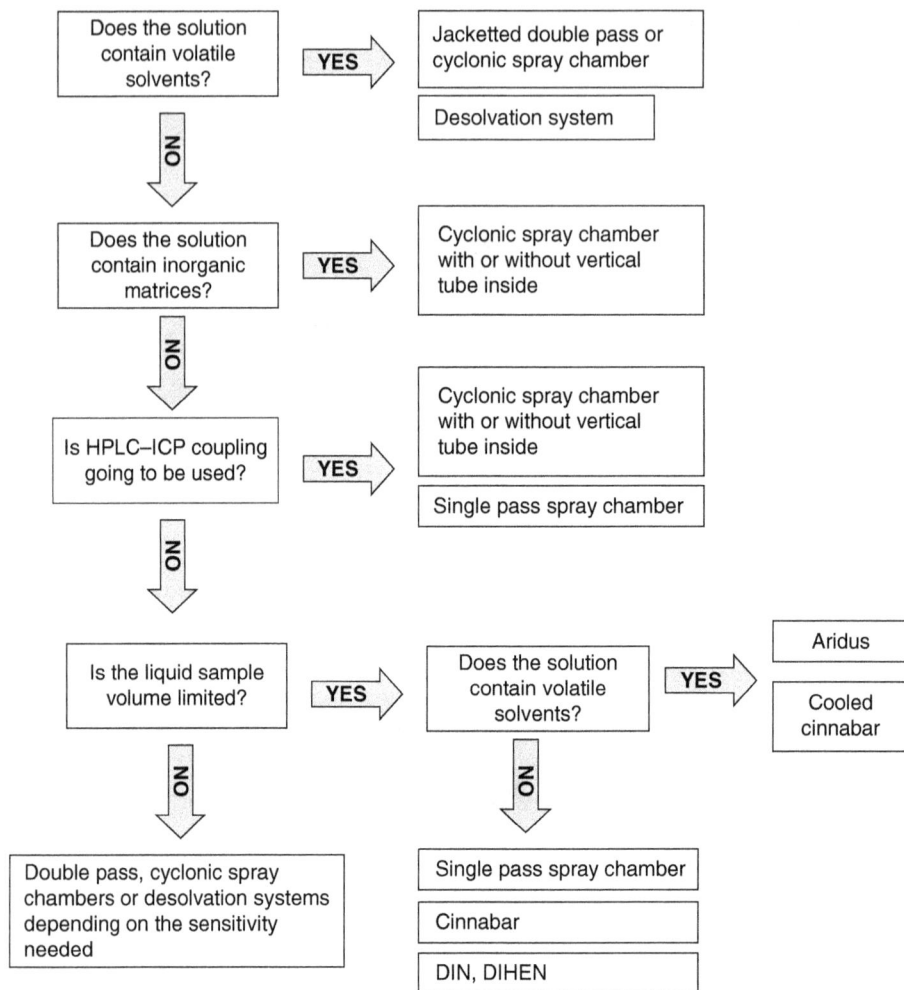

Figure 7.4 Guide for the selection of the best aerosol transport system.

7.6 PERISTALTIC PUMP

The default device to feed a nebulizer is the rotary peristaltic pump. The principle is based on periodic compressions of a flexible tubing by means of rotor with rollers. As the rotor turns, the fluid is forced to move through the tube. Such a pump has some advantages over alternative feed systems: (i) low dead volume, (ii) no contamination due to metallic parts, (iii) self-priming and (iv) several channels can be used, for example one for the sample delivery, another one for drainage and a last one for the addition of an internal standard. However, the pump suffers from some drawbacks, the main one being the discontinuous feeding leading to pulses in sample delivery, and, therefore, in aerosol formation. Pulse influence may be reduced by using several channels with phase-displaced rollers and by increasing the dumping effect of the spray chamber with enlarged dead volumes. In any case, it is advisable to ensure that several compressions will be obtained during the photon or ion acquisition time. Another limitation is the lifetime of the flexible tubing, which is typically of the order of 10^6–100×10^6 compressions.

Selection of pump tubing depends on three main parameters: chemical resistance to reagents (acids and organic solvents), service lifetime and inner diameter. This last parameter combined with the speed of the pump, usually expressed in revolutions per minute (RPM), will setup the liquid delivery rate. The material should also resist to adsorption of liquid fluids to avoid cross-contamination between samples. Note that xylene (dimethyl benzene) and kerosene are the most commonly used organic solvents.

Several materials are available for pump tubing: PVC, in particular, Solvent Flex PVC, Tygon® and Viton®. Tygon is a registered mark from Saint Gobain Performance Plastics, while Viton is a registered mark from Dupont Performance Elastomers. Some materials are clear, which makes possible to observe the path of an air bubble along the tubing, so as to verify the presence of blocked rollers.

PVC is satisfactory for acids, but is unsatisfactory for aqua regia and kerosene. Solvent Flex is yellow and is resistant not only to acids but also to aqua regia, but remains unsatisfactory for most organic solvents.

Tygon is clear, white and several qualities are available as pump tubing: Tygon LFL, Tygon R-3603 and Tygon 2075. Tygon LFL and Tygon R-3603 are excellent for acids and alcohols, but not for aromatic hydrocarbons. Tygon 2075 adds a resistance to ketones. Note that Tygon LFL has a long lifetime.

Viton is a fluorinated elastomer, opaque black and exhibits an excellent resistance to acids, oxidizing agents and to aromatic (xylene), kerosene or chlorinated solvents, but not to acetone. However, its lifetime is significantly reduced when compared to that of Tygon.

Table 7.5 indicates some theoretical tubing lifetime as a function of the number of rollers, the rotation speed (RPM) and the theoretical number of compressions that can accept the tubing.

According to the manufacturer, Tygon LFL and Tygon R-3603 will accept roughly 80×10^6 and 3×10^6 compressions, respectively. In the worst case, Tygon 3603 would work only a week. In any case, tubing is a consumable, and must be changed periodically.

A large selection of tubing inner diameter (i.d.) is available, making it possible to work from a few μL min^{-1} to several mL min^{-1}. The i.d. value is given by the colour of the two tags that are used to fix the tubing on the pump (Table 7.6). For conventional nebulizers, the i.d. is usually in

Table 7.5

Theoretical lifetime of pump tubing as a function of the number of rollers passing on the tubing, that is, the number of compressions per unit time

Pump roller number	RPM	min^{-1}	h^{-1}	Day^{-1} (8 h)	Life (days) Tygon LFL	Life (days) Tygon 3603
10	20	200	12×10^3	10^5	800	30
12	80	10^3	6×10^4	5×10^5	160	6

Table 7.6

Relation between the colour of the two tags and the
inner diameter of the pump tubing

Tag colors	i.d. mm
Orange/black	0.13
Orange/red	0.19
Orange/blue	0.25
Orange/green	0.38
Green/yellow	0.44
Orange/yellow	0.51
White/yellow	0.57
Orange/white	0.64
Black/black	0.76
Orange/orange	0.89
White/black	0.95
White/white	1.02
White/red	1.09
Red/red	1.14
Red/grey	1.22
Grey/grey	1.30
Yellow/yellow	1.42
Yellow/blue	1.52
Blue/blue	1.65
Blue/green	1.75
Green/green	1.85
Purple/purple	2.06
Purple/black	2.29
Purple/orange	2.54
Purple/white	2.79
Black/white	3.17

Table 7.7

Speed of the pump (RPM) to deliver $100\,\mu L\ min^{-1}$ with an orange/black tubing (0.13 mm i.d.)
(From www.geicp.com)

ICP system	Pump diameter (mm)	Number of rollers	Roller diameter (mm)	Pump RPM
ICAP 6000 (AES)	68.9	12	13.9	49.6
Varian 700 series (AES)	50.65	10	11.73	71.2
Agilent 7500 (MS)	44.0	10	10	82.2

the range 1–2 mm, while lower inner diameters may be used to feed a micronebulizer (e.g. orange/black tags). An example is given in Table 7.7 for a $100\,\mu L\ min^{-1}$ delivery rate.

To facilitate connection with capillary tubing from the sample and to the nebulizer, it may be useful to use dedicated tool so that the ends of the pump tubing is stretched.

An alternative is to buy tubing that are flared at the end to be able to insert a capillary tubing with an outer diameter (o.d.) slightly larger than the pump tubing i.d.

7.7 DIAGNOSIS

Users of an ICP system should plan systematic actions to verify that it satisfies the requirements for analytical performance and instrument maintenance. Based on previously published work on drift diagnostics [354], ionic-to-atomic line intensity ratio measurements [318], a procedure called QUID (quality control and identification of malfunctions) has been proposed [383, 422] to evaluate analytical performance, to control the major components of the ICP apparatus and to diagnose any possible deterioration of the system. A simplified version can be applied to the control of the sample introduction systems.

7.7.1 Use of Mg as a test element

The use of a single element is sufficient for such a diagnosis. Magnesium has been widely used for this purpose, selecting the Mg I 285.2 nm ($E_{exc} = 4.35$ eV) and the Mg II 280.2 nm lines ($E_{ion} = 7.65$ eV, $E_{exc} = 4.42$ eV). Practically, a 1 mg mL^{-1} test solution stabilized with a few percents of nitric acid is used. Net line intensities, background signals, relative standard deviations (RSD) of the intensities and ionic-to-atomic line intensity ratio will be measured. To avoid being limited by the shot noise when measuring the RSD, it is crucial to use an integration time that is long enough to be flicker-noise limited, that is by the fluctuations of the sample introduction system. This value can be obtained by plotting the RSD of the various lines as a function of the integration time and determining the time above which the RSD reaches a quasi-constant value. At least ten replicates should be carried out to obtain a good estimate of the RSD.

7.7.2 Measurement of the Mg II/Mg I ratio

In case of the use of photomultiplier detection, it is essential to use the same photomultiplier voltage (and the same amplifier gain when appropriate) when measuring the Mg II/Mg I line-intensity ratio. Otherwise, the ratio will be meaningless.

The two Mg lines were selected because of their wavelength closeness. Usually, there is no need to compensate for a different wavelength response curve as far as the same measurement conditions are used, for example the same high voltage in the case of a photomultiplier tube (PMT), and the same amplifier gain, if any. However, it may be necessary to use a correction factor in some cases. For instance, a 2400 line mm^{-1} grating can be used in the first order above 300 nm, and in the second order below 300 nm. Besides the use of an interference filter, the order selection can be performed by using a solar-blind PMT with a wavelength cut-off near 300 nm. Consequently, the wavelength response may exhibit a significant slope near the two Mg wavelengths. Another case is the use of an echelle grating. The two Mg lines may be located in adjacent orders, or at different locations within the same order. Because the diffraction efficiency is highly dependent on the location of the line within the order, it may also be necessary to compensate for a different wavelength response. A simple way to establish a correction factor is to assume that the continuum has a constant value in the range 280–285 nm. It is then sufficient to measure the background emission at 280.2 and 285.2 nm. If $B_{285}/B_{280} = \varepsilon$, the experimental Mg II/Mg I ratio has to be multiplied by ε. For instance, it was found that ε was equal to 1.85 and 1.8, for the Perkin-Elmer Optima 3000 and Varian Vista ICP systems, respectively. It should be mentioned that these values were obtained for systems running in our laboratory and have to be verified for other systems.

7.7.3 Procedure

The procedure to verify the energy transfer, the nebulizer efficiency, the nebulizer repeatability and the long-term reproducibility is summarized in Table 7.8. The SBR and %RSD tests on nebulizers can only be performed when the Mg II/Mg I test on the transfer of energy is valid. Otherwise, they will not be significant. An example of diagnosis over a period of 6 months is given in Figure 7.5 and 7.6 for a conventional concentric nebulizer associated with a double-pass spray chamber. The %RSD of the Mg II/Mg I was 3.5%, which is evidence of an excellent long-term stability of power delivery to the plasma. In marked contrast, the SBR was not as good. Actually, some problems were observed from 1 in Figure 7.6, because of leakage in gas delivery along with a degradation of the nebulizer. The nebulizer was changed in 2, and lasted for about 4 months, and was then replaced by another nebulizer in 3. Actually, following this diagnosis, it was decided to change the nebulizer every 4 months. Figure 7.6 shows that the %RSD varies over a large range: 0.4–1.2 with a mean of 0.7. Values below 1% were considered satisfactory. Possible malfunctions are summarized in Table 7.9.

Table 7.8

Summary of the procedure to verify the sample introduction system, based on the use of Mg as a test element

Energy transfer	The first step is to check that the transfer of energy is adequate by measuring the line intensity ratio of Mg II to Mg I, as this ratio is highly sensitive to the efficiency of energy transfer between the plasma and the sample
Nebulizer efficiency	The next step is to verify the nebulizer efficiency by measuring the SBR of the Mg I 285 nm line. For a given Mg II/Mg I line intensity ratio, the background remains constant. The SBR of the Mg I line is therefore directly proportional to the nebulizer efficiency
Nebulizer repeatability	Subsequently, the stability of the sample introduction system is checked by measuring the RSD. The average RSD for the two Mg lines is used
Long-term reproducibility	Long-term reproducibility may be evaluated by recording the Mg I line intensity

Table 7.9

Possible malfunctions to explain problems in the energy transfer, and the nebulizer efficiency and repeatability

Energy transfer	wrong power setup
	change in power display
	degradation in the generator (power tube, cooling)
	high reflected power
	variation in the coil shape
	recirculating water quality
	change in observation height
	change in plasma (outer) gas flow rate
	change in the carrier gas flow rate
	change in the torch position
	devitrification of the outer tube of the torch
	partial blocking of the injector
nebulizer efficiency	partial blocking of the nebulizer
	degradation of the tip(s) of the nebulizer
	partial blocking of the capillary tubing
	poor condition of the peristaltic pump tubing
	inadequate pump speed
	leak of the carrier gas
	change in the temperature of the spray chamber
	horizontal alignment of the torch
nebulizer repeatability	too low the integration time
	Pump roller quality
	poor condition of the peristaltic pump tubing
	poor cleaning of the inner part of the spray chamber
	too low a gas pressure
	driving of the matchbox capacitors
	drain instability
	high dark current because of a previous, intense signal (PMT)

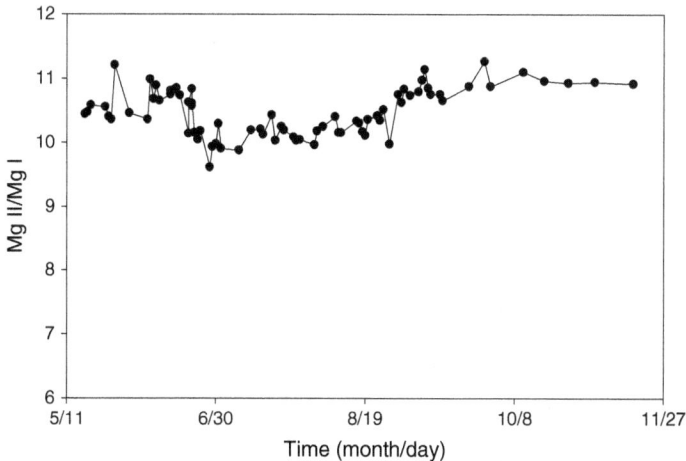

Figure 7.5 Example of Mg II/Mg I diagnosis over a period of 6 months. The %RSD of the ratio was 3.5.

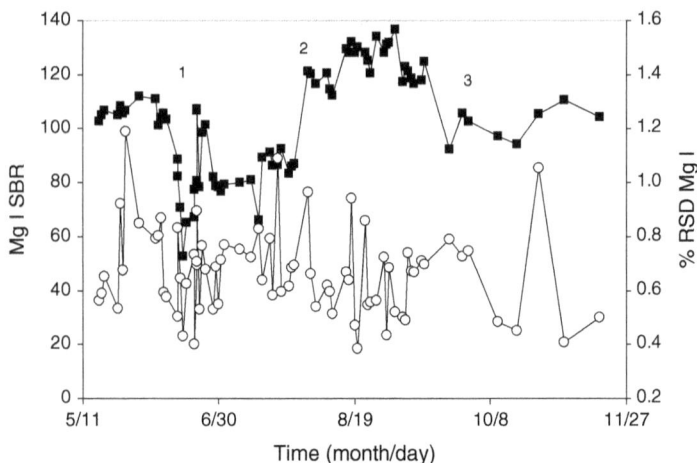

Figure 7.6 Example of Mg I SBR (■) diagnosis over a period of 6 months. In contrast to Figure 7.5, the %RSD was 19%. The % RSD was in the range 0.4–1.2 with a mean of 0.7.

7.8 OPERATION AND TROUBLESHOOTING OF CONCENTRIC NEBULIZERS

The different pneumatic concentric nebulizer manufacturers have made an effort to give their customers some guidelines regarding the proper use of the different devices to avoid the potential problems associated with working with these somewhat fragile accessories. In this section, these aspects have been considered. The information can be found in the websites of the manufacturers.

When working with a concentric nebulizer, some problems should be prevented. The nebulizer tip blocking is one of the most serious ones. Nebulizer blocking occurs because of the existence of constraints in its structure. It is worth a mention that, for example, the annular gas channel has a gap ranging from 10 to 30 μm. A more or less gradual drop in the signal magnitude and/or in its stability is normally a result of the nebulizer tip blocking.

Particles and fibres can originate from many sources. Some of these particles can be carried by the gas and, hence, one of the recommendations is to filter the carrier gas. The use of a Teflon tape to avoid gas leakage is one of the most prominent sources of particles that can deposit at the nebulizer gas exit area, thus degrading its performance. In other instances, turbid solutions have to be analysed. The obvious recommendation is to systematically filter all these solutions. When particles and colloids have an analytical interest, a high-solids nebulizer should be used.

If solutions having a moderate salt concentration are used, the nebulizer should be rinsed before turning off the gas. Even concentric nebulizers especially developed to work with saline solutions can be blocked if this point is neglected. In fact, it has been claimed that to use a nebulizer routinely, it should be flushed with a diluted acid solution (e.g. 2% nitric acid) for several minutes after its use. Then, demineralized water must be nebulized

Figure 7.7 Picture of a concentric C-type nebulizer after using it with an ammonium sulphate solution without nebulizer clearing. (Taken from ref. [127]).

for a couple of minutes. This ensures that samples deposits will not form on the inner walls of the tip once the solvent is evaporated.

Figure 7.7 demonstrates that if a C-type nebulizer is not cleaned after running it with a high salt content solution, crystallization occurs at the tip of the liquid capillary [127]. The rinsing solution should be compatible with the sample that has been analysed. Thus, for example, if a low pH saline sample has been analysed, an acidic solution should be used to clean the nebulizer tip. Then a pure water solution must be used to finally clean the nebulizer. Isopropyl alcohol can also be used at this stage. The nebulizer cleaning procedure finishes with a nebulizer drying step with argon. To avoid further solvent aspiration, the liquid supply must be switched off.

There are other sources of solid particles. Glass pneumatic nebulizers are mostly provided with plastic caps that prevent the deposition of airborne particles on the nebulizer tip during storage. The tip of the nebulizer should be neither knocked nor touched. Any deposit of body oils can degrade the performance of the nebulizer. Note again that once the nebulizer tip has been damaged, its behaviour cannot be restored.

To detect any nebulizer tip blocking, a continuous reading of pressure can be taken. When the nebulizer gas exit annulus blocks, the pressure required to maintain a given gas flow rate constant rises. Besides, if the blocking is produced at the liquid capillary tip, the liquid flow rate drops. A rapid way for checking the liquid flow rate variation has been suggested (http://www.geicp.com). It is possible to determine the velocity of an air bubble in the peristaltic pump uptake tube. At about 0.5 mL/min liquid flow rate, for 1.0 mm i.d. tubing, an air bubble moves at 1 cm s^{-1}. Any change in the liquid flow rate caused by, for example, the tubing deterioration can be detected in this way.

To preserve the nebulizer with an unaltered performance, nebulizer cleaning in ultrasonic baths should be avoided. This can cause capillary vibrations that can produce irreversible damages as it impacts against the inner walls of the nebulizer body.

If, after considering all the above mentioned points, a total or partial blocking in nebulizer occurs, there are some methods recommended by the manufacturers that can be tried to recover the concentric nebulizer. The nebulizer should first be removed and observed under microscope. A 20X or 30X magnification is recommended. Then several possibilities can be considered (http://www.meinhard.com):

1. There are some visible particles somewhere in the nebulizer tip. Several choices are available to try to eliminate the particle: (i) the nebulizer tip can be tapped gently against a wood surface, which would help the particle to go towards the zone of the capillary with higher diameter; (ii) a gas stream can be back flushed through the nebulizer tip. This can also be done by closing the liquid and gas inlets with the fingers, allowing the pressure to increase and to open suddenly at both inlets. The gas expansion helps the particle to be detached; (iii) a liquid solution can also be backflushed. In this case, isopropyl alcohol solution can solve the problem. Additionally, the nebulizer can be immersed into a surfactant solution for 24 h and then it can be flushed with warm water. To properly complete this step, the gas or the liquid inlets can be closed depending on the particle location; (iv) hot water and pressure applied to the nebulizer gas inlet are useful to eliminate Teflon shreds; (v) hydrofluoric acid is useful to eliminate silica particles. However, if the contact between the nebulizer and HF is prolonged, the glass walls can be degraded.
2. There is a solid deposit in the sample capillary and the chemical nature of the deposit is known. In this case, the nebulizer should be flushed with the most appropriate solvent to eliminate this deposit. To flush the nebulizer, either a bottle or a peristaltic pump can be used. If after observation under microscope it is verified that the deposit is gone, the nebulizer should be cleaned and dried. In case the deposit has not been eliminated, a general recommendation is to fill the nebulizer with a solvent and warm it with a heat lamp. Another choice is to use microwave heating. The solvent will boil and, hence, will carry out part or the entire residue. Furthermore, this heating will also dry the nebulizer. HF can be used in the case of siliceous deposits.
3. If the particle or deposit is of organic nature, hot (100 °C) nitric acid can be used to try to digest this solid matter. This solution can be forced to pass through the nebulizer conductions. If fusible solids such as waxes are present, the nebulizer must be carefully heated so as to avoid the formation of insoluble pyrolysis products.
4. Finally, if the particle is not removed even after following the steps mentioned above, a piano wire can be inserted into the liquid capillary and moved back and forth. This step must only be applied if any of the remedies mentioned before do not solve the blocking. If this is not carefully done, the nebulizer capillary can be irreversibly broken. In fact, some manufacturers recommend never using a wire to unblock the nebulizer tip.

The following scheme (Figure 7.8) has been proposed for cleaning a Meinhard®-type nebulizer.

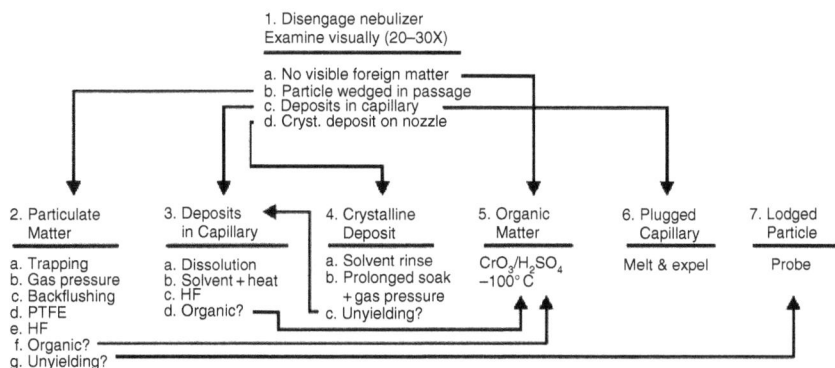

Figure 7.8 Cleaning strategy for a Meinhard® pneumatic concentric nebulizer. (Taken from http://www.meinhard.com).

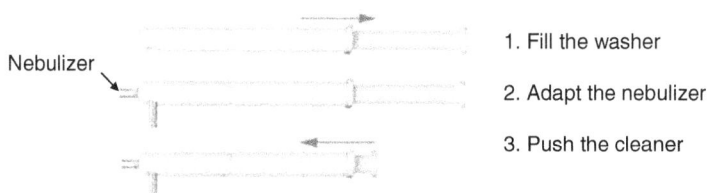

Figure 7.9 Pictures of a commercially available nebulizer washer and its operation (Adapted from http://www.geicp.com).

There are some cleaners commercially available to wash the nebulizer. Figure 7.9 shows how this system works.

Sometimes, the nebulizer is not cleaned even after using the device shown in Figure 7.9. In this case, it has been recommended to soak the nebulizer for 24 h and then to use the nebulizer washer with warm water. After a few flushes with warm water, methanol can be flushed to accelerate the final solvent evaporation.

The nebulizer performance is closely related to the operating conditions it has been designed for. Many manufacturers are currently selling concentric nebulizers specially conceived to work under clearly defined liquid flow rates. Thus, it is very important to remember that a given pneumatic concentric nebulizer will work best at liquid flow rates close to the 'design aspiration level' (http://www.geicp.com). This flow corresponds to the free liquid uptake rate. It is also worth to mention that a given nebulizer can be operated at a liquid flow rate higher than the best one. The liquid pressure must simply increase. For example, there are nebulizers that are able to work at liquid flow rates above 3 mL min^{-1}. However, it can be an impractical result to try to switch a micronebulizer at such a high-delivery flow rate. The liquid connections can burst because of the high pressure required. Note that there is a fourth power relationship between the applied pressure and the inner

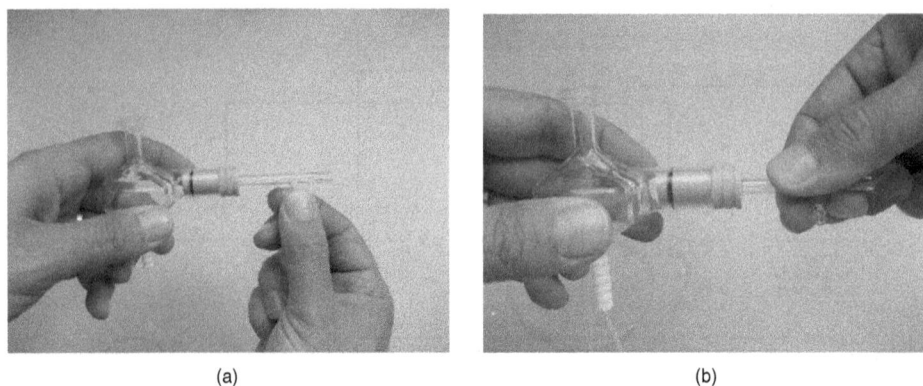

(a)   (b)

Figure 7.10 Illustrations indicating how to hold a nebulizer when it is removed or adapted at the spray chamber. The nebulizer must be held by the barrel (b) not by the side arm (a).

diameter of the sample conduction (i.e. that of the micronebulizer is lower than the i.d. of the capillary of a conventional concentric nebulizer).

Another important point to be considered is referred to the work with hydrofluoric acid. Solutions of this acid should not be used with glass or quartz nebulizers. This causes an irreversible damage of the nebulizer tip and sample capillary. For PFA concentric nebulizers, it is indicated that to carry out ppt levels analyses, it is necessary to clean the nebulizer with a 10% HCL + 10% HNO_3 solution for a period of time close to 30 min (http://www.elementalscientific.com). In case the nebulizer is strongly contaminated, it has been found that the use of hot (i.e. 80 °C) aqua regia solution can solve the problem.

Removing and assembling the nebulizer is a critical procedure. To avoid problems, the nebulizer must be held by the barrel and must be moved straight in or out. Figure 7.10 shows two pictures in which it is indicated what to do and what not to do when removing a nebulizer from the spray chamber.

7.9 OPERATION, MAINTENANCE AND TROUBLESHOOTING OF PARALLEL PATH NEBULIZERS

Working with a PPN is quite simple because it is robust and its behaviour is not modified for periods of operation time as long as three years. However, there are some instructions that have to be followed to properly work with these devices. Particularly, it has been recommended to wash the nebulizer only with water.

Because of the robustness of the PPN, when totally or partially blocked, it is possible to unblock it with a wire. However, it is a general recommendation to complete this cleaning under the observation of the nebulizer with a microscope and a 10 × magnification. This is mainly because if the gas orifice is touched with the wire, the nebulizer can be irreversibly damaged. Consequently, a filter must be used to remove particles coming with the argon stream.

For Teflon PPN, it is important to remember that the tip should never be touched with fingers or tissue, because this is a very soft material and the nebulizer can be damaged. This material is not wettable, thus it is not necessary to wash Teflon nebulizers with acid or surfactant solutions. Moreover, these solutions can induce a salt deposit on the inner walls of the gas exit and cause tip blocking.

7.10 OPERATION, MAINTENANCE AND TROUBLESHOOTING OF SPRAY CHAMBERS

Little information can be found in the literature concerning the instructions to use and properly maintain the different spray chambers. It is indicated that a spray chamber works at optimum when its walls are as close to the original state as possible. As a spray chamber is used, its surface becomes dirty and the signal stability can degrade. A clear sign of spray chamber surface degradation is the droplets collected on the inner walls of the chamber. Thus, a general procedure would be to use conditioning solutions to reach this goal (http://www.spray-chamber.com).

1. After each period of chamber use, it is worth washing it with distilled water for several minutes. In some cases, a diluted acid solution is also recommended for these purposes. This step is crucial to keep the chamber in a good shape as it prevents crystal deposits on the spray chamber once the solvent is evaporated. Figure 7.11 shows an example of cleaning efficiency for two conventional spray chambers: a double-pass (Figure 7.11a) and a cyclonic type one (Figure 7.11b). These pictures correspond to the chambers washed with distilled water for 1 min after being operated with a methylene blue solution. It can be observed that 1 min is not sufficient to completely clean the chamber and some blue droplets stay on the inner chambers walls.
2. The chamber must be thoroughly cleaned. The cleaning frequency recommended is once a week and the way of doing this is to immerse the chamber into either a basic

(a)   (b)

Figure 7.11 Pictures of a double-pass (a) and cyclonic (b) spray chambers when distilled water has been introduced for 1 min after running the system with a methylene blue solution.

detergent solution or a 1% nitric acid solution for 2 h. The former solution is used for cleaning glass chambers, whereas the latter one would be the chosen solutions for polymeric spray chambers.
3. To improve the performance of a glass spray chamber, it can be flushed monthly with a diluted hydrofluoric acid solution. Obviously, this step should be performed for short periods of time (less than 2 min).

To wash the spray chamber, scraping tools such as brushes must not be used, because these instruments cause surface erosion and future memory effects. If glass spray chambers are used, they must not be washed in an ultrasonic bath. There are solutions commercially available that can be used to clean the spray chamber surface. For example, the chamber can be soaked in a 25% strength solution RBS-25 (i.e. a detergent solution used for glass instruments in chemical, bacteriological and radiotechnical laboratories, http://www.sigmaaldrich.com) for one day or more. If this does not solve the problem, stronger acidic solutions (but not HF solutions) can be used. Hydrofluoric solutions are suitable for cleaning plastic chambers. This solution can also be used for the removal of eventual resident droplets on the spray chamber walls. A specific comment should be made concerning these acid solutions, because a monthly short (less than 2 min) flushing with 1% HF solutions can improve the stability of quartz and glass spray chambers (http://www.spray-chamber.com).

The plastic end cap of the spray chamber is a source of memory effects. Furthermore, it has a large connection surface and leakage can occur. Therefore, it is recommended to use spray chambers with no end caps.

Gas or sample leakage problems are a very important point that must be considered and avoided [807]. The usual connections present in a liquid sample introduction system are often the bottleneck for obtaining the best possible performance. As Figure 7.12 reveals, with a common sample introduction system, eight connections can be required. The analyst has to check for eventual leakages in every connection. If the sample connections

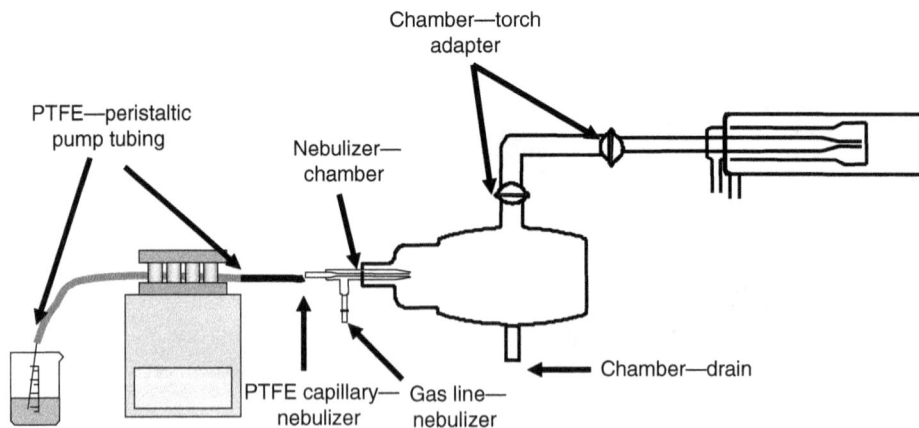

Figure 7.12 Connections found with a conventional ICP simple introduction system.

(PTFE–peristaltic pump and PTFE–nebulizer connections, Figure 7.12) are not tight enough, it can be easily verified, because either solution drops will form in the connection or air bubbles will appear in the liquid stream. In the former case, the sensitivity will be unusually low, whereas in the latter, the signal will be noisy.

Special mention has to be made about the gas leakage. The chamber and nebulizer assembly connections must be as tight as possible to avoid stability degradation and plasma ignition problems. Normally, with a glass pneumatic concentric nebulizer, ICP signal precisions lie between 0.2 and 0.5% RSD. If eventually, the RSD rises above this level, either there is a connection problem or the nebulizer gas flow is not the optimum one. To check for gas leaks, a soap solution can be carefully deposited on every connection, the bubble formation will confirm the problem.

There are additional advices that are given to check for possible sources of instability caused by the nebulizer. Thus, if the signal RSD is getting worse but the spray chamber seems to work properly, the nebulizer gas line should be explored. Tygon and other polymers can loose their flexibility with time. As a result, argon leakage can occur. Even a 1% loss of argon can lead to significant changes in the ICP sensitivity. An argon leakage can be easily detected using a silicone tube 1 m length as a stethoscope. If the signal is still instable, it is necessary to check for the nebulizer uptake. Sometimes, air bubbles entering together with the solution are responsible for stability degradation. To ensure that there are not gas losses, the gas nebulizer connection must be as tight as possible and the carrier line should be completely on the nebulizer (Figure 7.13). When the gas line is attached and detached many times, it can be degraded and its ability of sealing the gas conductions deteriorates. In this case, about 4 mm of the conduction can be cut to renew the gas line.

Another critical point is the spray chamber drain–waste tube connection. For plastic chambers, there is no way of observing the draining efficiency while the system is being operated. If the drain is not rapidly produced, the inner volume of the chamber will decrease and, hence the signal will drop. Furthermore, the signal stability will be compromised and the plasma will be difficult to ignite. One way to test the drain efficiency is to put some water in the spray chamber and determine if it drains without leaks [Gaines, http://www.ivstandards.com]. In the case of glass spray chambers, it is very important not to apply large mechanical forces when connecting drain and aerosol tubes or nebulizers,

Figure 7.13 Connecting the gas line to the nebulizer. Taken from ESI MicroFlow PFA Nebulizer installation guide.

(a)

(b)

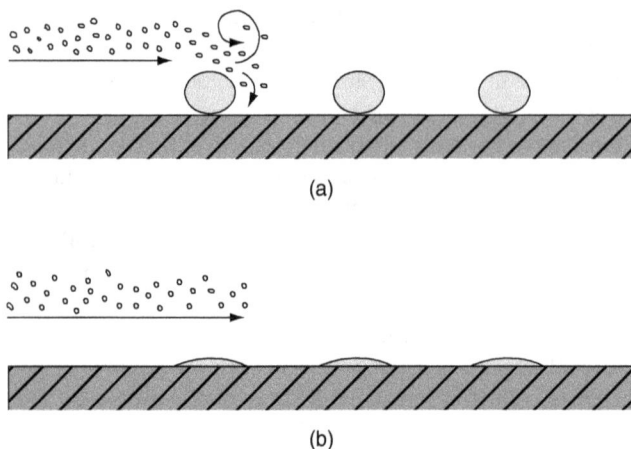

Figure 7.14 Effect of the chamber surface on the chamber performance. (Taken from ref. [702].)

otherwise, the glass can fracture. To protect the chamber drain exit, plastic caps are sometimes provided with the devices (http://www.geicp.com).

The status of the surface of the spray chamber plays a determinant role on its ability for draining the deposited solution. Thus, smooth surfaces promote the formation of coarse droplets on the chamber walls (Figure 7.14a). This fact also causes the loss of small droplets as well as the increase in the signal fluctuations. On the contrary, rough surfaces prevent this problem because the formation of a liquid film on the chamber inner walls is promoted (Figure 7.14b), hence decreasing the likelihood for small droplet trapping and enhancing the signal stability.

The spray chamber wall's temperature is a very important factor that can compromise the analytical results. It is sometimes observed that the ICP sensitivity changes with the time of the day or with the season. A 1 °C change in the chamber walls temperature can induce changes in the sensitivity higher than 1%. For this reason, jacketing the chamber has been suggested as a means for improving the long-term signal stability.

A table can be found in the ESI MicroFlow PFA Nebulizer installation guide, in which some of the points raised in the present section are summarized. This information is given in http://www.elementalscientific.com.

– 8 –

Applications

8.1 INTRODUCTION

Because of the characteristics of the ICP techniques, there has been an increased interest for applying them to many different fields. Trying to compile all the applications of ICP–AES and ICP–MS can lead to a large number of publications that are not within the scope of the present book. This chapter is intended to show selected recently issued examples of the applications of sample introduction systems.

There are different fields in which ICP techniques are increasingly used as it is indicated in some of the published reviews [747, 812] in which the spectacular growth of these techniques over the past decade is demonstrated. ICP–AES is a routine, mature technique for many applications, although some advances have been described that involve instrumental development focused on the sample introduction systems [811], fundamental studies and some methodological improvements [746]. ICP–MS is also used as a routine technique in many laboratories, and for this reason it has been claimed that few instrumental development works have appeared. Obviously, the advances produced in terms of instrumentation are of crucial importance to enlarge the field of application of these spectrometric methods. In geochemical and cosmochemical applications, the number of studies in which samples are analysed by means of ICP–MS has continuously increased. This has been possible because of the developments of this technique, mainly in the area of Sector Field ICP–MS and multi-collector ICP–MS [836]. The ability of ICP–AES and specially ICP–MS to determine extremely low analyte concentrations in limited-size samples makes this technique appropriate for forensic [837] or water [780] analyses, though in the second case many developments have been related with the use of an ICP–MS as a GC detection system. The ICP–MS technique has several advantages over other methods such as the radiochemical ones, because, unlike the latter method, it is not necessary to employ radiolabelled compounds. As a result, the sample preparation and handling procedures are less time-consuming in applications such as the determination of drug compounds [843]. Most of the developments produced in liquid sample introduction systems are related with low sample consumption systems. A description of some of these applications is given in the following section.

8.2 DESCRIPTION OF APPLICATIONS OF LOW SAMPLE CONSUMPTION SYSTEMS

There are many situations in which the amount of sample available for the analysis is very limited. Micronebulizers are ideal devices for performing analysis at low liquid flow rates. For example, with the MCN, it is possible to perform hair, biological samples and peptides analysis by ICP–MS. With this nebulizer, 100 μL of sample is enough to conduct a full analysis [495]. The MCN is also a good selection to carry out geological microsamples (i.e. 10 mg of sample) analysis by means of FIA–ICP–MS [644]. If the analyte concentration is very low, it is possible to achieve excellent signal stabilities by operating the nebulizer under free aspiration mode. This approach is especially interesting when conducting environmental analysis on low sample volumes (c.a. 1 mL) such as those obtained when dealing with Alpine snow and ice [556, 579].

As the MCN, the HEN has been successfully used as an interface between microscale flow injection (μFI)–HPLC and ICP–MS showing the following features [419]: (i) limits of detection in the femtogram range, (ii) less sample matrix is introduced into the plasma than in continuous mode and (iii) its coupling to a conventional spray chamber is simple, because its external dimensions are equal to those for conventional pneumatic concentric nebulizers. Other applications of the HEN include its coupling to a desolvation system for the analysis of corrosive samples [434] and its use as a nebulizer for the analysis of biological specimens [454].

Micronebulizers are also very interesting as interfaces between microbore columns and ICP techniques. A good example is the use of the MCN for the determination of bromate and bromide in water through ICP–MS [599] or some speciation studies [517]. Porous graphitic columns have also been used in conjunction with a MCN for the determination of boron in urine and blood plasma through ICP–AES and ICP–MS [673]. To avoid the detrimental effects caused by organic solvents, a Peltier cooled spray chamber can be used as it has been done for the chiral speciation of selenium compounds [740]. The MicroMist has also been used as an interface between HPLC and ICP–SFMS for arsenic speciation [759]. As regards the parallel path micronebulizer, it has been used as an interface for HPLC–ICP–MS for cobalt speciation [774].

Capillary electrophoresis also demands the participation of a micronebulizer as interface [511, 627, 635]. The MicroMist has proven to be useful for CE–ICP–MS for the speciation of arsenic in soils [692]. An example of the application of the HEN to speciation studies through capillary electrophoresis is its use for the analysis of metallothioneins [721]. Another good example is the use of the MCN for the separation and determination of $Cr_2O_7^{2-}$ and Cr^{3+} and Cu^{2+} and $Cu(EDTA)^{2-}$ [475]. The majority of the nebulizers used as interfaces between CZE–ICP techniques are self-aspirating systems and the separation can be degraded. The high-efficiency cross-flow micronebulizer (HECFMN) does not have suction effect, hence being very attractive for CE–ICP–MS coupling [645].

The DIHEN has been applied to the direct analysis of biological samples [590]. A microemulsion with Triton X-100 was prepared to carry out the analysis of gasoline samples by ID–ICP–MS in conjunction with a DIHEN [743]. This method was more appropriate and faster than the classical sample microwave-assisted digestion and subsequent analysis.

8.3 SELECTED APPLICATIONS

Table 8.1 summarizes a representative list with more than 80 applications of sample introduction systems. As regards conventional systems (i.e. those that are switched at liquid flow rates on the order of $1\,mL\,min^{-1}$), many different applications can be found. Table 8.1 shows new instances in which these systems are useful. On this subject, it is difficult to differentiate between applications of the conventional sample introduction systems and applications of ICP techniques.

In the case of micronebulizers, their most relevant applications are related to their use in topics in which the sample size is an important factor for several reasons (sample availability, toxicity, etc.). It can be verified that the sample volume consumed can be as low as several tens of nanolitres, but in most cases this variable is typically in the 0.1–1 mL range. In order to achieve this, either the sample is injected *via* a valve or stable pumping devices (i.e. HPLC, gas displacement pumps and syringe pumps) are used. Liquid flows as low as $0.5\,\mu L\,min^{-1}$ can be achieved.

Table 8.1

Selected recent applications of widely employed liquid sample introduction systems

Sample introduction system	Application/Technique	Comments	References
Pneumatic concentric micronebulizer coupled to a cooled Cinnabar chamber	Determination of gold in geological samples	The sample preparation step involves the use of HF, chromatographic separation of the matrix and a pre-concentration step. This leads to a detection limit of $2\,pg\,g^{-1}$	810
PTFE microflow nebulizer coupled to a tandem quartz glass spray chamber	Determination of the iron isotopic composition in Moon and Mars samples		784
Concentric nebulizer + single-pass spray chamber	Analysis of ink samples through ICP–MS with collision cell	An impact bead is used in order to remove the coarsest droplets produced by nebulizer	816
Cross-flow nebulizer coupled to a cyclonic spray chamber	Analysis of human cremated remains (cremains) by ICP–AES	Different samples of either cremains or concrete can be discriminated	825
Pneumatic nebulization	Determination of ^{129}I in environmental samples	More than 100 samples are analysed with satisfactory results as compared with X-ray spectrometry	753

(Continued)

Table 8.1

Selected recent applications of widely employed liquid sample introduction
systems—Continued

Sample introduction system	Application/Technique	Comments	References
Pneumatic nebulizer coupled to a Sturman-Masters spray chamber	Determination of the bioavailability of elements in foods	A continuous leaching-based system is used to study the progressive liberation of Zn and Pb caused by artificial saliva, gastric juice and intestinal juice	767
Cross-flow nebulizer coupled to a cooled double-pass spray chamber	Speciation of trace metals in cancerous thyroid tissues by ICP–MS	Size exclusion chromatography was coupled to ICP–MS. Some species of Cu, Zn, Cd and Pb are found in healthy thyroids that are not encountered in cancerous ones	744
Cross-flow nebulizer coupled to a cooled double-pass spray chamber	Determination of phytochelatins in plants cultivated under stress conditions	Ion pairing chromatography–ICP–MS was the used coupling	778
Pneumatic concentric + cyclonic spray chamber	Analysis of biological and environmental samples through ICP–AES	The samples have been on-line digested and a specially designed interface has been used for removing the generated gas	783
Cross-flow nebulizer coupled to a double-pass spray chamber	Analysis of olive, corn and sunflower oils through ICP–AES	The sample has been introduced as an emulsion and the quantification was successfully achieved using aqueous standards	782
V-groove pneumatic nebulizer coupled to a low-volume PTFE Sturman Masters spray chamber	Study of the stability of ceramic slurries	It is possible to carry out the determination of elements present in the smallest slurry particles with aqueous solutions	789
V-groove nebulizer coupled to a reduced inner volume Sturman-Masters spray chamber	Determination of impurities in fine TiN fine powders by ICP–AES	Calibration with aqueous standards provided good results compared by a method based on an alkaline fusion method	801

Table 8.1

Selected recent applications of widely employed liquid sample introduction
systems—Continued

Sample introduction system	Application/Technique	Comments	References
Either a cross-flow or a slurry nebulizer coupled to a cyclonic spray chamber	Analysis of gypsum and cement by ICP–AES	Particles of up to roughly 150 μm are present in the slurry. By using the generalized standard additions method, good recoveries were found for several CRMs	741
Pneumatic concentric nebulizer + cyclonic spray chamber	Elucidation of the exact stoichiometry of materials employed in the superconductor industry (YNi_2B_2C) through ICP–AES		683
Pneumatic concentric seaspray + double-pass spray chamber	Determination of Ca, Cl, K, Mg, Na and P in biodiesel samples through ICP–AES	Biodiesel samples were diluted with kerosene and the method yielded LODs for Na and K close to 1 and 7 ppb, respectively. Meanwhile, LODs for chlorine were of the order of 1 ppm	682
Pneumatic concentric nebulizer + Sturman–Masters spray chamber	Assessment of the mobility of specific elements in rocks	The system is reliable for carrying out analysis in a segmented flow injection modality	672
Pneumatic concentric nebulizer + cooled cyclonic spray chamber	Arsenic speciation through ICP–MS	A new arsenic metabolite is found in sheep's urine after arsenosugar ingestion: the dimethylarsinoylacetate (DMAA)	722
Pneumatic concentric nebulizer	Speciation in salmon eggs	The separation was carried out with surfactant-mediated HPLC–ICP what permitted the separation of big molecules such as proteins and anions such as PO_4^{3-}, Cl^- and Br^-	711

(Continued)

Table 8.1

Selected recent applications of widely employed liquid sample introduction
systems—Continued

Sample introduction system	Application/Technique	Comments	References
APEX desolvation system	Determination of low-abundance U isotopes in soils by ICP–MS	The U^+/UH^+ is improved by a factor of 106 with respect to a conventional sample introduction system	824
Pneumatic concentric nebulizer + cyclonic spray chamber	Determination of the S content in diesel fuels	Quantitative determinations indicated no significant differences between ICP–AES and EDXRFS results	833
Pneumatic concentric nebulizer + double-pass spray chamber	Classification of different marijuana samples through ICP–MS		841
	Estimation of the firing distance from the gunshot residue deposition pattern around a bullet entrance hole		835
Pneumatic nebulizer coupled to a double-pass spray chamber (http://www.varianinc.com)	Determination of stability constants of Co- and Zn-EDTA complexes	The method has been based on the use of ion chromatography coupled to ICP–MS	823
	Determination of Cd in polyethylene		815
MCN + single-pass spray chamber	Selenium speciation in selenized yeast/size exclusion chromatography and capillary zone electrophoresis–ICP–QMS	Five selenium compounds were baseline separated with LODs within the 7–$18 \, ng \, L^{-1}$ range with a 20 nL injection.	676
MCN	Main matrix elements and trace elements in plant reference samples/ICP–AES		478
MCN + double-pass spray chamber	Rh, Pd and Pt in snow and ice/ICP–DFMS	The LODs were 0.02, 0.08 and $0.008 \, pg \, g^{-1}$ for Rh, Pd and Pt, respectively, whereas the sample consumption volume was only 40–$80 \, \mu L \, min^{-1}$.	556

Table 8.1

Selected recent applications of widely employed liquid sample introduction
systems—Continued

Sample introduction system	Application/Technique	Comments	References
MCN + double-pass spray chamber	Trace elements in snow and ice/ICP–DFMS	All the procedures had to be carried out under ultra clean conditions	579
MCN	Total sulphur in soils, sediments, waters and a meteorite/ID ICP–SFMS	Sulphur is oxidized to sulphate. The determined concentrations were from 5.25 μg g^{-1} to 2% and the required sample size was just 13–40 mg.	644
MCN + double-pass spray chamber	Metals in milk powder (BCR CRM 63), bovine liver (BCR CRM 185) and mussel tissue (BCR CRM 278)/ICP–AES	The LODs were within the 0.13 to 174 μg L^{-1} range at 100 μL min^{-1} liquid flow rate	528
MCN + double-pass spray chamber	Metals in polymers/ ICP–QMS	The polymer samples were received as methanol solutions. Then they were dried and digested and the analysis was possible with the consumption of just 90 μL of sample	541
MCN	Cd and Zn in mouse and rabbit liver proteins/ CZE–ICP–QMS	The number of peaks appeared on the electropherogram depended on the Cd:Zn concentration ratio	653
MCN + double-pass spray chamber	Multi-element analysis in hair (NIES No. 5) and peptides	Analyses were carried out continuously with a sample consumption always lower than 100 μL	49
MCN	Bromate in drinking waters/flow injection ICP–QMS	With an anion exchanger and at 140 μL/min, the LOD was 0.13 μg L^{-1} (injected volume: 500 μL), the sample throughput was high (6 h^{-1}), and the precision was good (RSD < 2%)	599

(Continued)

Table 8.1

Selected recent applications of widely employed liquid sample introduction
systems—Continued

Sample introduction system	Application/Technique	Comments	References
MCN + double-pass spray chamber	Simultaneous speciation of two selenium and arsenic species in spiked groundwater/ IC–ICP–QMS	At a liquid flow rate of $100 \, \mu L \, min^{-1}$, the LODs were 1 and $4 \, ng \, mL^{-1}$ for As and Se, respectively	521
HEN	Cd, Cu and Zn speciation metallothioneins in eel liver cytosols/ CE–ICP–QMS		721
MMN + cyclonic spray chamber	Arsenic speciation in soils/CE–ICP–SFMS and high-performance ionic chromatography (HPIC)–ICP–SFMS	HPIC and CE allowed for the separation of 5 and 6 arsenic species, respectively. For HPIC, LODs were in the $0.04–0.08 \, ng \, g^{-1}$ range, whereas for CE they were 100 times higher	692
PFAN + Peltier cooled spray chamber	Absolute configuration of selenomethionine in Antarctic krill/ reversed-phase HPLC–ICP–QMS	The enantiomers of selenomethionine were derivatized into diastereomeric isoindole compounds. The achieved LOD for Se was $4 \, \mu g \, L^{-1}$	740
PFAN + double-pass spray chamber	Phosphorus in yeast lipid extracts/HPLC– ICP–QMS	Spray chamber cooling was used for the removal of interference. The absolute LODs for the different compounds ranged from 0.21 to 1.2 ng. The injected sample volume was $2 \, \mu L$ and seven different compounds were baseline resolved	757
PFAN + Teflon spray chamber	Phosphorylation degree of α and β-caseins/ ICP–QMS with dynamic reaction cell	The method can be used for determinations in the 10 to $1000 \, fmol \, \mu L^{-1}$ concentration range	670

Table 8.1

Selected recent applications of widely employed liquid sample introduction systems—Continued

Sample introduction system	Application/Technique	Comments	References
MM + Cinnabar	Iodide in food samples/ID ICP–MS	The short wash-out times allow for the determination of I in food samples without the need for adding oxidizing agents. The isotope dilution technique was applied by measuring the ^{129}I intensity	603
DCN + double-pass or cyclonic spray chamber	Trace elements in water (NIST SRM 1643c) and spinach (NIST SRM 1570)/ICP–QMS	Recoveries close to 100% were found for all the elements tested, except for Zn	762
Single-pass spray chamber	Sulphur in metalloproteins from bream liver/ CE–ICP–QMS with an octopole reaction cell	Limits of detection were $1.3\,\mu g\,L^{-1}$ and $3.2\,\mu g\,L^{-1}$ for ^{32}S and ^{34}S, respectively.	706
Single-pass spray chamber	Phosphorus in enzymatically digested calf thymus DNA/CE and HPLC–ICP–QMS with an octopole reaction cell	The four monophosphorylated deoxynucleotides present in the DNA chain were successfully resolved. Absolute limits of detection for P were 0.6 pg and 0.03 ng with CE and HPLC, respectively.	723
Single-pass spray chamber	Selenopeptide mapping in a selenium containing protein/reversed-Phase HPLC–ICP–QMS with and without collision cell	The liquid flow rate ranged from 0.5 to 7.5 $\mu L\,min^{-1}$ and just 200 nL of sample was injected. LODs were about 200 fg Se. Resolution was sharply increased with respect to conventional HPLC–ICP–MS	710
Single-pass spray chamber	Metallothioneins in human brain cytosols/ CZE–ICP–SFMS		713

(*Continued*)

Table 8.1

Selected recent applications of widely employed liquid sample introduction
systems—Continued

Sample introduction system	Application/Technique	Comments	References
MCN6000	Rear earth elements in marine particulate matter/ICP–SFMS	Detection limits were 1–40 ppq which, when combined with the low liquid flow rate ($100\,\mu L\,min^{-1}$) yielded 1–20 fg absolute LODs.	508
MCN6000	Arsenic in steel/ ICP–QMS	The background equivalent concentration was 25 times lower than for a conventional sample introduction system. At $60\,\mu L\,min^{-1}$ the limit of quantification was $0.12\,\mu g\,g^{-1}$.	564
MCN6000	P, V, Cr, Mn, Fe, Co, Ni, Cu, Zn, Y, Mo, Cd, Re, Tl and Pb in lake waters/ICP–SFMS	Elements were detected at levels below 1 ppb or ppt (Re).	719
MCN6000	^{239}Pu and ^{240}Pu in seawater/sequential injection ICP–SFMS	3–10 L of sample were pre-concentrated to 7 mL, thus affording LODs of 0.64 and 0.19 fg/mL for ^{239}Pu and ^{240}Pu, respectively.	671
MCN6000	V, Cr, Mn, Fe, Co, Ni, Cu, Zn, Cd and Pb in seawater samples	This analysis was carried out using only $50\,\mu L$ of sample without suffering from the interferences usually present in this kind of sample	578
MCN6000/MCN	Al, Sc, Ti, V, Cr, Mn, Fe, Ni, Co, Cu, As, Ag, Pt, Au, Pb in human milk/ ICP–SFMS		617
MCN6000	Isotopic analysis of uranium in tree bark/ ICP–QMS	Very low ^{238}U limits of detection ($0.004\,ng\,L^{-1}$) and good ^{235}U/^{238}U precisions (0.4–3.1% RSD) were achieved by operating the system at $80\,\mu L\,min^{-1}$	598

Table 8.1

Selected recent applications of widely employed liquid sample introduction
systems—Continued

Sample introduction system	Application/Technique	Comments	References
Desolvation system (MCN6000)	Determination of platinum group elements in carbonaceous chondrite	Excellent sensitivities and long-term stabilities	785
Aridus	Determination of Pb isotopes in lavas	The system was washed with 2% nitric acid solution during 1–2 min with good sample throughput. To lower the background, the nebulizer was removed and dipped in a beaker with Milli-Q water	754
Desolvation system (MCN6000)	Determination of Am in environmental samples though ICP–MS	The absolute limit of detection is 0.86 fg	838
Aridus	Determination of Am and Pu in soils and sediments though ICP–MS	The desolvation system reduces the interference of UH^+ on $^{239}Pu^+$. The LODs achieved for Pu are included within the 10 to 23 fg g^{-1} range.	839, 849
AridusTM	Protactinium in silicate rock/multi-collector ICP–MS	A few tens of femtograms of Pa are sufficient to yield good precisions and accuracies	748
AridusTM/MMN	Anionic and cationic arsenic compounds in freshwater fish/HPLC–ICP–SFMS	Conventional HPLC (1.5 mL min^{-1}) was coupled to a micronebulizer system by applying a 1:7.5 splitting factor. 100 μL of sample were injected	759
AridusTM	Ir and Pt/pre-concentration ICP–SFMS	Limits of detection achieved were 0.02 ppg and 0.08 ppq for Ir and Pt, respectively	764
AridusTM	Mo isotope composition/multi-collector ICP–MS	Mo isotope variations can be determined to a precision of 0.2‰	643
AridusTM	Fe isotopic composition/multi-collector ICP–MS	The isotopic composition of Fe can be measured with a precision of 0.2 (parts per 10,000)	749

(Continued)

Table 8.1

Selected recent applications of widely employed liquid sample introduction systems—Continued

Sample introduction system	Application/Technique	Comments	References
AridusTM/DIN	Drug substances containing chlorine, bromine and iodine/ Revered phase HPLC– ICP–MS	DIN was less sensitive to analyte structure than the AridusTM	726
AridusTM	Uranium isotopic ratios in geological samples/ multi-collector ICP–MS	A precise $^{234}U/^{238}U$ analysis was carried out with 200 ng of sample. The precision (2σ) of the isotopic measurements was 0.8‰	714
AridusTM	Uranium and plutonium in soils/ICP–SFMS	By working at a 100 µL/min liquid flow rate, limits of detection were $0.2\,pg\,L^{-1}$ and $0.04\,pg\,g^{-1}$ for uranium and plutonium, respectively	693
AridusTM	Si isotope ratios in silica, diatomite and sponges/ multi-collector ICP–MS	Variations in the $^{28}Si/^{29}Si$ are determined with precisions better than 0.1‰. This procedure requires only 3 µg Si.	717
AridusTM	$^{236}U/^{238}U$ ratios in uranium minerals/ ICP–SFMS	With a $100\,\mu L\,min^{-1}$ sample consumption rate, the uranium isotope ratio was measured down to the level of 10–7.	761
AridusTM with a PFAN	^{239}Pu in urine/ICP–SFMS	For a sample with a Pu concentration as low as $7.6\,pg\,L^{-1}$, the error with respect to the expected value was lower than 2%.	725
AridusTM/MMN/ DIHEN	$^{236}U/^{238}U$ isotope ratio in soil simples/ICP–QMS with an hexapole reaction cell, ICP– SFMS and multi- collector ICP–MS	These experiments can be carried out with uranium levels as low as $fg\,g^{-1}$	691
AridusTM	Thorium isotopic measurements in volcanic rocks/multi- collector ICP–MS	A liquid flow rate moderately low (i.e. $0.5\,mL\,min^{-1}$) was used.	647

Table 8.1

Selected recent applications of widely employed liquid sample introduction
systems—Continued

Sample introduction system	Application/Technique	Comments	References
Demountable DIN	Nickel and bismuth in sediments/ preconcentration lab-on-valve ICP–MS	With a 2.0 mL sample volume consumed, the LODs obtained were 15 and 4 ng L^{-1} for Ni and Bi, respectively.	658
DIHEN/MMN	Uranium, thorium and plutonium isotope ratios in radioactive wastes and uranium isotope ratios in soils/ ICP–QMS	The ^{240}Pu/^{239}Pu ratio was measured for concentrations as low as 12 ng L^{-1}.	555
DIHEN/MM	Concentration and isotope ratios for long-lived radionuclides in radioactive wastes	Polyatomic interferences were removed by extracting the analytes (e.g. U, Th and ^{99}Tc) through a liquid extractant or by means of a solid exchanger	580
DIHEN	Sn and Cr isotopes in human lung fibroblast cells/ICP–TOF–MS	Precise isotope ratios can be obtained by measuring simultaneously fast (12.75 ms) transient signals for several isotopes	689
Demountable DIN	Selenium speciation in human urine samples/ ion pairing ICP– Quadrupole MS	Four identified selenium compounds and an unidentified one were separated and detected working at 50 μL min^{-1} and injecting 3 μL of sample.	688
Demountable DIN/ MMN + cyclonic spray chamber	Selenium speciation in human urine samples/ reversed phase, ion pairing LC– and CE–ICP–Quadrupole MS	Two selenium metabolites were identified at 50 μL min^{-1} and injecting 3 μL of sample	758
DIN	Boron in undigested blood plasma and urine samples/isotope dilution ICP– Quadrupole MS	The total sample volume required was 1 mL and the injected volume was 50 μL. About 94% of the carbon was removed by protein precipitation	769

(Continued)

Table 8.1

Selected recent applications of widely employed liquid sample introduction
systems—Continued

Sample introduction system	Application/Technique	Comments	References
MCN6000/DIN/ MCN + cyclonic spray chamber	Enriched [77]Se yeast samples/reversed phase LC–ICP–Quadrupole MS	More than 30 selenium containing compounds were separated by consuming just 3 µL of sample and working at a 50 µL min^{-1} liquid flow rate	790
dDIHEN/DIHEN	Trace metals in urine samples/ICP–QMS	No nebulizer clogging was observed for 1:5 diluted urine samples and good recoveries are achieved	607
LB-DIHEN	Trace metals in herbal extracts/ICP–QMS	The sample volume required was 20 µL. A standard additions methodology was applied	593
DIHEN	Cd2 + in seawater samples/anodic stripping voltammetry– ICP–QMS	The obtained concentration $(26 \pm 4\,ppt)$ was in accordance with the actual one $(25 \pm 3\,ppt)$	708
Demountable DIHEN	Ni and Bi in river sediments (CRM 320) and urine samples	Matrix elimination and analyte pre-concentration were carried out by means of a 'lab-on-valve' methodology. The aspirated sample volume of 2.0 mL was reduced to 60 µL through the pre-concentration step. Limits of detection were on the order of ng L^{-1}	658
Demountable DIHEN	Cd and Pb in urine samples/sequential injection ICP–QMS	Matrix was removed and the analytes pre-concentrated by means of a suspension of PTE beads. 3 mL of sample were aspirated and the retained analyte was eluted with 40 µl of a nitric acid solution. The LODs were 2.9 and 6.0 ng L^{-1} for Cd and Pb, respectively	695

Table 8.1

Selected recent applications of widely employed liquid sample introduction
systems—Continued

Sample introduction system	Application/Technique	Comments	References
DIHEN	31P in a phosphopeptide mixture and 127I in a synthetic tryoxine/ micro- and nano-LC– ICP–SFMS	Eleven phosphorus compounds were separated and detected by injecting 5 µL of sample the liquid flow rate being just $4\,\mu L\,min^{-1}$. Absolute LOD for 127I was 40 fmol	677
DIHEN	Hg in freeze-dried urine (NIST SRM 2670), I and B in bovine muscle (NIST SRM 8414), Hg and I in seahorse genital tonic pills and B in rodent liver samples/ flow injection ICP–QMS	Good recoveries were found for the determination of these memoryprone elements	675
DIHEN	Na, Mg, K, Ca, Fe, Mg in homeopathic nerve tonic tablets/ICP–QMS under cool conditions	Unlike under normal conditions, when operated under cool conditions, the background levels were low enough to allow the precise analysis of these kinds of samples	608
DIHEN	^{56}Fe, ^{52}Cr, ^{59}Co, ^{64}Cu, ^{208}Pb, ^{27}Al, ^{55}Mn, ^{65}Zn, ^{108}Ag, ^{88}Sr in silicon wafer surfaces/ICP– QMS with hexapole collision cell	A 100 µL drop scanned the sample surface and dissolved the contaminants. It was then diluted to 2 mL and analysed by the method of standard additions. The analytes surface concentration range was 0.49 to $6.5 * 109$ atoms/ cm^{-2}	718
DIHEN	Sn and Cr in human fibroblast cells/ ICP–TOFMS	Only 10 µL of sample allowed for a precise determination of isotopic ratios	689

(Continued)

Table 8.1

Selected recent applications of widely employed liquid sample introduction
systems—Continued

Sample introduction system	Application/Technique	Comments	References
DIN	Ba, Cu, Pb and Zn in undigested honey samples/ID ICP–QMS	A 50 µL diluted sample aliquot was injected into the system and the results were in good agreement with those obtained with a conventional digestion-based method	659
DIN	Lead speciation in rainwater/ICP–QMS	At a $40\,\mu L\,min^{-1}$ delivery liquid flow rate, limits of detection for inorganic lead and triethyllead were 90 and $200\,ng\,L^{-1}$, respectively.	656
DIN/MCN	Palladium in road dust and car exhaust fumes/ICP–QMS	The injected sample volume ranged from 1.5 to 5 µL, the Pd concentration being on the order of $20\,\mu g\,L^{-1}$	651

Acronyms

AAS	atomic absorption spectrometry
AES	atomic emission spectrometry
ALC	(height) above load coil
AIR	aerosol ionic redistribution
CCN	concentric capillary nebulizer
CE	capillary electrophoresis
CFN	cross-flow nebulizer
CZE	capillary zone electrophoresis
DCN	demountable concentric nebulizer
DIHEN	direct injection high-efficiency nebulizer
DIN	direct injection nebulizer
EIE	easily ionized element
EPPN	enhanced parallel-path nebulizer
FI	flow injection
GC	gas chromatography
HEN	high-efficiency nebulizer
hf	high-frequency
HPLC	high performance liquid chromatography
IP	ionization potential
IRZ	initial radiation zone
IS	internal standard
MCN	microconcentric nebulizer
MED	matrix energy demand
MLR	multiple linear regression
MMN	MicroMist nebulizer
MTSC	Myers-Tracy compensation (method)
MW	microwave
NAZ	normal analytical zone
OCN	oscillating-capillary nebulizer
PCR	principal component regression
PCTFE	polychlorotrifluoroethylene
PEEK	polyetheretherketone
PFA	perfluoroalkoxy
PFAN	PFA nebulizer
PHZ	preheating zone
PI	polyimide
PLS	partial least square (regression)

PMT	photomultiplier tube
PP	polypropylene
PPMN	parallel-path micronebulizer
PPN	parallel-path nebulizer
PPS	polyphenylene sulfide
PRISM	parameter-related internal standard method
PTFE	polytetrafluoroethylene
PVC	polyvinyl chloride
PVDF	polyvinylidene difluoride
QMS	quadrupole mass spectrometer
SBR	signal-to-background ratio
SFMS	sector-field mass spectrometer
SSN	sonic-spray nebulizer
TISIS	torch-integrated sample introduction system
REE	rare earth element
RPM	revolution per minute
RSD	relative standard deviation
VC	volume concentration

References

[1] G.L. Gouy, Photometric research on colored flames, Ann. Chim., 18 (1879) 5–101.

[2] S. Nukiyama, Y. Tanasawa, Experiments on the atomization of liquids in air stream, Trans. Soc. Mech. Eng. Jpn., 5 (1939) 68.

[3] T.B. Reed, Induction coupled plasmas, J. Appl. Phys., 32 (1961) 821–824.

[4] T.B. Reed, Growth of refractory crystals using the induction plasma torch, J. Appl. Phys., 32 (1961) 2534–2535.

[5] R.H. Wendt, V.A. Fassel, Induction coupled plasma spectrometric excitation source, Anal. Chem., 37 (1965) 920–922.

[6] S. Greenfield, I. Ll. Jones, C.T. Berry, High-pressure plasmas as spectroscopic emission sources, Analyst, 89 (1964) 713–720.

[7] S. Greenfield, I.L.W. Jones, C.T. Berry, L.G. Bunch, The high-frequency torch: some facts, figures and thoughts, Proc. Soc. Anal. Chem., 2 (1965) 111–113.

[8] R.H. Wendt, V.A. Fassel, Atomic absorption spectroscopy with induction coupled plasmas, Anal. Chem., 38 (1966) 337–338.

[9] H.C. Hoare, R.A. Mostyn, Emission spectrometry of solutions and powders with a high-frequency plasma source, Anal. Chem., 39 (1967) 1153–1155.

[10] J.B. Willis, Atomisation problems in atomic-absorption spectroscopy. I. Operation of a typical nebuliser, spray chamber and burner system, Spectrochim. Acta, 23 (1967) 811–830.

[11] A. Hell, W.F. Ulrich, N. Shifrin, J. Ramirez-Muñoz, Laminar flow burner system with infrared heated spray chamber, Appl. Opt., 7 (1968), 1317.

[12] A. Veillon, M. Margoshes, An evaluation of the induction-coupled, radio frequency plasma torch for atomic emission and atomic absorption spectrometry, Spectrochim. Acta, 23B (1968) 503–512.

[13] R.S. Babington, A.A. Yetman, W.R. Slivka, Method of atomizing liquid in a mono-disperse spray, U.S. Patent 3421692 (1969).

[14] R.S. Babington, A.A. Yetman, W.R. Slivka, Apparatus for spraying liquids in mono-dispersed form, U.S. Patent 3421699 (1969).

[15] R.W. Park, Behavior of water drops colliding in humid nitrogen, Ph. D Thesis, Univeristy of Wisconsin (1970).

[16] K. Kitagawa, T. Takeuchi, Spectroscopic studies of microwave-excited plasma, Anal. Chim. Acta, 60 (1971) 309–318.

[17] M.J. Matteson, The separation of charge at the gas-liquid interface by dispersion of various electrolyte solutions, J. Colloid Interface Sci., 37 (1971) 879–890.

[18] G. Uny, J.N. Lottin, J.P. Tardif, J. Spitz, Etude des performances d'un brûleur préchauffé en spectrométrie d'absorption atomique, Spectrochim. Acta, 26 (1971) 151–156.

[19] G.F. Kirkbright, A.F. Ward, T.S. West, The determination of sulphur and phosphorus by atomic emission spectrometry with an induction coupled high-frequency plasma source, Anal. Chim. Acta, 59 (1972) 241–251.

[20] G.F. Kirkbright, A.F. Ward, T.S. West, Atomic emission spectrometry with an induction-coupled high-frequency plasma source. The determination of iodine, mercury, arsenic and selenium, Anal. Chim. Acta, 64 (1973) 353–362.

[21] R.H. Scott, V.A. Fassel, R.N. Kniseley, D.E. Nixon, Inductively coupled plasma-optical emission analytical spectrometry, Anal. Chem., 46 (1974) 75–80.

[22] R.N. Kniseley, H. Amenson, C.C. Butler, V.A. Fassel, An improved pneumatic nebulizer for use at low nebulizing gas flow, Appl. Spectrosc., 28 (1974) 285–286.

[23] G.F. Larson, V.A. Fassel, R.H. Scott, R.N. Kniseley, Inductively coupled plasma optical emission analytical spectrometry. A study of some interelement effects, Anal. Chem., 47 (1975) 238–243.

[24] R.H. Scott, A. Strasheim, Determination of trace elements in plant materials by inductively coupled plasma optical emission spectrometry, Anal. Chim. Acta, 75 (1975) 71–78.

[25] J.M. Mermet, J. Robin, Etude des interférences dans un plasma induit par haute fréquence, Anal. Chim. Acta, 75 (1975) 271–279.

[26] J.M. Mermet, Sur les mécanismes d'excitation des éléments introduits dans un plasma HF d'argon, C.R. Acad. Sci. Paris, Série B, 281 (1975) 273–275.

[27] P.W.J.M. Boumans, F.J. DeBoer, Studies of an inductively coupled high-frequency argon plasma for optical emission spectrometry. II. Compromise conditions for simultaneous multielement analysis, Spectrochim. Acta, 30B (1975) 309–334.

[28] J.M. Mermet, Comparaison des températures et des densités électroniques mesurées sur le gaz plasmagène et sur des éléments excités dans un plasma HF, Spectrochim. Acta, 30B (1975) 383–396.

[29] P.W.J.M. Boumans, F.J. DeBoer, F.J. Dahmen, H. Hoelzel, A. Meyer, A comparative investigation of some analytical performance characteristics of an inductively coupled plasma and a capacitively coupled microwave plasma for solution analysis by emission spectrometry, Spectrochim. Acta, 30B (1975) 449–469.

[30] S. Greenfield, H. McD. McGeachin, P.B. Smith, Nebulization effects with acid solutions in ICP spectrometry, Anal. Chim. Acta, 84 (1976) 67–78.

[31] M.H. Abdallah, J.M. Mermet, C. Trassy, Etude spectrométrique d'un plasma induit par haute fréquence. II. Etude de différents types d'effets interéléments observés, Anal. Chim. Acta, 84 (1976) 329–339.

[32] F.E. Lichte, S.R. Koirtyohann, Inductively coupled plasma emission from a different angle, ICP Inf. Newsl., 2 (1976) 192–200.

[33] P.W.J.M. Boumans, F.J. DeBoer, Studies of a radio frequency inductively coupled argon plasma for optical emission spectrometry. III. Interference effects under compromise conditions for simultaneous multielement analysis, Spectrochim. Acta, 31B (1976) 355–375.

[34] M. Franklin, C. Baber, S.R. Koirtyohann, Spectral source profiling with a photodiode array, Spectrochim. Acta, 31B (1976) 589–597.

[35] D.J. Kalnicky, V.A. Fassel, R.N. Kniseley, Excitation temperatures and electron number densities experienced by analyte species in inductively coupled plasma with and without the presence of an easily ionized element, Appl. Spectrosc., 31 (1977) 137–150.

[36] K. Ohls, K.H. Koch, Praktische Anwendung der simultanen ICP-Emissionsspektroskopie, ICP Inf. Newsl., 5 (1977) 192–200.

[37] J. Posterdörfer, J. Gebhart, G. Röbig, Effect of evaporation on the size distribution of nebulized aerosols, J. Aerosol Sci., 8 (1977) 371–380.

[38] C.I.M. Beenakker, Evaluation of a microwave-induced plasma in helium at atmospheric pressure as an element-selective detector for gas chromatography, Spectrochim. Acta, 32B (1977) 173–187.

[39] P.W.J.M. Boumans, F.J. DeBoer, An experimental study of a 1-kW, 50 MHz RF inductively coupled plasma with a pneumatic nebulizer, and a discussion of experimental evidence for a nonthermal mechanism, Spectrochim. Acta, 32B (1977) 365–395.

[40] G.R. Kornblum, L. de Galan, A study of the interference of cesium and phosphate in the low-power inductively coupled radio frequency argon plasma using spatially resolved emission and absorption measurements, Spectrochim. Acta, 32B (1977) 455–478.

[41] R.F. Suddendorf, K.W. Boyer, Nebulizer for analysis of high salt content samples with inductively coupled plasma emission spectrometry, Anal. Chem., 50 (1978) 1769–1771.

[42] R.L. Dahlquist, J.W. Knoll, Inductively coupled plasma-atomic emission spectrometry: Analysis of biological materials and soils for major, trace and ultra-trace elements, Appl. Spectrosc., 32 (1978) 1–30.

[43] R.K. Skogerboe, K.W. Olsen, Aerosols, aerodynamics, and atomic analysis, Appl. Spectrosc., 32 (1978) 181–187.

[44] B. Bogdain, An improved concentric glass nebulizer from J.E. Meinhard Associates – Preliminary results, ICP Inf. Newsl., 3 (1978) 491–493.

[45] G.R. Kornblum, W. Van der Waa, L. de Galan, Reduction of argon consumption by a water cooled torch in inductively coupled plasma emission spectrometry, Anal. Chem., 51 (1979) 2378–2381.

[46] P. Schutyser, E. Janssens, Evaluation of spray chambers for use in inductively coupled plasma atomic emission spectrometry, Spectrochim. Acta, 34B (1979) 443–449.

[47] J.E. Meinhard, Pneumatic nebulizers, present and future, Applications of Plasma Emission Spectroscopy, R.M. Barnes, Ed., Heyden, London (1979) 1–14.

[48] P.W.J.M. Boumans, Inductively coupled plasma-atomic emission spectroscopy: Its present and future position in analytical chemistry, Fresenius Z. Anal. Chem., 299 (1979) 337–361.

[49] S.S. Berman, J.W. McLaren, S.N. Willie, Simultaneous determination of five trace metals in seawater by inductively coupled plasma emission spectrometry with ultrasonic nebulization, Anal. Chem., 52 (1980) 488–492.

[50] J.W. Novak, Jr, D.E. Lillie, A.W. Boorn, R.F. Browner, Fixed crossflow nebulizer for use with inductively coupled plasmas and flames, Anal. Chem., 52 (1980) 576–579.

[51] J.A. Borowiec, A.W. Boorn, J.H. Dillard, M.S. Cresser, R.F. Browner, M.J. Matteson, Interference effects from aerosol ionic redistribution in analytical atomic spectrometry, Anal. Chem., 52 (1980) 1054–1059.

[52] R.N. Savage, G.M. Hieftje, Vaporization and ionization interferences in a miniature inductively coupled plasma, Anal. Chem., 52 (1980) 1267–1272.

[53] S.R. Koirtyohann, J.S. Jones, D.A. Yates, Nomenclature system for the low-power argon inductively coupled plasma, Anal. Chem., 52 (1980) 1965–1966.

[54] R.S. Houk, V.A. Fassel, G.D. Flesch, H.J. Svec, A.L. Gray, C.E. taylor, Inductively coupled argon plasmas as ion sources for mass spectrometric determination of trace elements, Anal. Chem., 52 (1980) 2283–2289.

[55] A. Sugimae, Determination of trace elements in sea water by inductively coupled plasma emission spectrometry, Anal. Chim. Acta, 121 (1980) 331–336.

[56] P.K. Hon, O.W. Lau, C.S. Mok, Direct determination of metals in lubricating oils using atomic-absorption spectrometry and aqueous inorganic standards, Analyst, 105 (1980) 919–921.

[57] M.S. Cresser, R.F. Browner, Observations on the effects of impact beads, mixer paddles, and auxiliary oxidant on droplet distributions in analytical flame spectroscopy, Appl. Spectrosc., 34 (1980) 364–368.

[58] J.R. Garbarino, H.E. Taylor, A Babington-type nebulizer for use in the analysis of natural water samples by inductively coupled plasma spectrometry, Appl. Spectrosc., 34 (1980) 584–590.

[59] M.W. Blades, G. Horlick, Photodiode array measurement system for implementing Abel inversion on emission from an inductively coupled plasma, Appl. Spectrosc., 34 (1980) 696–699.

[60] S.M.R. El Shanawany, A.H. Lefebvre, Airblast atomization: The effect of linear scale on mean drop size, J. Energy, 4 (1980) 184–189.

[61] S. Nikdel, J.D. Winefordner, Interference of potassium on barium measurements in the inductively coupled plasma, Microchem. J., 25 (1980) 254–256.

[62] M.S. Cresser, R.F. Browner, A method for investigating size distributions of aqueous droplets in the range 0.5–1 μm produced by pneumatic nebulizers, Spectrochim. Acta, 35B (1980) 73–79.

[63] J.F. Alder, R.M. Bombelka, G.F. Kirkbright, Electronic excitation and ionization temperature measurements in a high-frequency inductively coupled plasma source and the influence of water vapor on plasma parameters, Spectrochim. Acta, 35B (1980) 163–175.

[64] H. Kawaguchi, T. Ito, K. Ota, A. Mizuike, Effects of matrix on spatial profiles of emission from an inductively coupled plasma, Spectrochim. Acta, 35B (1980) 199–206.

[65] A.W. Boorn, M.S. Cresser, R.F. Browner, Evaporation characteristics of organic solvent aerosols used in analytical atomic spectrometry, Spectrochim. Acta, 35B (1980) 823–832.

[66] R.F. Suddendorf, K.W. Boyer, Mechanical device to produce a finely disperse aerosol, U.S. Patent 4206160 (1980).

[67] D.W. Hausler, L.T. Taylor, Nonaqueous on-line simultaneous determination of metals by size exclusion chromatography with inductively coupled plasma atomic emission spectrometric detection, Anal. Chem., 53 (1981) 1223–1227.

[68] H. Uchida, Y. Nojiri, H. Haraguchi, K. Fuwa, Simultaneous multi-element analysis by inductively coupled plasma emission spectrometry utilizing micro-sampling techniques with internal standard, Anal. Chim. Acta, 123 (1981) 57–63.

[69] A.D. Weiss, R.N. Savage, G.M. Hieftje, Development and characterization of a 9-mm inductively coupled argon plasma source for atomic emission spectrometry, Anal. Chim. Acta, 124 (1981) 245–258.

[70] A. Montaser, V.A. Fassel, J. Zalewski, A critical comparison of Ar and Ar-N$_2$ inductively coupled plasmas as excitation sources for atomic emission spectrometry, Appl. Spectrosc., 35 (1981) 292–302.

[71] R.S. Houk, H.J. Svec, V.A. Fassel, Mass spectrometric evidence for suprathermal ionization in an inductively coupled argon plasma, Appl. Spectrosc., 35 (1981) 380–384.

[72] S.R. Koirtyohann, J.S. Jones, C.P. Jester, D.A. Yates, Use of spatial emission profiles and a nomenclature system as aids in interpretating matrix effects in the low-power argon inductively coupled plasma, Spectrochim. Acta, 36B (1981) 49–59.

[73] R.M. Belchamber, G. Horlick, Pressure measurements in the nebulizer spray chamber of an inductively coupled plasma, Spectrochim. Acta, 36B (1981) 581–583.

[74] H. Kawaguchi, T. Ito, A. Mizuike, Axial profiles of excitation and gas temperatures in an inductively coupled plasma, Spectrochim. Acta, 36B (1981) 615–623.

[75] M.W. Blades, G. Horlick, The vertical spatial characteristics of analyte emission in the inductively coupled plasma, Spectrochim. Acta, 36B (1981) 861–880.

[76] M.W. Blades, G. Horlick, Interferences from easily ionizable element matrices in inductively coupled plasma emission spectroscopy – a spatial study, Spectrochim. Acta, 36B (1981) 881–900.

[77] J. Kragten, A. Parczewski, Factorial analysis of matrix effects in ICP-OES and AAS, Talanta, 28 (1981) 901–907.

[78] H. Anderson, H. Kaiser, B. Meddings, High precision (<0.50%RSD) in routine analysis by ICP using a high-pressure (200 psig) cross-flow nebulizer, Developments in Atomic Plasma Spectrochemical Analysis, R.M. Barnes, Ed., Heyden, London (1981) 251–277.

[79] A. Miyazaki, A. Kimura, K. Bansho, Y. Umezaki, Simultaneous determination of heavy metals in waters by inductively coupled plasma atomic emission spectrometry after extraction into diisobutyl ketone, Anal. Chim. Acta, 144 (1982) 213–221.

[80] B.S. Whaley, K.R. Snable, R.F. Browner, Spray chamber placement and mobile phase flow rate effects in liquid chromatography/inductively coupled plasma atomic emission spectrometry, Anal. Chem., 54 (1982) 162–165.

[81] D.D. Smith, R.F. Browner, Measurement of aerosol transport efficiency in atomic spectrometry, Anal. Chem., 54 (1982) 533–537.

[82] L.R. Layman, F.E. Fricke, Glass frit nebulizer for atomic spectrometry, Anal. Chem., 54 (1982) 638–642.

[83] A.W. Boorn, R.F. Browner, Effects of organic solvents in inductively coupled plasma atomic emission spectrometry, Anal. Chem., 54 (1982) 1402–1410.

[84] R.F. Browner, A.W. Boorn, D.D. Smith, Aerosol transport model for atomic spectrometry, Anal. Chem., 54 (1982) 1411–1419.

[85] P.D. Goulden, D.H.J. Anthony, Determination of trace metals in freshwaters by inductively coupled argon plasma atomic emisison spectrometry with a heated spray chamber and desolvation, Anal. Chem., 54 (1982) 1678–1681.

[86] J.P. Rybarczyk, C.P. Jester, D.A. Yates, S.R. Koirtyohann, Spatial profiles of interelement effects in the inductively coupled plasma, Anal. Chem., 54 (1982) 2162–2170.

[87] G.J. Schmidt, W. Slavin, Inductively coupled plasma emission spectrometry with internal standardization and subtraction of plasma background fluctuations, Anal. Chem., 54 (1982) 2491–2495.

[88] L. Ebdon, M.R. Cave, A study of pneumatic nebulization systems for inductively coupled plasmas emission spectrometry, Analyst, 107 (1982) 172–178.

[89] M. Thompson, M.H. Ramsey, B.J. Coles, Interactive matrix matching: a new method of correcting interference effects in inductively coupled plasma spectrometry, Analyst, 107 (1982) 1286–1288.

[90] B. Capelle, J.M. Mermet, J. Robin, Influence of the generator frequency on the spectral characteristics of inductively coupled plasma, Appl. Spectrosc., 36 (1982) 102–106.

[91] J.E. Roederer, G.J. Bastiaans, M.A. Fernandez, K.J. Fredeen, Spatial distribution of interference effects in ICP emission analysis, Appl. Spectrosc., 36 (1982) 383–389.

[92] J.F. Wolcott, C. Sobel, Fabrication of a Babington-type nebulizer for ICP sources, Appl. Spectrosc., 36 (1982) 685–686.

[93] A. Batal, J.M. Mermet, Influence of the collision processes on the energy range of analytical lines in ICP-AES, Can. J. Spectrosc., 27 (1982) 37–45.

[94] M.S. Cresser, Theoretical aspects of organic solvent enhancement effects in atomic spectroscopy, Prog. Anal. At. Spectrosc., 5 (1982) 35–62.

[95] P.W.J.M. Boumans, M.Ch. Lux-Steiner, Modification and optimization of a 50 MHz inductively coupled argon plasma with special reference to analyses using organic solvents, Spectrochim. Acta, 37B (1982) 97–126.

[96] M.W. Blades, G.M. Hieftje, On the significance of radiation trapping in the inductively coupled plasma, Spectrochim. Acta, 37B (1982) 191–197.

[97] F.J.M.J. Maessen, J. Balke, J.L.M. De Boer, Preservation of accuracy and precision in the analytical practice of low power ICP-AES, Spectrochim. Acta, 37B (1982) 517–526.

[98] W.H. Gunter, K. Visser, P.B. Zeeman, Some aspects of matrix interference caused by elements of low ionization potential in inductively coupled plasma atomic emission spectrometry, Spectrochim. Acta, 37B (1982) 571–581.

[99] M.W. Blades, Some considerations regarding temperature, electron density, and ionization in the argon inductively coupled plasma, Spectrochim. Acta, 37B (1982) 869–879.

[100] R.J. Lovett, A rate model of inductively coupled argon plasma analyte spectra, Spectrochim. Acta, 37B (1982) 969–985.

[101] W.C. Hinds, Aerosol Technology, John Wiley & Sons, New York, chap.13 (1982) 271.

[102] N.K. Rizk, A.H. Lefebvre, Influence of atomizer design features on mean drop size, AIAA J., 21 (1983) 1139–1142.

[103] R.F. Smith, R.F. Browner, Reply to exchange of comments on the measurement of aerosol transport efficiencies in atomic spectrometry, Anal. Chem., 55 (1983) 373–374.

[104] M.H. Ramsey, M. Thompson, B.J. Coles, Modified concentric glass nebulizer for reduction of memory effects in inductively coupled plasma spectrometry, Anal. Chem., 55 (1983) 1626–1629.

[105] J. Xu, H. Kawaguchi, A. Mizuike, Effects of organic acids and solvents in inductively coupled plasma emission spectrometry, Anal. Chim. Acta, 152 (1983) 133–139.

[106] J. Lee, Calcium matrix effects in multi-element analysis of animal bone by inductively-coupled plasma emission spectrometry, Anal. Chim. Acta, 152 (1983) 141–147.

[107] A.L. Gray, A.R. Date, Inductively coupled plasma mass spectrometry using continuum flow ion extraction, Analyst, 108 (1983) 1033–1050.

[108] K.D. Summerhays, P.J. Lamothe, T.L. Fries, Volatile species in inductively coupled plasma atomic emission spectrometry: implications for enhanced sensitivity, Appl. Spectrosc., 37 (1983) 25–28.

[109] L.M. Faires, C.T. Apel, T.N. Niemczyk, Intra-alkali matrix effects in the inductively coupled plasma emission spectroscopy, Appl. Spectrosc., 37 (1983) 558–565.

[110] Z. Zadgorska, H. Nickel, M. Mazurkiewicz, G. Wolff, Contribution to the quantitative analysis of oxide layers formed on high-temperature alloys using inductively coupled plasma atomic emission spectrometry. II. Optimization of the ICP working parameters and typical analysis of oxide layers, Fresenius Z. Anal. Chem., 314 (1983) 356–361.

[111] D.J. Douglas, E.S.K. Quan, R.G. Smith, Elemental analysis with an atmospheric pressure plasma (MIP, ICP)/quadrupole mass spectrometer system, Spectrochim. Acta, 38B (1983) 39–48.

[112] S. Xi-en, C. Qui-Ian, Effect of a small amount of phosphoric acid in inductively coupled plasma atomic spectroscopy, Spectrochim. Acta, 38B (1983) 115–121.

[113] I.J.M.M. Raaijmakers, P.W.J.M. Boumans, B. van der Sijde, D.C. Schram, A theoretical study and experimental investigation of non-LTE phenomena in an inductively-coupled argon plasma-1. Characterization of the discharge, Spectrochim. Acta, 38B (1983) 697–706.

[114] P.A.M. Ripson, L. de Galan, Empirical power balances for conventional and externally cooled inductively-coupled argon plasmas, Spectrochim. Acta, 38B (1983) 707–726.

[115] S.D. Olsen, A. Strasheim, Correlation of the analytical signal to the characterized nebulizer spray, Spectrochim. Acta, 38B (1983) 973–975.

[116] S.A. Myers, D.H. Tracy, Improved performance using internal standardization in inductively-coupled plasma emission spectroscopy, Spectrochim. Acta, 38B (1983) 1227–1253.

[117] H. Ishii, K. Satoh, Determination of rare earths in lanthanum oxide by inductively coupled plasma emission derivative spectrometry, Talanta, 30 (1983) 111–115.

[118] A. Lorber, Z. Golbart, Generalized internal reference method for simultaneous multichannel analysis, Anal. Chem., 56 (1984) 37–43.

[119] A. Montaser, G.R. Huse, R.A. Wax, S. Chan, D.W. Golightly, J.S. Kane, A.F. Dorrzapf, Analytical performance of a low-gas flow-torch optimized for inductively coupled plasma atomic emission spectroscopy, Anal. Chem., 56 (1984) 283–288.

[120] K.E. Lawrence, G.W. Rice, V. Fassel, Direct liquid sample introduction for flow injection analysis and liquid chromatography with inductively coupled, argon plasma spectrometric detection, Anal. Chem., 56 (1984) 289–292.

[121] P. Barrett, E. Pruszkowska, Use of organic solvents for inductively coupled plasma analyses, Anal. Chem., 56 (1984) 1927–1930.

[122] D.D. Smith, R.F. Browner, Influence of aerosol drop size on signals and interferences in flame atomic absorption spectrometry, Anal. Chem., 56 (1984) 2702–2708.

[123] J.L. Imbert, J.M. Mermet, Utilisation de la spectrométrie d'émission avec un plasma induit par haute fréquence pour l'analyse du thorium, de l'yttrium et de l'uranium dans l'acide phosphorique contenant des teneurs élevées en fer, chrome et zirconium, Analusis, 12 (1984) 209–213.

[124] M.A.E. Wandt, M.A.B. Pougnet, A.L. Rodgers, Determination of calcium, magnesium, and phosphorus in human stones by inductively coupled plasma-atomic emission spectroscopy, Analyst, 109 (1984) 1071–1074.

[125] H.M. Ramsey, M. Thompson, Communication. Improved precision in inductively coupled plasma atomic-emission spectrometry by a parameter-related internal standard method, Analyst, 109 (1984) 1625–1626.

[126] M.P. Doherty, G.M. Hieftje, Jet-impact nebulization for sample introduction in ICP spectrometry, Appl. Spectrosc., 38 (1984) 405–411.

[127] B.R. Baginsky, J.E. Meinhard, Some effects of high-solids matrices on the sample delivery system and the Meinhard concentric nebulizer during ICP emission analyses, Appl. Spectrosc., 38 (1984) 568–572.

[128] M.H. Miller, D. Eastwood, M.S. Hendrick, Excitation of analytes and enhancement of emission intensities in a d.c. plasma jet: a critical review leading to proposed mechanistic models, Spectrochim. Acta, 39B (1984) 13–56.

[129] A. Gustavsson, A tutorial review of the basic theory for and practical aspects of aerosol chambers, Spectrochim. Acta, 39B (1984) 85–94.

[130] Y. Fujishiro, M. Kubota, R. Ishida, A study of designs of a cross flow nebulizer for ICP atomic emission spectrometry, Spectrochim. Acta, 39B (1984) 617–620.

[131] J.W. Mills, G.M. Hieftje, A detailed consideration of resonance radiation in the argon inductively coupled plasma, Spectrochim. Acta, 39B (1984) 859–866.

[132] F.J.M.J. Maessen, P.J.H. Seeverens, G. Kreuning, Analytical aspects of organic solvent load reduction in normal-power ICPs by aerosol thermostatting at low temperature, Spectrochim. Acta, 39B (1984) 1171–1180.

[133] J.L. Fabec, M.L. Ruschak, Determination of nickel, vanadium, and sulfur in crudes and heavy crude fractions by inductively coupled argon plasma/atomic emission spectrometry and flame atomic absorption spectrometry, Anal. Chem., 57 (1985) 1853–1863.

[134] C.A. Monnig, S.R. Koirtyohann, Mie scattering from sample aerosol inside the inductively coupled plasma, Anal. Chem., 57 (1985) 2533–2536.

[135] J.A. Olivares, R.S. Houk, Ion sampling for inductively coupled plasma mass spectrometry, Anal. Chem., 57 (1985) 2674–2679.

[136] D.A. Wilson, A. Yuen, G.M. Hieftje, Comparison of the helium/oxygen/acetylene and air/acetylene flames as atom sources for continuum-source atomic fluorescence spectrometry, Anal. Chim. Acta, 171 (1985) 241–249.

[137] M. Thompson, The capabilities of inductively coupled plasma atomic-emission spectrometry-some conjectures and refutations, Analyst, 110 (1985) 443–449.

[138] M.II. Ramsey, M. Thompson, Correlated variance in simultaneous inductively coupled plasma atomic emission spectrometry: its cause and correlation by a parameter-related internal standard method, Analyst, 110 (1985) 519–530.

[139] M. Thompson, M.H. Ramsey, Matrix effects due to calcium in inductively coupled plasma atomic emission spectrometry: their nature, source and remedy, Analyst, 110 (1985) 1413–1422.

[140] L.M. Faires, T.M. Bieniewski, C.T. Apel, T.M. Niemscyk, Top-down versus side-on viewing of the inductively coupled plasma, Appl. Spectrosc., 39 (1985) 5–9.

[141] R.K. Skogerboe, S.J. Freeland, Effects of solution composition on the physical characteristics of aerosols produced by nebulization, Appl. Spectrosc., 39 (1985) 925–930.

[142] A. Delijska, M. Vounchkov, Possibilities for correcting the influence of mineral acids and of drift in ICP Emission Spectral Analysis, Fresenius Z. Anal. Chem., 321 (1985) 448–452.

[143] W. Nisamaneepong, D.L. Haas, J.A. Caruso, The use of organic solvents with a glass frit nebulizer for sample introduction in inductively coupled plasma emission spectrometry, Spectrochim. Acta, 40B (1985) 3–10.

[144] R. Rezaaiyaan, J.W. Olesik, G.M. Hieftje, Interferences in a low-flow, low-power inductively coupled plasma, Spectrochim. Acta, 40B (1985) 73–83.

[145] R.I. Botto, Method for correcting for acid and salt matrix interferences in ICP-AES, Spectrochim. Acta, 40B (1985) 397–412.

[146] S.E. Long, R.D. Snooks, R.F. Browner, Some observations on electrothermal vaporisation for sample introduction into the inductively coupled plasma, Spectrochim. Acta, 40B (1985) 553–568.

[147] M.W. Blades, B.L. Caughlin, Excitation temperature and electron density in the inductively coupled plasma – aqueous vs organic solvent introduction, Spectrochim. Acta, 40B (1985) 579–591.

[148] M.H. Miller, E. Keating, D. Eastwood, M.S. Hendrick, Measured and modeled enhancement of transition metal emissions in the d.c. plasma jet, Spectrochim. Acta, 40B (1985) 593–616.

[149] W. Gunter, K. Visser, P.B. Zeeman, Ionization interferences under various operating conditions in a 9, 27 and 50 MHz ICP, and a study of shifts in level populations of calcium through simultaneous absorption-emission measurements in a 9 MHz ICP, Spectrochim. Acta, 40B (1985) 617–629.

[150] B.L. Caughlin, M.W. Blades, Spatial profiles of electron number density in the inductively coupled plasma, Spectrochim. Acta, 40B (1985) 987–993.

[151] N. Furuta, Spatial profile measurement of ionization and excitation temperatures in an inductively coupled plasma, Spectrochim. Acta, 40B (1985) 1013–1022.

[152] L.J. Prell, C. Monning, R.E. Harris, S.R. Koirtyohann, The role of the electrons in the emission enhancement by easily ionized elements low in the inductively coupled plasma, Spectrochim. Acta, 40B (1985) 1401–1410.

[153] S.K. Chan, A. Montaser, A helium inductively coupled plasma for atomic emission spectrometry, Spectrochim. Acta, 40B (1985) 1467–1472.

[154] K.E. LaFrenière, G.W. Rice, V.A. Fassel, Flow injection analysis with inductively coupled plasma atomic emission spectroscopy: critical comparison of conventional pneumatic, ultrasonic and direct injection nebulization, Spectrochim. Acta, 40B (1985) 1495–1504.

[155] J.A. Olivares, R.S. Houk, Suppression of analyte signal by various concomitant salts in inductively coupled plasma mass spectrometry, Anal. Chem., 58 (1986) 20–25.

[156] D.R. Luffer, E.D. Salin, Rapid throughput nebulizer-spray chamber system for inductively coupled plasma atomic emission spectrometry, Anal. Chem., 58 (1986) 654–656.

[157] J.A. Koropchak, D.H. Winn, Thermospray interfacing for flow injection analysis with inductively coupled plasma atomic emission spectrometry, Anal. Chem., 58 (1986) 2558–2561.

[158] D.E. Nixon, G.A. Smith, Comparison of Jarrell-Ash, Perkin-Elmer, and modified Perkin-Elmer nebulizers for inductively coupled plasma analysis, Anal. Chem., 58 (1986) 2886–2888.

[159] R.S. Houk, Mass spectrometry of inductively coupled plasmas, Anal. Chem., 58 (1986) 97A–105A.

[160] M.D. Palmieri, J.S. Fritz, J.J. Thomson, R.S. Houk, Separation of trace rare earths and other metals from uranium by liquid-liquid extraction with quantitation by inductively-coupled plasma/mass spectrometry, Anal. Chim. Acta, 184 (1986) 187–196.

[161] O. Kujirai, K. Yamada, M. Kohri, H. Okochi, Analysis of heat-resistant alloys by inductively coupled plasma/atomic emission spectrometry with hydrofluoric acid-resistant sample introduction systems, Appl. Spectrosc., 40 (1986) 962–968.

[162] P.A. Vieira, H. ZhiZhuang, S.-K. Chan, A. Montaser, Evaluation of recycling cyclone spray chamber for ICP-AES, Appl. Spectrosc., 40 (1986) 1141–1146.

[163] J.B. Willis, Aerosols in flame atomic spectroscopy, Fresenius Z. Anal. Chem., 324 (1986) 846–854.

[164] L.S. Dale, S.J. Buchanan, A comparison of cloud chambers for use in inductively coupled plasma nebulisation systems, J. Anal. At. Spectrom., 1 (1986) 59–62.

[165] M.H. Ramsey, M. Thompson, A predictive model of plasma matrix effects in inductively coupled plasma atomic emission spectrometry, J. Anal. At. Spectrom., 1 (1986) 185–193.

[166] H. Matusiewicz, F.L. Fricke, R.M. Barnes, Modification of a commercial electrothermal vaporisation system for inductively coupled plasma spectrometry: evaluation and matrix effects, J. Anal. At. Spectrom., 1 (1986) 203–209.

[167] F.J.M.J. Maessen, G. Kreuning, J. Balke, Experimental control of the solvent load of inductively coupled argon plasmas and effects of the chloroform plasma load on their analytical performance, Spectrochim. Acta, 41B (1986) 3–25.

[168] Y.Q. Tang, C. Trassy, Inductively coupled plasma: the role of water in axial excitation temperatures, Spectrochim. Acta, 41B (1986) 143–150.

[169] C.J. Pickford, R.M. Brown, Comparison of ICP-MS with ICP-ES: detection power and interference effects experienced with complex matrices, Spectrochim. Acta, 41B (1986) 183–187.

[170] J. Lee, J.R. Sedcole, M.W. Pritchard, Matrix interactions in an inductively coupled argon plasma optimised for simultaneous multi-element analysis by atomic emission spectrometry, Spectrochim. Acta, 41B (1986) 217–225.

[171] J.R. Sedcole, J. Lee, M.W. Pritchard, Internal standard selection in the presence of matrix interactions in an inductively coupled plasma optimised for simultaneous multi-element analysis by atomic emission spectrometry, Spectrochim. Acta, 41B (1986) 227–235.

[172] R.M. Belchamber, D. Betteridge, A.P. Wade, A.J. Cruickshank, P. Davison, Removal of matrix effect in ICP-AES multi-element analysis by simplex optimisation, Spectrochim. Acta, 41B (1986) 503–505.

[173] G. Gillson, G. Horlick, An atomic fluorescence study of easily ionizable element interferences in the inductively coupled plasma, Spectrochim. Acta, 41B (1986) 619–624.

[174] S.E. Long, R.F. Browner, Influence of water on the spatial excitation behaviour of selected elements in the inductively coupled plasma, Spectrochim. Acta, 41B (1986) 639–649.

[175] G.D. Rayson, G.M. Hieftje, A steady-state approach to evaluation of proposed excitation mechanisms in the analytical ICP, Spectrochim. Acta, 41B (1986) 683–697.

[176] A. Goldwasser, J.M. Mermet, Contribution of the charge-transfer process to the excitation mechanisms in inductively coupled plasma atomic emission spectroscopy, Spectrochim. Acta, 41B (1986) 725–739.

[177] Z. Walker, M.W. Blades, Measurement of excited state level populations for atomic and ionic iron in the inductively coupled plasma, Spectrochim. Acta, 41B (1986) 761–775.

[178] C. Zicai, R.M. Barnes, Characterization of a recycling nebulization system for inductively coupled plasma spectrometry-I. Stability and humidified argon carrier gas, Spectrochim. Acta, 41B (1986) 979–989.

[179] N. Furuta, Spatial emission distribution of YO, Y I, Y II and Y III radiation in an inductively coupled plasma for the elucidation of excitation mechanisms, Spectrochim. Acta, 41B (1986) 1115–1129.

[180] C. Zicai, R.M. Barnes, Characterization of a recycling nebulization system for inductively coupled plasma spectrometry-II. Matrix and memory effects, Spectrochim. Acta, 41B (1986) 1151–1155.

[181] J.W. Elgersma, F.J.M.J. Maessen, W.M.A. Niessen, A low consumption thermospray nebulizer with a fused silica vaporizer for inductively coupled plasma atomic emission spectrometry, Spectrochim. Acta, 41B (1986) 1217–1220.

[182] S.A. Schwartz, G.A. Meyer, Characterization of aerosols generated by thermospray nebulization for atomic spectroscopy, Spectrochim. Acta, 41B (1986) 1287–1298.

[183] M.A. Sharaf, D.L. Illman, B.R. Kowalski, Chemometrics, John Wiley and Sons Inc., New York, chap. 4 (1986) 119–147.

[184] M.W. Routh, J.E. Goulter, D.B. Tasker, S.D. Arellano, Investigation of ICP-AES nebulization techniques for the analysis of microsamples, Am. Lab., 19 (1987) 102–104.

[185] K.E. LaFreniere, V.A. Fassel, D.E. Eckels, Elemental speciation via high-performance liquid chromatography combined with inductively coupled plasma atomic emission spectroscopic detection: application of a direct injection nebulizer, Anal. Chem., 59 (1987) 879–887.

[186] K.I. Mahan, T.A. Forderaro, T.L. Garza, R.M. Martinez, G.A. Maruney, M.R. Trivisonno, E.M. Willging, Microwave digestion techniques in the sequential extraction of calcium, iron, chromium, manganese, lead and zinc in sediments, Anal. Chem., 59 (1987) 938–945.

[187] J. Farino, J.R. Miller, D.D. Smith, R.F. Browner, Influence of solution uptake rate on signals and interferences in inductively coupled plasma optical emission spectrometry, Anal. Chem., 59 (1987) 2303–2309.

[188] K.R. Beebe, B.R. Kowalski, An introduction to multivariate calibration and analysis, Anal. Chem., 59 (1987) 1007A–1017A.

[189] H. Kawaguchi, T. Tanaka, T. Nakamura, M. Morishita, Matrix effects in inductively coupled plasma mass spectrometry, Anal. Sci., 3 (1987) 305–308.

[190] H.B. Fannin, C.J. Seliskar, D.C. Miller, Studies of energy transfer and ionization processes in a helium ICP, Appl. Spectrosc., 41 (1987) 621–624.

[191] J.J. Thompson, R.S. Houk, A study of internal standardization in inductively coupled plasma-mass spectrometry, Appl. Spectrosc., 41 (1987) 801–806.

[192] D.A. Wilson, G.H. Vickers, G.M. Hieftje, Ionization temperatures in the inductively coupled plasma determined by mass spectrometry, Appl. Spectrosc., 41 (1987) 875–880.

[193] J.A. Koropchak, D.H. Winn, Fundamental characteristics of thermospray aerosols and sample introduction for atomic spectrometry, Appl. Spectrosc., 41 (1987) 1311–1318.

[194] M. Thompson, M.H. Ramsey, B.J. Coles, C.M. Du, Correction of matrix effects in inductively coupled plasma atomic emission spectrometry by interactive power adjustment, J. Anal. At. Spectrom., 2 (1987) 185–188.

[195] K.A. Vermeiren, P.D.P. Taylor, R. Dams, Use of thermospray nebuliser for organic solvent introduction into the inductively coupled plasma, J. Anal. At. Spectrom., 2 (1987) 383–387.

[196] T. Brotherton, B. Barnes, N. Vela, J. Caruso, Evaluation of the grid-type nebuliser for organic solvent introduction into the inductively coupled plasma, J. Anal. At. Spectrom., 2 (1987) 389–396.

[197] M.H. Ramsey, M. Thompson, High accuracy analysis by inductively coupled plasma atomic emission spectrometry using the parameter-related internal standard method, J. Anal. At. Spectrom., 2 (1987) 497–502.

[198] R.F. Browner, G. Zhu, Sample introduction in plasma emission and mass spectrometry, J. Anal. At. Spectrom., 2 (1987) 543–547.

[199] R.C. Hutton, A.N. Eaton, Role of aerosol water vapour loading in inductively coupled plasma mass spectrometry, J. Anal. At. Spectrom., 2 (1987) 595–598.

[200] T. Brotherton, J. Caruso, Evaluation of the grid-type nebuliser for the introduction of high dissolved salt and high solids content solutions into the inductively coupled plasma, J. Anal. At. Spectrom., 2 (1987) 695–698.

[201] S.H. Tan, G. Horlick, Matrix-effect observations in inductively coupled plasma mass spectrometry, J. Anal. At. Spectrom., 2 (1987) 745–763.

[202] M.A. Vaughan, G. Horlick, S.H. Tan, Effect of operating parameters on analyte signals in inductively coupled plasma mass spectrometry, J. Anal. At. Spectrom., 2 (1987) 765–772.

[203] P.L. Kempster, J.F. van Staden, H.R. van Vliet, Investigation of small volume cloud chambers for use in inductively coupled plasma nebulisation, J. Anal. At. Spectrom., 2 (1987) 823–828.

[204] B.T.G. Ting, M. Janghorbani, Application of inductively coupled plasma-mass spectrometry to accurate isotopic analysis for human metabolic studies, Spectrochim. Acta, 42B (1987) 21–27.

[205] H.P. Longerich, B.J. Fryer, D.F. Strong, C.J. Kantipuly, Effects of operating conditions on the determination of the rare earth elements by inductively coupled plasma-mass spectrometry, Spectrochim. Acta, 42B (1987) 75–92.

[206] H.P. Longerich, B.J. Fryer, D.F. Strong, Trace analysis of natural alloys by inductively coupled plasma-mass spectrometry: application to archeological native silver artifacts, Spectrochim. Acta, 42B (1987) 101–109.

[207] R.I. Botto, Matrix interferences in the analysis of organic solutions by inductively coupled plasma-atomic emission spectrometry, Spectrochim. Acta, 42B (1987) 181–199.

[208] B.L. Caughlin, M.W. Blades, Effect of wet and dry nebulizer gas on the spatial distribution of the electron density, Spectrochim. Acta, 42B (1987) 353–360.

[209] D. Beauchemin, J.W. McLaren, S.S. Berman, Study of the effects of concomitant elements in inductively coupled plasma mass spectrometry, Spectrochim. Acta, 42B (1987) 467–490.

[210] G.C. Turk, O. Axner, N. Omenetto, Optical detection of laser-induced ionization in the inductively coupled plasma for the study of ion-electron recombination and ionization equilibrium, Spectrochim. Acta, 42B (1987) 873–881.

[211] D.C. Gregoire, The effect of easily ionizable concomitant elements on non-spectroscopic interferences in inductively coupled plasma-mass spectrometry, Spectrochim. Acta, 42B (1987) 895–907.

[212] M. Murillo, J.M. Mermet, Transfer of energy between the surrounding plasma and the central channel in inductively coupled plasma- atomic emission spectrometry, Spectrochim. Acta, 42B (1987) 1151–1162.

[213] R.F. Browner, Fundamental aspects of aerosol generation and transport, Inductively Coupled Plasma Emission Spectrometry, Part II Applications and Fundamentals, P.W.J.M. Boumans Ed., John Wiley, New York, (1987) 244–288.

[214] A.W. Boorn, R.F. Browner, Applications:organics, Inductively Coupled Plasma Emission Spectrometry, Part II Applications and Fundamentals, P.W.J.M. Boumans Ed., John Wiley, New York, (1987) 151–216.

[215] A.G.T. Gustavsson, Liquid sample introduction into plasmas, Inductively Coupled Plasmas in Analytical Atomic Spectrometry, A. Montaser, D.W. Golightly Eds., VCH Publishers, New York, Part III (1987) 399–430.

[216] S.J. Jiang, R.S. Houk, M.A. Stevens, Alleviation of overlap interferences for determination of potassium isotope ratios by inductively coupled plasma mass spectrometry, Anal. Chem., 60 (1988) 1217–1221.

[217] G.R. Gillson, D.J. Douglas, J.E. Fulford, K.W. Halligan, S.D. Tanner, Non spectroscopic interelement interferences in inductively coupled plasma mass spectrometry, Anal. Chem., 60 (1988) 1472–1474.

[218] R.D. Satzger, Evaluation of inductively coupled plasma mass spectrometry for the determination of trace elements in food, Anal. Chem., 60 (1988) 2500–2504.

[219] C. Vandecasteele, M. Nagels, H. Vanhoe, R. Dams, Suppression of analyte signal in inductively coupled plasma mass spectrometry and the use of an internal standard, Anal. Chim. Acta, 211 (1988) 91–98.

[220] K. Kitagawa, H. Kikuchi, Laser Fraunhofer method for the correction of physical interference in an atomic-absorption nebulizer, Anal. Sci., 4 (1988) 53–57.

[221] T. Cecconie, S. Muralidharan, H. Freiser, Characterization of micro-Babington and cross-flow type nebulizers for inductively coupled plasma atomic emission spectrometry, Appl. Spectrosc., 42 (1988) 177–179.

[222] P.C. Hauser, M.W. Blades, Atomization of organic compounds in the inductively coupled plasma, Appl. Spectrosc., 42 (1988) 595–598.

[223] J. Marshall, G. Rodgers, W.C. Campbell, Myers-Tracy signal compensation in inductively coupled plasma atomic emission spectrometry with high dissolved solids solutions, J. Anal. At. Spectrom., 3 (1988) 241–244.

[224] J.R. Dean, L. Ebdon, H.M. Crews, R.C. Massey, Characteristics of flow injection inductively coupled plasma mass spectrometry for trace metal analysis, J. Anal. At. Spectrom., 3 (1988) 349–354.

[225] R.C. Hutton, A.N. Eaton, Analysis of solutions containing high levels of dissolved solids by inductively coupled plasma mass spectrometry, J. Anal. At. Spectrom., 3 (1988) 547–550.

[226] B.L. Sharp, Pneumatic nebulisers and spray chambers for inductively coupled plasma spectrometry. Part I. Nebulisers, J. Anal. At. Spectrom., 3 (1988) 613–652.

[227] D.W. Koppenaal, L.F. Quinton, Development and assessment of a helium inductively coupled plasma ionisation source for inductively coupled plasma mass spectrometry, J. Anal. At. Spectrom., 3 (1988) 667–672.

[228] D.J. Douglas, J.B. French, Gas dynamics of the inductively coupled plasma mass spectrometry interface, J. Anal. At. Spectrom., 3 (1988) 743–747.

[229] D.J. Douglas, L.A. Kerr, Study of solids deposition on inductively coupled plasma mass spectrometry samplers and skimmers, J. Anal. At. Spectrom., 3 (1988) 749–752.

[230] D. Beauchemin, J.W. McLaren, S.S. Berman, Use of external calibration for the determination of trace metals in biological materials by inductively coupled plasma mass spectrometry, J. Anal. At. Spectrom., 3 (1988) 775–780.

[231] B.L. Sharp, Pneumatic nebulisers and spray chambers for inductively coupled plasma spectrometry. A review. Part II. Spray chambers, J. Anal. At. Spectrom., 3 (1988) 939–963.

[232] C. Shuyin, L. Yunlong, A novel type of concentric glass nebulizer for an analysis of solutions with high salt content by inductively coupled plasma atomic emission spectroscopy, Spectrochim. Acta, 43B (1988) 287–291.

[233] P.E. Walters, C.A. Barnardt, The role of desolvation and hydrogen addition on the excitation features on the inductively coupled plasma, Spectrochim. Acta, 43B (1988) 325–337.

[234] L. Ebdon, A.R. Collier, Particle size effects on kaolin slurry analysis by inductively coupled plasma-atomic emission spectrometry, Spectrochim. Acta, 43B (1988) 355–369.

[235] S. Dahai, Z. Zhanxia, Q. Haowen, C. Mingxiang, Matrix effects due to metallic elements in inductively coupled plasma emission spectrometry, Spectrochim. Acta, 43B (1988) 391–403.

[236] J.W. McLaren, D. Beauchemin, S.S. Berman, Analysis of the marine sediment reference materials PACS-1 by inductively coupled plasma mass spectrometry, Spectrochim. Acta, 43B (1988) 413–420.

[237] A. Gustavsson, Characterization of a membrane interface for sample introduction into atom reservoirs for analytical atomic spectrometry, Spectrochim. Acta, 43B (1988) 917–922.

[238] S. Greenfield, M.S. Salman, J.F. Tyson, Atomiser, source, inductively coupled plasmas in atomic fluorescence spectrometry: a study of chemical and ionisation interference effects, Spectrochim. Acta, 43B (1988) 1087–1092.

[239] S. Nowak, J.A.M. van der Mullen, D.C. Schram, Electron density and temperature determination in an inductively coupled plasma using a non-equilibrium concept, Spectrochim. Acta, 43B (1988) 1235–1245.

[240] A. Canals, J. Wagner, R.F. Browner, V. Hernandis, Empirical model for estimating drop size distributions of aerosols generated by inductively coupled plasma nebulizers, Spectrochim. Acta, 43B (1988) 1321–1335.

[241] J.S. Crain, R.S. Houk, F.G. Smith, Matrix interferences in inductively coupled plasma-mass spectrometry: some effects of skimmer orifice diameter and ion lens voltages, Spectrochim. Acta, 43B (1988) 1355–1364.

[242] S.E. Long, R.F. Browner, Influence of water on conditions in the inductively coupled argon plasma, Spectrochim. Acta, 43B (1988) 1461–1471.

[243] M.N.A. Noordermeer, One-piece high-solids nebulizer, US Patent 4880164 (1988).

[244] J.W. Olesik, L.J. Smith, E.J. Williamsen, Signal fluctuations due to individual droplets in inductively coupled plasma atomic emission spectrometry, Anal. Chem., 61 (1989) 2002–2008.

[245] R.H. Clifford, A. Montaser, S.A. Dolan, S.G. Capar, Thimble glass frit nebulizer for atomic spectrometry, Anal. Chem., 61 (1989) 2777–2784.

[246] T. Adachi, H. Takeishi, Y. Sasaki, K. Motojima, Enhancement of ruthenium determination by inductively coupled plasma atomic emission spectrometry by addition of periodic acid, Anal. Chim. Acta, 218 (1989) 77–84.

[247] H. Uchida, W.R. Masamba, T. Uchida, B.W. Smith, J.D. Winefordner, A new desolvation system for use with capacitatively coupled microwave plasma and inductively coupled plasma-atomic emission spectrometry, Appl. Spectrosc., 43 (1989) 425–430.

[248] J.W. Olesik, E.J. Williamsen, Easily and noneasily ionizable element matrix effects in inductively coupled plasma optical spectrometry, Appl. Spectrosc., 43 (1989) 1223–1232.

[249] E.G. Chudinov, I.I. Ostroukhova, G.V. Varvanina, Acid effects in ICP-AES, Fresenius Z. Anal. Chem., 335 (1989) 25–33.

[250] H. Berndt, B. Schaldach, Improvement of power of detection in ICP-OES by a new way of sample introduction (hydraulic high-pressure nebulization), Fresenius Z. Anal. Chem., 335 (1989) 367–369.

[251] N. Kovacic, B. Budic, V. Hudnik, Matrix effects from magnesium and lithium in inductively coupled plasma atomic emission spectrometry, J. Anal. At. Spectrom., 4 (1989) 33–37.

[252] T.J. Brotherton, W.L. Shen, J.A. Caruso, Use of Hildebrand grid nebulizer for analysis of high matrix solutions containing easily ionisable elements with inductively coupled plasma mass spectrometry, J. Anal. At. Spectrom., 4 (1989) 39–44.

[253] A.J. Ambrose, L. Ebdon, M.E. Foulkes, P. Jones, Direct atomic spectrometric analysis by slurry atomisation. Part 8. Flow injection inductively coupled plasma atomic emission spectrometry, J. Anal. At. Spectrom., 4 (1989) 219–222.

[254] D. Beauchemin, K.W.M. Siu, J.W. McLaren, S.S. Berman, Determination of arsenic species by high-performance liquid chromatography-inductively coupled plasma mass spectrometry, J. Anal. At. Spectrom., 4 (1989) 285–289.

[255] H. Isoyama, T. Uchida, T. Niwa, C. Iida, G. Nakagawa, Small spray chamber for inductively coupled plasma atomic emission spectrometry and its evaluation by a digital signal sampling technique, J. Anal. At. Spectrom., 4 (1989) 351–355.

[256] L. Ebdon, E. Hywel-Evans, N.W. Barnett, Simple optimisation of experimental conditions in inductively coupled plasma atomic emission spectrometry with organic solvent introduction, J. Anal. At. Spectrom., 4 (1989) 505–508.

[257] Y. Igarashi, H. Kawamura, K. Shiraishi, Y. Takaku, Determination of thorium and uranium in biological samples by inductively coupled plasma mass spectrometry using internal standardisation, J. Anal. At. Spectrom., 4 (1989) 571–576.

[258] H.P. Longerich, Effect of nitric acid , acetic acid and ethanol in inductively coupled plasma mass spectrometric ion signals as a function of nebulizer gas flow with implications on matrix suppression and enhancements, J. Anal. At. Spectrom., 4 (1989) 665–667.

[259] X. Wang, A. Lasztity, M. Viczian, R.M. Barnes, Sample analysis by on-line isotope dilution inductively coupled plasma mass spectrometry, J. Anal. At. Spectrom., 4 (1989) 761–766.

[260] N. Bradshaw, E.F.H. Hall, N.E. Sanderson, Inductively coupled plasma as an ion source for high-resolution mass spectrometry, J. Anal. At. Spectrom., 4 (1989) 801–803.

[261] I. Steffan, G. Vujicic, A new nebulizer for inductively coupled plasma analysis of solutions with high salt content, Spectrochim. Acta, 44B (1989) 229–233.

[262] W. Doherty, An internal standardization procedure for the determination of yttrium and the rare elements in geological materials by inductively coupled plasma-mass spectrometry, Spectrochim. Acta, 44B (1989) 263–280.

[263] G. Kreuning, F.J.M.J. Maessen, Effects of the solvent plasma load of various solvents on the excitation conditions in medium power inductively coupled plasmas, Spectrochim. Acta, 44B (1989) 367–384.

[264] S. Nowak, J.A.M. van der Mullen, A.C.A.P. van Lammeren, D.C. Schram, On the influence of water on the electron density in an argon inductively coupled plasma, Spectrochim. Acta, 44B (1989) 411–418.

[265] P. Yang, R.M. Barnes, J. Mostaghimi, M.I. Boulos, Application of a two-dimentional model in the simulation of an analytical inductively coupled plasma discharge, Spectrochim. Acta, 44B (1989) 657–666.

[266] M.T. Cicerone, P.B. Farnsworth, A simple, non-invasive method for the measurement of gas flow velocities in the inductively coupled plasma, Spectrochim. Acta, 44B (1989) 897–907.

[267] K. Backstrom, A. Gustavsson, P. Hietala, A membrane interface for organic solvent sample introduction into inductively coupled plasmas, Spectrochim. Acta, 44B (1989) 1041–1048.

[268] J.M. Mermet, Ionic to atomic line intensity ratio and residence time in inductively coupled plasma-atomic emission spectrometry, Spectrochim. Acta, 44B (1989) 1109–1116.

[269] E.G. Chudinov, G.V. Varvanina, Matrix effects of acids in inductively coupled plasma atomic emission spectrometry. Comparison of different instruments, Zh. Anal. Khim., 44 (1989) 814–826.

[270] A. Lefebvre, Atomisation and Sprays, Hemisphere Publishing Corporation, New York (1989).

[271] Z. Sulcek, P. Povondra, Methods of decomposition in inorganic analysis, CRC Press, Boca Raton, Fl (1989).

[272] D.C. Grégoire, Sample introduction techniques for the determination of osmium isotope ratios by inductively coupled plasma mass spectrometry, Anal. Chem., 62 (1990) 141–146.

[273] J.W. Olesik, A.W. Moore, Influence of small amounts of organic solvents in aqueous samples on argon inductively coupled plasma spectrometry, Anal. Chem., 62 (1990) 840–845.

[274] D.R. Wiederin, R.S. Houk, R.K. Winge, A.P. D'Silva, Introduction of organic solvents into inductively coupled plasmas by ultrasonic nebulization with cryogenic desolvation, Anal. Chem., 62 (1990) 1155–1160.

[275] M.E. Ketterer, Determination of rhenium in groundwater by inductively coupled plasma mass spectrometry with on-line cation exchange membrane sample cleanup, Anal. Chem., 62 (1990) 2522–2526.

[276] D.B. Ward, M. Bell, Determination of rubidium by isotope dilution inductively coupled plasma mass spectrometry as an alternative to thermal ionization mass spectrometry, Anal. Chim. Acta, 229 (1990) 157–162.

[277] G.P. Klinkhammer, L.H. Chan, Determination of barium in marine waters by isotope dilution inductively coupled plasma mass spectrometry, Anal. Chim. Acta, 232 (1990) 323–329.

[278] C.A. Johnson, Rapid ion-exchange technique for the separation and preconcentration of chromium (VI) and chromium (III) in fresh waters, Anal. Chim. Acta, 238 (1990) 273–278.

[279] E. Yoshimura, H. Suzuki, S. Toda, Interference by mineral acids in inductively coupled plasma atomic emission spectrometry, Analyst, 115 (1990) 167–171.

[280] L.K. Olson, J.A. Caruso, J.T. Creed, K.A. Wolnik, F.L. Fricke, Evaluation of a cooled recirculating nebulizer for inductively coupled plasma spectrometry, Appl. Spectrosc., 44 (1990) 716–721.

[281] T.W. Avery, C. Chakrabarty, J.J. Thompson, Characterization and optimization of a direct injection nebulizer for introduction of organic solvents and volatile analyte species into an inductively coupled plasma, Appl. Spectrosc., 44 (1990) 1690–1698.

[282] A. Canals, V. Hernandis, R.F. Browner, Experimental evaluation of the Nukiayama-Tanasawa equation for pneumatic nebulisers used in plasma atomic emission spectrometry, J. Anal. At. Spectrom., 5 (1990) 61–66.

[283] L.C. Bates, J.W. Olesik, Effect of sample aerosol transport rate on inductively coupled plasma atomic emission and fluorescence, J. Anal. At. Spectrom., 5 (1990) 239–247.

[284] H. Isoyama, T. Uchida, C. Iida, G. Nakagawa, Small spray chamber with a concentric nebuliser for inductively coupled plasma atomic emission spectrometry, J. Anal. At. Spectrom., 5 (1990) 307–310.

[285] B.S. Ross, D.M. Chambers, G.H. Vickers, P. Yang, G.M. Hieftje, Characterization of a 9-mm torch for inductively coupled plasma mass spectrometry, J. Anal. At. Spectrom., 5 (1990) 351–358.

[286] G.E. Hall, J.C. Pelchat, J. Loop, Determination of zirconium, niobium, hafnium and tantalum at low levels in geological materials by inductively coupled plasma mass spectrometry, J. Anal. At. Spectrom., 5 (1990) 339–349.

[287] J.C. Ivaldi, W. Slavin, Cross-flow nebulizers and testing procedures for inductively coupled plasma nebulisers, J. Anal. At. Spectrom., 5 (1990) 359–364.

[288] H. Isoyama, T. Uchida, C. Iida, G. Nakagawa, Recycling nebulisation system with exchangeable spray chamber for inductively coupled plasma atomic emission spectrometry, J. Anal. At. Spectrom., 5 (1990) 365–369.

[289] E.H. Evans, L. Ebdon, Effect of organic solvents and molecular gases on polyatomic ion interferences in inductively coupled plasma mass spectrometry, J. Anal. At. Spectrom., 5 (1990) 425–430.

[290] J.R. Pretty, E.H. Evans, E.A. Blubaugh, W.-L. Shen, J.A. Caruso, T.M. Davidson, Minimisation of sample matrix effects and signal enhancement for trace analytes using anodic stripping voltammetry with detection by inductively coupled plasma atomic emission spectrometry and inductively coupled plasma mass spectrometry, J. Anal. At. Spectrom., 5 (1990) 437–443.

[291] J. Wang, W.-L. Shen, B.S. Sheppard, E.H. Evans, J.A. Caruso, F.L. Fricke, Effect of ion-lens tuning and flow injection on non-spectroscopic matrix interferences in inductively coupled plasma mass spectrometry, J. Anal. At. Spectrom., 5 (1990) 445–449.

[292] D.E. Nixon, Excitation modulation by water: effects of desolvatation on line intensities, temperatures and ion-atom ratios produced by inductively coupled plasmas, J. Anal. At. Spectrom., 5 (1990) 531–536.

[293] C. Pan, G. Zhu, R.F. Browner, Comparison of desolvation effects with aqueous and organic (carbon tetrachloride) sample introduction for inductively coupled plasma atomic emission spectrometry, J. Anal. At. Spectrom., 5 (1990) 537–542.

[294] D.C. Grégoire, Determination of boron in fresh and saline waters by inductively coupled plasma mass spectrometry, J. Anal. At. Spectrom., 5 (1990) 623–626.

[295] M. Thompson, M.H. Ramsey, Extrapolation to infinite dilution: a method for overcoming matrix effects, J. Anal. At. Spectrom., 5 (1990) 701–704.

[296] N. Ashgriz, J.Y. Poo, Coalescence and separation in binary collisions of liquid drops, J. Fluid Mech., 221 (1990) 183–204.

[297] T.D. Hettipathirana, A.P. Wade, M.W. Blades, Effects of organic acids in low power inductively coupled argon plasma-optical emission spectroscopy, Spectrochim. Acta, 45B (1990) 271–280.

[298] J.J. Tiggelman, F.J. Oukes, M.T.C. de Loos-Vollebregt, Signal compensation for the inductively coupled plasma-atomic emission spectrometry analysis of high-solid solutions, Spectrochim. Acta, 45B (1990) 927–932.

[299] Y.-S. Kim, H. Kawaguchi, T. Tanaka, A. Mizuike, Non-spectroscopic matrix interferences in inductively coupled plasma-mass spectrometry, Spectrochim. Acta, 45B (1990) 333–339.

[300] M. Huang, P.Y. Yang, D.S. Hanselman, C.A. Monnig, G.M. Hieftje, Verification of a Maxwellian electron-energy distribution in the inductively coupled plasma, Spectrochim. Acta, 45B (1990) 511–520.

[301] R. Tsukahara, M. Kubota, Studies with desolvation in inductively coupled plasma-mass spectrometry, Spectrochim. Acta, 45B (1990) 581–589.

[302] A. Canals, V. Hernandis, R.F. Browner, Evolution of drop size distributions for pneumatically generated aerosols in inductively coupled plasma-atomic emission spectrometry, Spectrochim. Acta, 45B (1990) 591–601.

[303] D.G.J. Weir, M.W. Blades, An electronic device for control of solvent-plasma load for inductively coupled plasma spectroscopy, Spectrochim. Acta, 45B (1990) 615–618.

[304] J.W. Olesik, S.J. Den, Effect of central gas flow rate and water on an argon inductively coupled plasma: spatially resolved emission, ion-atom intensity ratios and electron number densities, Spectrochim. Acta, 45B (1990) 731–752.

[305] J.W. McLaren, J.W. Lam, A. Gustavsson, Evaluation of a membrane interface sample introduction system for inductively coupled plasma mass spectrometry, Spectrochim. Acta, 45B (1990) 1091–1094.

[306] A. Gustavsson, P. Hietala, A membrane interface for aqueous sample introduction into inductively coupled plasma, Spectrochim. Acta, 45B (1990) 1103–1108.

[307] M. Marichy, M. Mermet, J.M. Mermet, Some effects of low acid concentrations in inductively coupled plasma atomic emission spectrometry, Spectrochim. Acta, 45B (1990) 1195–1201.

[308] M.A. Vaughan, G. Horlick, Effect of sampler and skimmer orifice size on analyte oxide signals in inductively coupled plasma mass spectrometry, Spectrochim. Acta, 45B (1990) 1289–1299.

[309] G.H. Vickers, D.A. Wilson, G.M. Hieftje, Spatial dependence of ion measurements in inductively coupled plasma-mass pectrometry, Spectrochim. Acta, 45B (1990) 499–509.

[310] J. Sneddon, Sample Introduction in Atomic Spectroscopy, Elsevier, Amsterdam, (1990).

[311] J.W. Olesik, Elemental analysis using ICP-OES and ICP-MS. An evaluation and assessment of remaining problems, Anal. Chem., 63 (1991) 12A–21A.

[312] D.R. Wiederin, F.G. Smith, R.S. Houk, Direct injection nebulization for inductively coupled plasma mass spectrometry, Anal. Chem., 63 (1991) 219–225.

[313] P. Allain, L. Jaunault, Y. Mauras, J.M. Mermet, T. Delaporte, Signal enhancement of elements due to the presence of carbon-containing compounds in inductively coupled plasma mass spectrometry, Anal. Chem., 63 (1991) 1497–1498.

[314] C.J. Lord III, Determination of trace metals in crude oil by inductively coupled plasma mass spectrometry with microemulsion sample introduction, Anal. Chem., 63 (1991) 1594–1599.

[315] D.R. Wiederin, R.E. Smyczek, R.S. Houk, On-line standard additions with direct injection nebulization for inductively coupled plasma mass spectrometry, Anal. Chem., 63 (1991) 1626–1631.

[316] A. Montaser, H. Tan, I. Ishii, S.-H. Nam, M. Cai, Argon inductively coupled plasma mass spectrometry with thermospray, ultrasonic, and pneumatic nebulizers, Anal. Chem., 63 (1991) 2660–2665.

[317] J. Toole, K. McKey, M. Baxter, Determination of uranium in marine sediment pore waters by isotope dilution inductively coupled plasma mass spectrometry, Anal. Chim. Acta, 245 (1991) 83–88.

[318] J.M. Mermet, Use of magnesium as test element for inductively coupled plasma atomic emission spectrometry diagnostics, Anal. Chim. Acta, 250 (1991) 85–94.

[319] S.C. Hight, J.I. Rader, Use of the Hildebrand grid nebulizer for inductively coupled plasma atomic emission spectrometric analysis of foodware leach solutions and rodent soft tissues and femurs, Analyst, 116 (1991) 1013–1017.

[320] B.S. Ross, P. Yang, G.M. Hieftje, The investigation of a 13-mm torch for use in inductively coupled plasma mass spectrometry, Appl. Spectrosc., 45 (1991) 190–197.

[321] G.D. Rayson, D.Y. Shen, Impact of scattering on axial viewing absorption measurements within an inductively coupled plasma, Appl. Spectrosc., 45 (1991) 706–708.

[322] S.E. Hobbs, J.W. Olesik, Laser-excited fluorescence studies of matrix-induced errors in inductively coupled plasma spectrometry: implications for ICP-Mass spectrometry, Appl. Spectrosc., 45 (1991) 1395–1407.

[323] D.R. Wiederin, R.S. Houk, Measurements of aerosol particle sizes from a direct injection nebulizer, Appl. Spectrosc., 45 (1991) 1408–1412.

[324] M.A. Tarr, Guangxuan Zhu, R.F. Browner, Fundamental aerosol studies with an ultrasonic nebulizer, Appl. Spectrosc., 45 (1991) 1424–1432.

[325] S.V. Mani, T. Vesala, J.A. Raper, G.J. Jameson, Evaporation of polydisperse organic aerosols at ambient conditions, J. Aerosol Sci., 22 (1991) S81–S84.

[326] L.M. Garden, J. Marshall, D. Littlejohn, Correction of mineral acid interferences in inductively coupled plasma optical emission spectrometry on copper and manganese using internal standardization, J. Anal. At. Spectrom., 6 (1991) 159–163.

[327] J. Mora, V. Hernandis, A. Canals, Influence of solvent physical properties on drop size distribution, transport and sensitivity in Flame Atomic Absorption Spectrometry with pneumatic nebulization, J. Anal. At. Spectrom., 6 (1991) 573–579.

[328] J. Marshall, J. Franks, Matrix interferences from methacrylic acid solutions in inductively coupled plasma mass spectrometry, J. Anal. At. Spectrom., 6 (1991) 591–600.

[329] J.S. Crain, R.S. Houk, High speed photographic study of plasma fluctuations and intact aerosol particles or droplets in inductively coupled plasma mass spectrometry, J. Anal. At. Spectrom., 6 (1991) 601–604.

[330] B.S. Ross, D.M. Chambers, G.M. Hieftje, Fundamental and applied investigations in plasma-source mass spectrometry for elemental analysis, Mikrochim. Acta, 104 (1991) 287–297.

[331] B. Huang, J. Yang, A. Pei, X. Zeng, P.W.J.M. Boumans, Studies on inductively coupled plasma sample introduction -2. Use of ethanol-water solutions and desolvation, Spectrochim. Acta, 46B (1991) 407–416.

[332] D.M. Chambers, J. Poehlman, P. Yang, G.M. Hieftje, Fundamental studies of the sampling process in an inductively coupled plasma mass spectrometer -1. Langmuir probe measurements, Spectrochim. Acta, 46B (1991) 741–760.

[333] D.M. Chambers, G.M. Hieftje, Fundamental studies of the sampling process in an inductively coupled plasma mass spectrometer -2. Ion kinetic energy measurements, Spectrochim. Acta, 46B (1991) 761–784.

[334] D.M. Chambers, B.S. Ross, G.M. Hieftje, Fundamental studies of the sampling process in an inductively coupled plasma mass spectrometer -3. Monitoring the ion beam, Spectrochim. Acta, 46B (1991) 785–804.

[335] J.W. Olesik, J.C. Fister 3, Incompletely desolvated droplets in argon inductively coupled plasmas: their number, original size and effect on emission intensities, Spectrochim. Acta, 46B (1991) 851–868.

[336] J.C. Fister 3, J.W. Olesik, Vertical and radial emission profiles and ion-atom intensity ratios in inductively coupled plasmas: the connection to vaporizing droplets, Spectrochim. Acta, 46B (1991) 869–883.

[337] I. Ishii, H. Tan, S.K. Chan, A. Montaser, Helium ICP-AES: effects of induction frequency and forward power on plasma formation and analytical and fundamental properties, Spectrochim. Acta, 46B (1991) 901–916.

[338] Zicheng Peng, H. Klinkenberg, T. Beeren, W. van Borm, Determination of germanium, palladium and platinum in highly concentrated solutions of phosphoric acid and ammonium nitrate by flow injection inductively coupled plasma mass spectrometry (FI-ICP-MS), Spectrochim. Acta, 46B (1991) 1051–1061.

[339] J.C. Ivaldi, J. Vollmer, W. Slavin, The conespray nebulizer for inductively coupled plasma atomic emission spectrometry, Spectrochim. Acta, 46B (1991) 1063–1072.

[340] J.W. Elgersma, J. Balke, F.J.MJ. Maessen, The performance of a low consumption thermospray nebulizer for specific use in micro-HPLC and general use in FI with ICP-AES detection, Spectrochim. Acta, 46B (1991) 1073–1088.

[341] F. Laborda, M.T.C de Loos-Vollebregt, L. de Galan, Coupling of HPLC and ICP-AES for speciation, Spectrochim. Acta, 46B (1991) 1089–1098.

[342] B.S. Ross, G.M. Hieftje, Alteration of the ion-optic lens configuration to eliminate mass-dependent matrix interference effects in inductively coupled plasma mass spectromerty, Spectrochim. Acta, 46B (1991) 1263–1273.

[343] S.E. Hobbs, J.W. Olesik, Inductively coupled plasma mass spectrometry signal fluctuations due to individual aerosol droplets and vaporizing particles, Anal. Chem., 64 (1992) 274–282.

[344] A.P. Weber, R. Keil, L. Tobler, U. Baltensperger, Sensitivities of inductively coupled plasma optical emission spectrometry for dry and wet aerosols, Anal. Chem., 64 (1992) 672–676.

[345] M.B. Shabani, T. Akagi, A. Masuda, Preconcentration of trace rare-earth elements in seawater by complexation with bis(2-ethylhexyl) hydrogen phosphate and 2-ethylexyl dihydrogen phosphate adsorbed on a C_{18} cartridge and determination by inductively coupled plasma mass spectrometry, Anal. Chem., 64 (1992) 737–742.

[346] M.J. Powell, E.S.K. Quan, D.W. Boomer, D.R. Wiederin, Inductively coupled plasma mass spectrometry with direct injection nebulization for mercury analysis of drinking water, Anal. Chem., 64 (1992) 2253–2257.

[347] W.T. Corns, L. Ebdon, S.J. Hill, P.B. Stockwell, Effects of moisture on the cold vapour determination of mercury and its removal by use of membrane dryer tubes, Analyst, 117 (1992) 717–720.

[348] C. Robles, J. Mora, A. Canals, Experimental evaluation of the Nukiyama-Tanasawa equation for pneumatically generated aerosols in flame atomic spectrometry, Appl. Spectrosc., 46 (1992) 669–676.

[349] B.A. Meinhard, D.K. Brown, J.E. Meinhard, The effect of nebulizer structure on flame emission, Appl. Spectrosc., 46 (1992) 1134–1139.

[350] M. Wu, G.M. Hieftje, A new spray chamber for inductively coupled plasma spectrometry, Appl. Spectrosc., 46 (1992) 1912–1918.

[351] S.J. Hill, J. Hartley, L. Ebdon, Determination of trace metals in volatile organic solvents using inductively coupled plasma atomic emission spectrometry and inductively coupled plasma mass spectrometry, J. Anal. At. Spectrom., 7 (1992) 23–28.

[352] G.M. Hieftje, P.J. Galley, M. Glick, D.S. Hanselman, New developments and final frontiers in inductively coupled plasma mass spectrometry, J. Anal. At. Spectrom., 7 (1992) 69–79.

[353] X. Jian, L. Qingyan, L. Wenchong, Q. Haowen, T. Jingyuan, Z. Zhanxia, Matrix effects of easily ionized elements on the spatial distribution of electron number densities in an inductively coupled plasma using an optical fibre probe and a photodiode array spectrometer, J. Anal. At. Spectrom., 7 (1992) 131–134.

[354] M. Carré, E. Poussel, J.M. Mermet, Drift diagnostics in inductively coupled plasma atomic emission spectrometry, J. Anal. At. Spectrom., 7 (1992) 791–797.

[355] M.A. Tarr, G. Zhu, R.F. Browner, Transport effects with dribble and jet ultrasonic nebulizers, J. Anal. At. Spectrom., 7 (1992) 813–817.

[356] Z. Chen, H.P. Longerich, B.J. Fryer, Recycling nebulization system with a disposable spray chamber for analysis of sub-milligram samples of geological materials using inductively coupled plasma mass spectrometry, J. Anal. At. Spectrom., 7 (1992) 905–914.

[357] A. Stroh, U. Völlkopf, E.R. Denoyer, Analysis of samples containing large amounts of dissolved solids using microsampling flow injection inductively coupled plasma mass spectrometry, J. Anal. At. Spectrom., 7 (1992) 1201–1205.

[358] N. Jakubowski, I. Feldmann, D. Stuewer, Analytical improvement of pneumatic nebulization in ICP-MS by desolvation, Spectrochim. Acta, 47B (1992) 107–118.

[359] R.F. Browner, A. Canals, V. Hernandis, Effect on analyte and solvent transport on signal intensities in inductively coupled plasma atomic emission spectrometry, Spectrochim. Acta, 47B (1992) 659–673.

[360] S.D. Tanner, Space charge in inductively coupled plasma mass spectrometry: calculation and implications, Spectrochim. Acta, 47B (1992) 809–823.

[361] J.A. Koropchak, M. Veber, J. Herries, Fused silica aperture thermospray sample introduction to inductively coupled plasma atomic emission spectrometry, Spectrochim. Acta, 47B (1992) 825–834.

[362] E.H. Evans, J.A. Caruso, Optimization strategies for the reduction of non-spectroscopic interferences in inductively coupled plasma mass spectrometry, Spectrochim. Acta, 47B (1992) 1001–1012.

[363] F. Vankaecke, H. Vanhoe, R. Dams, C. Vandecasteele, The use of internal standards in ICP-MS, Talanta, 39 (1992) 737–742.

[364] J.H. Kalivas, Practical Guide to Chemometrics, S.J. Haswell Ed., Marcel Dekker, Inc. New York, chap. 4, Calibration (1992) 99–149.

[365] T.J. Overcamp, S.E. Scarlett, Effect of Reynolds number on the Stokes number of cyclones, Aerosol Sci. Tech., 19 (1993) 362–370.

[366] H. Tan, B.A. Meinhard, J.E. Meinhard, Recent investigations of Meinhard concentric nebulizers, Federation of Analytical Chemistry and Spectroscopy Societies, Philadelphia, PA, (1993).

[367] S.-H. Nam, W.R.L. Masamba, A. Montaser, Investigation of helium inductively coupled plasma mass spectrometry for the detection of metals and nonmetals in aqueous solutions, Anal. Chem., 65 (1993) 2784–2790.

[368] X.-J. Yang, J.-S. Guan, End-on viewed inductively coupled plasma for the determination of trace impurities in high-purity scandium oxide by extraction chromatography, Anal. Chim. Acta, 279 (1993) 261–272.

[369] J. Goossens, F. Vanhaecke, L. Moens, R. Dams, Elimination of interferences in the determination of arsenic and selenium in biological samples by inductively coupled plasma mass spectrometry, Anal. Chim. Acta, 280 (1993) 137–143.

[370] B. Walczak, W. Wegscheider, Non-linear modelling of chemical data by combinations of linear and neural net methods, Anal. Chim. Acta, 283 (1993) 508–517.

[371] S.C.K. Shum, S.K. Johnson, H.-M. Pang, R.S. Houk, Spatially resolved measurements of size and velocity distributions of aerosol droplets from a direct injection nebulizer, Appl. Spectrosc., 47 (1993) 575–583.

[372] K. Hu, R.S. Houk, Inductively coupled plasma mass spectrometry with an enlarged sampling orifice and offset ion lens. II. Polyatomic ion interferences and matrix effects, J. Am. Soc. Mass Spectrom., 4 (1993) 28–37.

[373] E.H. Evans, J.J. Giglio, Interferences in inductively coupled plasma mass spectrometry. A review, J. Anal. At. Spectrom., 8 (1993) 1–18.

[374] A.I. Ruiz, A. Canals, V. Hernandis, Effect of surfactants in Flame Atomic Absorption Spectrometry with pneumatic nebulization: influence of hydrophobic chain length, J. Anal. At. Spectrom., 8 (1993) 109–113.

[375] A.R. Eastgate, R.C. Fry, G.H. Gower, Radiation versus conduction in heated spray chamber desolvation for inductively coupled plasmas, J. Anal. At. Spectrom., 8 (1993) 305–308.

[376] I.D. Holclajtner-Antunovic, M.R. Tripkovic, Study of the matrix effect of easily and non-easily ionizable elements in an inductively coupled argon plasma. Part II. Equilibrium plasma composition, J. Anal. At. Spectrom., 8 (1993) 359–365.

[377] F. Vanhaecke, R. Dams, C. Vandecasteele, Zone model as an explanation for signal behaviour and non spectral interferences in inductively coupled plasma mass spectrometry, J. Anal. At. Spectrom., 8 (1993) 433–438.

[378] A. Miyazaki, R.A. Reimer, Determination of lead isotope ratios and concentrations in sea-water by inductively coupled plasma mass spectrometry after preconcentration using Chelex-100, J. Anal. At. Spectrom., 8 (1993) 449–452.

[379] I.B. Brenner, E. Dorfman, Application of ultrasonic nebulization for the determination of rare earth elements in phosphates and related sedimentary rocks using inductively coupled plasma atomic emission spectrometry with comments on dissolution procedures, J. Anal. At. Spectrom., 8 (1993) 833–838.

[380] P. Richner, Detection limits versus matrix effects: analysis of solutions with high amounts of dissolved solids by flow injection inductively coupled plasma mass spectrometry, J. Anal. At. Spectrom., 8 (1993) 927–931.

[381] G.E.M. Hall, J.C. Pelchat, Determination of palladium and platinum in fresh waters by inductively coupled plasma mass spectrometry and activated charcoal preconcentration, J. Anal. At. Spectrom., 8 (1993) 1059–1065.

[382] D.E. Nixon, Comparison of two Meinhard nebulizers operating at the same argon flow but different pressures, Spectrochim. Acta, 48B (1993) 447–459.

[383] E. Poussel, J.M. Mermet, O. Samuel, Simple experiments for the control, the evaluation and the diagnosis of inductively coupled plasma sequential systems, Spectrochim. Acta, 48B (1993) 743–755.

[384] P.J. Galley, M. Glick, G.M. Hieftje, Easily ionizable element interferences in inductively coupled plasma atomic emission spectrometry-I. Effect on radial analyte emission patterns, Spectrochim. Acta, 48B (1993) 769–788.

[385] S.E. Hobbs, J.W. Olesik, The effect of desolvating droplets and vaporizing particles on ionization and excitation in Ar inductively coupled plasmas, Spectrochim. Acta, 48B (1993) 817–833.

[386] R.H. Clifford, H. Tan, H. Liu, A. Montaser, F. Zarrin, P.B. Keady, Particle size measurements in the submicron range by the differential electromobility technique: comparison of aerosols from thermospray, ultrasonic, pneumatic and frit-type nebulizers, Spectrochim. Acta, 48B (1993) 1221–1235.

[387] J.B. French, B. Etkin, R. Jong, Monodispersed dried microparticulate injector for analytical instrumentation, Anal. Chem., 66 (1994) 685–691.

[388] J.W. Olesik, J.F. Kinzer, B. Harkleroad, Inductively coupled plasma optical emission spectrometry using nebulizers with widely different sample consumption rates, Anal. Chem., 66 (1994) 2022–2030.

[389] H. Liu, A. Montaser, Phase-Doppler diagnostic studies of primary and tertiary aerosols produced by a high-efficiency nebulizer, Anal. Chem., 66 (1994) 3233–3242.

[390] J.W. Olesik, S.E. Hobbs, Monodispersed dried microparticulate injector: a new tool for studying fundamental processes in inductively coupled plasmas, Anal. Chem., 66 (1994) 3371–3378.

[391] M. Wu, Y. Madrid, G.M. Hieftje, New spray chamber for use in flow injection plasma emission spectrometry, Anal. Chim. Acta, 286 (1994) 155–167.

[392] R. Cidu, L. Fanfani, P. Shand, W.M. Edmunds, L. Van't dack, R. Gijbels, Determination of gold at the ultratrace level in natural waters, Anal. Chim. Acta, 296 (1994) 295–304.

[393] G. Légère, E. Salin, Fast-clearing spray chamber for ICP-AES, Appl. Spectrosc., 48 (1994) 761–765.

[394] S.D. Tanner, L.M. Cousins, D.J. Douglas, Reduction of space charge effects using a three-aperture gas dynamic vacuum interface for inductively coupled plasma mass spectrometry, Appl. Spectrosc., 48 (1994) 1367–1372.

[395] S.D. Tanner, D.J. Douglas, J.B. French, Gas and ion dynamics of a three-aperture vacuum interface for inductively coupled plasma mass spectrometry, Appl. Spectrosc., 48 (1994) 1373–1378.

[396] S. Roncevic, M. Siroki, Effects of low acetic acid concentration in inductively coupled plasma atomic emission spectrometry, J. Anal. At. Spectrom., 9 (1994) 99–104.

[397] B. Budic, V. Hudnik, Matrix effects of potassium chloride and phosphoric acid in argon inductively coupled plasma atomic emission spectrometry, J. Anal. At. Spectrom., 9 (1994) 53–57.

[398] E. Poussel, J.M. Mermet, D. Deruaz, Dissociation of analyte oxide ions in inductively coupled plasma mass spectrometry, J. Anal. At. Spectrom., 9 (1994) 61–66.

[399] H. Yanping, Z. Zhanxia, Z. Jainguo, Simulation of the vaporization process in inductively coupled plasma atomic emission spectrometry with a modified model using the Monte Carlo technique, J. Anal. At. Spectrom., 9 (1994) 213–216.

[400] A. Fernandez, M. Murillo, N. Carrion, J.M. Mermet, Influence of operating conditions on the effects of acids in inductively coupled plasma atomic emission spectrometry, J. Anal. At. Spectrom., 9 (1994) 217–221.

[401] L.C. Alves, M.G. Minnich, D.R. Wiederin, R.S. Houk, Removal of organic solvents by cryogenic desolvation in inductively coupled plasma mass spectrometry, J. Anal. At. Spectrom., 9 (1994) 399–403.

[402] G. Xiao, D. Beauchemin, Reduction of matrix effects and mass discrimination in inductively coupled plasma mass spectrometry with optimized argon-nitrogen plasmas, J. Anal. At. Spectrom., 9 (1994) 509–518.

[403] Y. Nakamura, K. Takahashi, O. Kujirai, H. Okochi, C.W. McLeod, Evaluation of an axially and radially viewed inductively coupled plasma using an echelle spectrometer with wavelength modulation and second-derivative detection, J. Anal. At. Spectrom., 9 (1994) 751–757.

[404] R.I. Botto, J.J. Zhu, Use of an ultrasonic nebulizer with membrane desolvation for analysis of volatile solvents by inductively coupled plasma atomic emission spectrometry, J. Anal. At. Spectrom., 9 (1994) 905–912.

[405] M.J. Bloxham, S.J. Hill, P.J. Worsfold, Determination of trace metals in sea-water and the on-line removal of matrix interferences by flow injection with inductively coupled plasma mass spectrometric detection, J. Anal. At. Spectrom., 9 (1994) 935–938.

[406] I. Feldmann, W. Tittes, N. Jakubowski, D. Stüwer, G. Tölg, J.A.C. Broekaert, Performance characteristics of inductively coupled plasma mass spectrometry with high mass resolution, J. Anal. At. Spectrom., 9 (1994) 1007–1014.

[407] E.H. Larsen, S. Stürup, Carbon-enhanced inductively coupled plasma mass spectrometric detection of arsenic and selenium and its application to arsenic speciation, J. Anal. At. Spectrom., 9 (1994) 1099–1105.

[408] D.G. Weir, M.W. Blades, Characteristics of an inductively coupled argon plasma operating with organic aerosols. Part I. Spectral and spatial observations, J. Anal. At. Spectrom., 9 (1994) 1311–1322.

[409] D.G. Weir, M.W. Blades, Characteristics of an inductively coupled argon plasma operating with organic aerosols. Part II. Axial profiles of solvent and analyte species in a chloroform-loaded plasma, J. Anal. At. Spectrom., 9 (1994) 1323–1334.

[410] S.-H. Nam, J.-S. Lim, A. Montaser, High-efficiency nebulizer for argon inductively coupled plasma mass spectrometry, J. Anal. At. Spectrom., 9 (1994) 1357–1362.

[411] M. Wu, G. Hieftje, The effect of easily ionized elements on analyte emission efficiency in inductively coupled plasma spectrometry, Spectrochim. Acta, 49B (1994) 149–161.

[412] S.K. Luo, H. Berndt, Sample introduction in inductively coupled plasma spectrometry by hydraulic high-pressure nebulization, Spectrochim. Acta, 49B (1994) 485–492.

[413] D.S. Hanselman, N.N. Sesi, M. Huang, G.M. Hieftje, The effect of sample matrix on electron density, electron temperature and gas temperature in the argon inductively coupled plasma examined by Thomson and Rayleigh scattering, Spectrochim. Acta, 49B (1994) 495–526.

[414] D. Pollman, C. Pilger, R. Hergenröder, F. Leis, P. Tschöpel, J.A.C. Broekaert, Noise power spectra of inductively coupled plasma mass spectrometry using a cooled spray chamber, Spectrochim. Acta, 49B (1994) 683–690.

[415] P.J. Galley, G.M. Hieftje, Easily ionisable element (EIE) interferences in inductively couped plasma atomic emission spectrometry-II. Minimization of EIE effects by choice of observation volume, Spectrochim. Acta, 49B (1994) 703–724.

[416] M. Cai, I. Ishii, R.H. Clifford, A. Montaser, B.A. Palmer, L.R. Layman, Fundamental properties of helium inductively coupled plasmas measured by high-resolution Fourier transform spectrometry, Spectrochim. Acta, 49B (1994) 1081–1095.

[417] D.G.J. Weir, M.W. Blades, The response of the inductively coupled plasma argon plasma to solvent plasma load: spatially resolved maps of electron density obtained from the intensity of one argon line, Spectrochim. Acta, 49B (1994) 1231–1250.

[418] Y. Liu, V. Lopez-Avila, J.J. Zhu, D.R. Wiederin, W.F. Beckert, Capillary electrophoresis coupled on-line with inductively coupled plasma mass spectrometry for elemental speciation, Anal. Chem., 67 (1995) 2020–2025.

[419] S.A. Pergantis, E.M. Heithmar, T.A. Hinners, Microscale flow injection and microbore high-performance liquid chromatography coupled with inductively coupled plasma mass spectrometry with a high-efficiency nebulizer, Anal. Chem., 67 (1995) 4530–4535.

[420] K.G. Heumann, S. Eisenhut, S. Gallus, E.H. Hebeda, R. Nusko, A. Vengosh, T. Walczyk, Recent developments in thermal ionization mass spectrometric techniques for isotope analysis. A review, Analyst, 120 (1995) 1291–1299.

[421] J.J. Giglio, J. Wang, J.A. Caruso, Evaluation of a direct injection nebulizer (DIN) for liquid sample introduction in helium microwave-induced plasma mass spectrometry (MIP-MS), Appl. Spectrosc., 49 (1995) 314–319.

[422] J.M. Mermet, E. Poussel, ICP emission spectrometers: 1995 Analytical figures of merit, Appl. Spectrosc., 49 (1995) (10) 12A–18A.

[423] E.R. Denoyer, D. Jacques, E. Debrah, S.D. Tanner, Determination of trace elements in uranium: practical benefits of a new ICP-MS lens system, At. Spectrosc., 16 (1995) 1–6.

[424] H. Tao, A. Miyazaki, Decrease of solvent water loading in inductively coupled plasma mass spectrometry by using a membrane separator, J. Anal. At. Spectrom., 10 (1995) 1–5.

[425] P.D.P. Taylor, P. De Bièvre, A.J. Walder, A. Entwistle, Validation of the analytical linearity and mass discrimation correction model exhibited by a multiple collector inductively coupled plasma mass spectrometer by means of a set of synthetic uranium isotope mixtures, J. Anal. At. Spectrom., 10 (1995) 395–398.

[426] R.I. McCrindle, C.J. Rademeyer, Excitation temperature and analytical parameters for an ethanol-loaded inductively coupled plasma atomic emission spectrometer, J. Anal. At. Spectrom., 10 (1995) 399–404.

[427] M. Bettinelli, S. Spezia, U. Baroni, G. Bizzari, Determination of trace elements in fuel oils by inductively coupled plasma mass spectrometry after acid mineralization of the sample in a microwave oven. J. Anal. At. Spectrom., 10 (1995) 555–560.

[428] L. Moens, F. Vanhaecke, J. Riondato, R. Dams, Some figures of merit of a new double focusing inductively coupled plasma mass spectrometer, J. Anal. At. Spectrom., 10 (1995) 569–574.

[429] H. Vanhoe, S. Saverwijns, M. Parent, L. Moens, R. Dams, Analytical characteristics of an inductively coupled plasma mass spectrometer coupled with a thermospray nebulization system, J. Anal. At. Spectrom., 10 (1995) 575–581.

[430] G. Zoorob, M. Tomlinson, J.-S. Wang, J. Caruso, Evaluation of the direct injection nebulizer in the coupling of high-performance liquid chromatography to inductively coupled plasma mass spectrometry, J. Anal. At. Spectrom., 10 (1995) 853–858.

[431] R. Muñoz Olivas, C.R. Quétel, O.F.X. Donard, Sensitive determination of selenium by inductively coupled plasma mass spectrometry with flow injection and hydride generation in the presence of organic solvents, J. Anal. At. Spectrom., 10 (1995) 865–870.

[432] S.D. Tanner, Characterization of ionization and matrix suppression in inductively coupled cold plasma mass spectrometry, J. Anal. At. Spectrom., 10 (1995) 905–921.

[433] G. Li, Y. Duan, G.M. Hieftje, Space-charge effects and ion distribution in plasma source mass spectrometry, J. Mass. Spectrom., 30 (1995) 841–848.

[434] H. Wildner, G. Wunsch, Comparison of nebulizer efficiencies with thermal trace-matrix-separation for the analysis of corrosive samples by ICP-MS on the semi-micro-scale, J. Prakt. Chem., 337 (1995) 542–547.

[435] R.F.J. Dams, J. Goossens, L. Moens, Spectral and non-spectral interferences in inductively coupled plasma mass spectrometry, Mikrochim. Acta, 119 (1995) 277–286.

[436] J.W. Olesik, L.C. Bates, Characterization of aerosols produced by pneumatic nebulizers for inductively coupled plasma sample introduction: effect of liquid and gas flow rates on volume based drop size distributions, Spectrochim. Acta, 50B (1995) 285–303.

[437] M. Carré, K. Lebas, M. Marichy, M. Mermet, E. Poussel, J.M. Mermet, Influence of the sample introduction system on acid effects in inductively coupled plasma atomic emission spectrometry, Spectrochim. Acta, 50B (1995) 271–283.

[438] A. Canals, V. Hernandis, J.L. Todolí, R.F. Browner, Fundamental studies on pneumatic generation and aerosol transport in atomic spectrometry: effect of mineral acids on emission intensity in inductively coupled plasma atomic emission spectrometry, Spectrochim. Acta, 50B (1995) 305–321.

[439] I.B. Brenner, J.M. Mermet, I. Segal, G.L. Long, Effect of nitric and hydrochloric acids on rare earth element (REE) intensities in inductively coupled plasma emission spectrometry, Spectrochim. Acta, 50B (1995) 323–331.

[440] I.B. Brenner, I. Segal, M. Mermet, J.M. Mermet, Study of the depressive effect of nitric acid on the line intensities of rare earth elements in inductively coupled plasma atomic emission spectrometry, Spectrochim. Acta, 50B (1995) 333–340.

[441] J.C. Ivaldi, J.F. Tyson, Performance evaluation of an axially viewed horizontal inductively coupled plasma for optical emission spectrometry, Spectrochim. Acta, 50B (1995) 1207–1226.

[442] J.A. Burgener, Parallel path induction pneumatic nebulizer, US Patent, 5411208 (1995).

[443] M.P. Dziewatkoski, L.B. Daniels, J.W. Olesik, Time-resolved inductively coupled plasma mass spectrometry measurements with individual, monodisperse drop sample introduction, Anal. Chem. 68 (1996) 1101–1109.

[444] Yu Vin Yi, A. Masuda, Simultaneous determination of ruthenium, palladium, iridium, and platinum at ultratrace levels by isotope dilution inductively coupled plasma mass spectrometry in geological samples, Anal. Chem., 68 (1996) 1444–1450.

[445] P.A. Patterson, R.J. Munz, Gas and particles flow patterns in cyclones at room and elevated temperature, Can. J. Chem. Eng., 74 (1996) 213–221.

[446] W.R.L. Cairns, L. Ebdon, S.J. Hill, A high performance liquid chromatography-inductively coupled plasma-mass spectrometry interface employing desolvation for speciation studies of platinum in chemotherapy drugs, Fresenius J. Anal. Chem., 355 (1996) 202–208.

[447] R.I. McCrindle, C.J. Rademeyer, Ethanol loading in an inductively coupled plasma, Fresenius J. Anal. Chem., 355 (1996) 264–266.

[448] F. Vanhaecke, J. Riondato, L. Moens, R. Dams, Non-spectral interferences encountered with a commercially available high resolution ICP-mass spectrometer, Fresenius J. Anal. Chem., 355 (1996) 397–400.

[449] S.H. Nam, H. Zhang, M. Cai, J.S. Lim, A. Montaser, A status report on helium inductively coupled plasma mass spectrometry, Fresenius J. Anal. Chem., 355 (1996) 510–520.

[450] I.B. Brenner, J. Zhu, A. Zander, Evaluation of an ultrasonic nebulizer-membrane separation interface (USN-MEMSEP) with ICP-AES for the determination of trace elements by solvent extraction, Fresenius J. Anal. Chem., 355 (1996) 774–777.

[451] J.L.M. de Boer, M. Velterop, Empirical procedure for the reduction of mixed-matrix effects in inductively coupled plasma atomic emission spectrometry using an internal standard and proportional correction, Fresenius J. Anal. Chem., 356 (1996) 362–370.

[452] J.W. Olesik, M.P. Dziewatkoski, Time-resolved measurements of individual ion cloud signals to investigate space charge effects in plasma mass spectrometry, J. Am. Soc. Mass Spectrom., 7 (1996) 362–367.

[453] M. Murillo, J. Chirinos, Study of the effect of organic emulsified samples in inductively coupled plasma atomic emission spectrometry, J. Anal. At. Spectrom., 11 (1996) 253–257.

[454] H. Liu, A. Montaser, S.P. Dolan, R.S. Schwartz, Evaluation of low sample consumption, high-efficiency nebulizer for elemental analysis of biological samples using inductively coupled plasma mass spectrometry, J. Anal. At. Spectrom., 11 (1996) 307–311.

[455] F. Vanhaecke, M. Van Holderbeke, L. Moens, R. Dams, Evaluation of a commercially available microconcentric nebulizer for inductively coupled plasma mass spectrometry, J. Anal. At. Spectrom., 11 (1996) 543–548.

[456] R.I. Botto, J.J. Zhu, Universal calibration for analysis of organic solutions by inductively coupled plasma atomic emission spectrometry, J. Anal. At. Spectrom., 11 (1996) 675–681.

[457] S. Augagneur, B. Médina, J. Szpunar, R. Lobinski, Determination of rare earth elements in wine by inductively coupled plasma mass spectrometry using a microconcentric nebulizer, J. Anal. At. Spectrom., 11 (1996) 713–721.

[458] Zhongxing Chen, B.J. Fryer, H.P. Longerich, S.E. Jackson, Determination of precious metals in milligram samples of sulfides and oxides using inductively coupled plasma mass spectrometry after ion exchange preconcentration, J. Anal. At. Spectrom., 11 (1996) 805–809.

[459] A. Risnes, W. Lund, Comparison of systems for eliminating interferences in the determination of arsenic and antimony by hydride generation inductively coupled plasma atomic emission spectrometry, J. Anal. At. Spectrom., 11 (1996) 943–948.

[460] J.L. Todolí, A. Canals, V. Hernandis, Behaviour of single-bore high pressure pneumatic nebulizer operating with alcohols in inductively coupled plasma atomic emission spectrometry, J. Anal. At. Spectrom., 11 (1996) 949–956.

[461] D.G. Weir, M.W. Blades, Characteristics of an inductively coupled argon plasma operating with organic aerosols. Part 4. Noise power spectra, J. Anal. At. Spectrom., 11 (1996) 1011–1018.

[462] L.S. Zhang, S.M. Combs, Using the installed spray chamber as a gas-liquid separator for the determination of germanium, arsenic, selenium, tin, antimony, tellurium and bismuth by hydride generation inductively coupled plasma mass spectrometry, J. Anal. At. Spectrom., 11 (1996) 1043–1048.

[463] L. Wang, S.W May, R.F. Browner, S.H. Pollock, Low-flow interface for liquid chromatography-inductively coupled plasma mass spectrometry speciation using an oscillating capillary nebulizer, J. Anal. At. Spectrom., 11 (1996) 1137–1146.

[464] C. Rivas, L. Ebdon, S.J. Hill, Effect of different spray chambers on the determination of organotin compounds by high-performance liquid chromatography-inductively coupled plasma mass spectrometry, J. Anal. At. Spectrom., 11 (1996) 1147–1150.

[465] H. Liu, R.H. Clifford, S.P. Dolan, A. Montaser, Investigation of a high-efficiency nebulizer and a thimble glass frit nebulizer for elemental analysis of biological materials by inductively coupled plasma atomic emission spectrometry, Spectrochim. Acta, 51B (1996) 27–40.

[466] X. Chen, R.S. Houk, Spatially resolved measurements of ion density behind the skimmer of an inductively coupled plasma mass spectrometer, Spectrochim. Acta, 51B (1996) 41–54.

[467] H. Klinkenberg, W. Van Borm, F. Souren, A theoretical adaptation of the classical isotope dilution technique for practical routine analytical determinations by means of inductively coupled plasma mass spectrometry, Spectrochim. Acta, 51B (1996) 139–153.

[468] H. Niu, R.S. Houk, Fundamentals aspects of ion extraction in inductively coupled plasma mass spectrometry, Spectrochim. Acta, 51B (1996) 779–815.

[469] J. Yang, T.S. Conver, J.A. Koropchak, Use of a multi-tube Nafion membrane dryer for desolvation with thermospray sample introduction to inductively coupled plasma atomic emission spectrometry, Spectrochim. Acta, 51B (1996) 1491–1503.

[470] I. Novotny, J.C. Fariñas, W. Jia-liang, E. Poussel, J.M. Mermet, Effect of power and carrier gas flow rate on the tolerance to water loading in inductively coupled plasma atomic emission spectrometry, Spectrochim. Acta, 51B (1996) 1517–1526.

[471] N.N. Sesi, G.M. Hieftje, Studies into the interelement matrix effect in inductively coupled plasma spectrometry, Spectrochim. Acta, 51B (1996) 1601–1628.

[472] J.A. Koropchak, S. Sadain, B. Szotek, Dispersion of discrete sample signals within aerosol spray chambers: preliminary investigations, Spectrochim. Acta, 51B (1996) 1733–1745.

[473] A. Eastgate, W. Vogel, Sample nebulizer and evaporation chamber for ICP and MIP emission or mass spectrometry and spectrometers comprising the same, US Patent 5534998 (1996).

[474] T.W. Burgoyne, G.M. Hieftje, R.A. Hites, Space charge evaluation in a plasma source mass spectrograph, Anal. Chem., 69 (1997) 485–489.

[475] E. Mei, H. Ichihashi, W. Gu, S. Yamasaki, Interface for coupling capillary electrophoresis to inductively coupled plasma and on-column concentration technique, Anal. Chem., 69 (1997) 2187–2192.

[476] L.A. Allen, J.J. Leach, R.S. Houk, Spatial location of the space charge effect in individual ion clouds using monodisperse dried microparticulate injection with a twin quadrupole inductively coupled plasma mass spectrometer, Anal. Chem., 69 (1997) 2384–2391.

[477] A.C. Lazar, P.B. Farnsworth, Characterization of an inductively coupled plasma with xylene solutions introduced as monodisperse aerosols, Anal. Chem., 69 (1997) 3921–3929.

[478] M. Hoenig, H. Baeten, S. Vanhentenrijk, G. Ploegaerts, T. Bertholet, Evaluation of various commercially available nebulization devices for inductively coupled plasma atomic emission spectromtry, Analusis, 25 (1997) 13–19.

[479] J.W. Olesik, Investigating the fate of individual sample droplets in inductively coupled plasmas, Appl. Spectrosc., 51 (1997) 158A–175A.

[480] J.W. Olesik, J.A. Kinzer, G.J. McGowan, Observation of atom and ion clouds produced from single droplets of sample in inductively coupled plasmas by optical emission and laser-induced fluorescence imaging, Appl. Spectrosc., 51 (1997) 607–616.

[481] A.C. Lazar, P.B. Farnsworth, Investigation of the analytical performance of an MDMI-ICP-AES system, Appl. Spectrosc., 51 (1997) 617–624.

[482] G. Lidén, A. Gudmundsson, Semi-empirical modelling to generalise the dependence of cyclone collection efficiency on operating conditions and cyclone design, J. Aerosol. Sci., 28 (1997) 853–874.

[483] D. Günther, H. Cousin, B. Magyar, I. Leopold, Calibration studies on dried aerosols for laser-ablation inductively coupled plasma mass spectrometry, J. Anal. At. Spectrom., 12 (1997) 165–170.

[484] I.B. Brenner, A. Zander, M. Plantz, J. Zhu, Characterization of an ultrasonic nebulizer-membrane separation interface with inductively coupled plasma mass spectrometry for the determination of trace elements by solvent extraction, J. Anal. At. Spectrom., 12 (1997) 273–279.

[485] C. Dubuisson, E. Poussel, J.M. Mermet, Comparison of axially and radially viewed inductively coupled plasma atomic emission spectrometry in terms of signal-to-background ratio and matrix effects, J. Anal. At. Spectrom., 12 (1997) 281–286.

[486] K. O'Hanlon, L. Ebdon, M. Foulkes, Effect of ionizable elements on the mass transport efficiency of solutions and slurries used in plasma emission spectrometry, J. Anal. At. Spectrom., 12 (1997) 329–331.

[487] E.H. Larsen, M.B. Ludwigsen, Determination of iodine in food-related certified reference materials using wet ashing and detection by inductively coupled plasma mass spectrometry, J. Anal. At. Spectrom., 12 (1997) 435–439.

[488] J. Mora, J.L. Todolí, A. Canals, V. Hernandis, Comparative study of several nebulizers in inductively coupled plasma atomic emission spectrometry: low-pressure versus high-presure nebulization, J. Anal. At. Spectrom., 12 (1997) 445–451.

[489] A. Tangen, R. Trones, T. Greibrokk, W. Lund, Microconcentric nebulizer for the coupling of micro liquid chromatography and capillary zone electrophoresis with inductively coupled plasma mass spectrometry, J. Anal. At. Spectrom., 12 (1997) 667–670.

[490] I.B. Brenner, A. Zander, M. Cole, A. Wiseman, Comparison of axially and radially viewed inductively coupled plasmas for multi-element analysis: effect of sodium and calcium, J. Anal. At. Spectrom., 12 (1997) 897–906.

[491] T.U. Probst, N.G. Berryman, P. Lemmen, L. Weissfloch, T. Auberger, D. Gabel, J. Carlsson, Comparison of inductively coupled plasma atomic emission spectrometry and inductively coupled plasma mass spectrometry with quantitative neutron capture radiography for the determination of boron in biological samples from cancer therapy, J. Anal. At. Spectrom., 12 (1997) 1115–1122.

[492] R.S. Houk, R.K. Winge, X. Chen, High speed photographic study of wet droplets and solid particles in the inductively coupled plasma, J. Anal. At. Spectrom., 12 (1997) 1139–1148.

[493] W. Branagh, C. Whelan, E.D. Salin, System for automatic selection of operating conditions for inductively coupled plasma atomic emission spectrometry, J. Anal. At. Spectrom., 12 (1997) 1307–1315.

[494] M.G. Minnich, R.S. Houk, M.A. Woodin, D.C. Christiani, Method to screen urine samples for vanadium by inductively coupled plasma mass spectrometry with cryogenic desolvation, J. Anal. At. Spectrom., 12 (1997) 1345–1350.

[495] S.D. Lofthouse, G.M. Greenway, S.C. Stephen, Microconcentric nebuliser for the analysis of small sample volumes by inductively coupled plasma mass spectrometry, J. Anal. At. Spectrom., 12 (1997) 1373–1376.

[496] D.A. Sadler, F. Sun, S.E. Howe, D. Littlejohn, Comparison of procedures for correction of matrix interferences in the multi-element analysis of soils by ICP-AES with a CCD detection system, Mikrochim. Acta, 126 (1997) 301–311.

[497] J.L. Todolí, J. Mora, J.M. Cano, A. Canals, Organic sample introduction in inductively coupled plasma atomic emission spectrometry by hydraulic high-pressure nebulization, Quim. Anal. (Barcelona), 16 (1997) 177–189.

[498] Y. Chen, P.B. Farnsworth, Ion deposition experiments as a tool for the study of the spatial distribution of analyte ions in the second vacuum stage of an inductively coupled plasma mass spectrometer, Spectrochim. Acta, 52B (1997) 231–239.

[499] X. Romero, E. Poussel, J.M. Mermet, Influence of the operating conditions on the efficiency of internal standardization in inductively coupled plasma atomic emission spectrometry, Spectrochim. Acta, 52B (1997) 487–493.

[500] X. Romero, E. Poussel, J.M. Mermet, The effect of sodium on analyte ionic line intensities in inductively coupled plasma atomic emission spectrometry, Spectrochim. Acta, 52B (1997) 495–502.

[501] M. Huang, S.A. Lehn, E.J. Andrews, G.M. Hieftje, Comparison of electron concentrations, electron temperatures, gas kinetic temperatures, and excitation temperatures in argon ICPs operated at 27 and 40 MHz, Spectrochim. Acta, 52B (1997) 1173–1193.

[502] L. Gras, J. Mora, J.L. Todolí, V. Hernandis, A. Canals, Behaviour of a desolvation system based on microwave radiation heating for use in inductively coupled plasma atomic emission spectrometry, Spectrochim. Acta, 52B (1997) 1201–1213.

[503] I. Llorente, M. Gómez, C. Cámara, Improvement in selenium determination in water by inductively coupled plasma mass spectrometry through use of organic compounds as matrix modifiers, Spectrochim. Acta, 52B (1997) 1825–1838.

[504] M. Rehkämper, A.N. Halliday, Development and application of new iondashexchange techniques for the separation of the platinum group and other siderophile elements from geological samples, Talanta, 44 (1997) 663–672.

[505] J. Szpunar, J. Bettmer, M. Robert, H. Chassaigne, K. Cammann, R. Lobinski, O.F.X. Donard, Validation of the determination of copper and zinc in blood plasma and urine by ICP-MS with cross-flow and direct injection nebulization, Talanta, 44 (1997) 1389–1396.

[506] H.M. Kingston, S.J. Haswell, Eds, Microwave Enhanced Chemistry. Fundamentals, Sample Preparation and Applications, ACS, Washington DC (1997).

[507] J.A. McLean, H. Zhang, A. Montaser, A direct injection high-efficiency nebulizer for inductively coupled plasma mass spectrometry, Anal. Chem., 70 (1998) 1012–1020.

[508] M.P. Field, R.M. Sherrell, Magnetic sector ICPMS with desolvating micronebulization: interference-free sub-picogram determination of rare earth elements in natural samples, Anal. Chem., 70 (1998) 4480–4486.

[509] M. Rupprecht, T. Probst, Development of a method for the systematic use of bilinear multivariate calibration methods for the correction of interferences in inductively coupled plasma-mass spectrometry, Anal. Chim. Acta, 358 (1998) 205–225.

[510] A.G. González, Two level factorial experimental designs based on multiple linear regression models: a tutorial digest illustrated by case studies, Anal. Chim. Acta, 360 (1998) 227–241.

[511] B. Michalke, P. Schramel, The coupling of capillary electrophoresis to ICP-MS, Analusis, 26 (1998) M51–M56.

[512] J. Mora, J.L. Todolí, I. Rico, A. Canals, Aerosol desolvation studies with a thermospray nebulizer coupled to inductively coupled plasma atomic emission spectrometry, Analyst, 123 (1998) 1229–1234.

[513] S.H. Nam, A. Montaser, E.F. Cromwell, Exploration of a helium inductively coupled plasma/Mattauch-Herzog mass spectrometer for simultaneous multielement analysis, Appl. Spectrosc., 52 (1998) 161–167.

[514] O.Y. Akinbo, J.W. Carnahan, Effects of a membrane desolvator on the analytical performance of a 120 W helium microwave-induced plasma: Aqueous desolvation, Appl. Spectrosc., 52 (1998) 1079–1085.

[515] A.S. Al-Ammar, R. Barnes, Correction for drift in ICP -OES measurements by internal standardization using spectral lines of the same analyte as internal reference, At. Spectrosc., 19 (1998) 18–22.

[516] J.L.M. de Boer, W. van Leeuwen, U. Kohlmeyer, P.M. Breugem, The determination of chromium, copper and nickel in groundwater using axial plasma inductively coupled plasma atomic emission spectrometry and proportional correction matrix effect reduction, Fresenius J. Anal. Chem., 360 (1998) 213–218.

[517] S. Saverwyns, K. Van Hecke, F. Vanhaecke, L. Moens, R. Dams, Evaluation of a commercially available microbore anion exchange column for chromium speciation with detection by ICP-mass spectrometry and hyphenation with microconcentric nebulization, Fresenius J. Anal. Chem., 363 (1999) 490–494.

[518] N. Fitzgerald, J.F. Tyson, D.A. Leighty, Reduction of water loading effects in inductively coupled plasma mass spectrometry by a Nafion membrane dryer device, J. Anal. At. Spectrom., 13 (1998) 13–16.

[519] J.L. Todolí, J.M. Mermet, A. Canals, V. Hernandis, Acid effects in inductively coupled plasma atomic emission spectrometry with different nebulizers operated at very low sample consumption rates, J. Anal. At. Spectrom., 13 (1998) 55–62.

[520] C. Dubuisson, E. Poussel, J.M. Mermet, J.L. Todolí, Comparison of the effect of acetic acid with axially and radially viewed inductively coupled plasma atomic emission spectrometry: influence of the operating conditions, J. Anal. At. Spectrom., 13 (1998) 63–67.

[521] A. Woller, H. Garraud, J. Boisson, A.M. Dorthe, P. Fodor, O.F.X. Donard, Simultaneous speciation of redox species of arsenic and selenium using an anion-exchange microbore column coupled with a micro-nebulizer and an inductively coupled plasma mass spectrometer as detector, J. Anal. At. Spectrom., 13 (1998) 141–149.

[522] I. Rodushkin, T. Ruth, D. Klockare, Non-spectral inteferences caused by a saline water matrix in quadrupole and high resolution inductively coupled plasma mass spectrometry, J. Anal. At. Spectrom., 13 (1998) 159–166.

[523] M.G. Minnich, R.S. Houk, Comparison of cryogenic and membrane desolvation for attenuation of oxide, hydride and hydroxide ions and ions containing chlorine in inductively coupled plasma mass spectrometry, J. Anal. At. Spectrom., 13 (1998) 167–174.

[524] D. Pozebon, V.L. Dressler, A.J. Curtius, Determination of copper, cadmium, lead, bismuth and selenium (IV) in sea-water by electrothermal vaporization inductively coupled plasma mass spectrometry after on-line separation, J. Anal. At. Spectrom., 13 (1998) 363–369.

[525] J.M. Mermet, Revisitation of the matrix effects in inductively coupled plasma atomic emission spectrometry: the key role of the spray chamber, J. Anal. At. Spectrom., 13 (1998) 419–422.

[526] T.D. Hettipathirana, D.E. Davey, Evaluation of a microconcentric nebuliser-cyclonic spray chamber for flow injection simultaneous multielement inductively coupled plasma optical emission spectrometry, J. Anal. At. Spectrom., 13 (1998) 483–488.

[527] J. Hamier, E.D. Salin, Tandem calibration methodology: dual nebulizer sample introduction system for inductively coupled plasma atomic emission spectrometry, J. Anal. At. Spectrom., 13 (1998) 497–505.

[528] M. De Wit, R. Blust, Determination of metals in saline and biological matrices by axial inductively coupled plasma atomic emission spectrometry using microconcentric nebulization, J. Anal. At. Spectrom., 13 (1998) 515–520.

[529] J.L. Todolí, J.M. Mermet, Minimization of acid effects at low consumption rates in an axially viewed inductively coupled plasma atomic emission spectrometer by using micronebulizer-based sample introduction systems, J. Anal. At. Spectrom., 13 (1998) 727–734.

[530] L.B. Allen, P.H. Siitonen, H.C. Thompson, Jr Impact of membrane desolvation on the effects of sodium on response in inductively coupled plasma atomic emission spectrometry with ultrasonic sample introduction, J. Anal. At. Spectrom., 13 (1998) 735–741.

[531] J.A. McLean, M.G. Minnich, L.A. Iacone, H. Liu, A. Montaser, Nebulizer diagnostics: fundamental parameters, challenges, and techniques on the horizon, J. Anal. At. Spectrom., 13 (1998) 829–842.

[532] I.I. Stewart, J.W. Olesik, Transient acid effects in inductively coupled plasma optical emission spectrometry and inductively coupled plasma mass spectrometry, J. Anal. At. Spectrom., 13 (1998) 843–854.

[533] C. B'Hymer, K.L. Sutton, J.A. Caruso, Comparison of four nebulizer-spray chamber interfaces for the high-performance liquid chromatographic separation of arsenic compounds using inductively coupled plasma mass spectrometric detection, J. Anal. At. Spectrom., 13 (1998) 855–858.

[534] B. Budic, Matrix effects in inductively coupled plasma atomic emission spectrometry using an ultrasonic nebulizer, J. Anal. At. Spectrom., 13 (1998) 869–874.

[535] J. Chirinos, A. Fernandez, J. Franquiz, Multi-element optimization of the operating parameters for inductively coupled plasma atomic emission spectrometry with a charge injection device detector for the analysis of samples dissolved in organic solvents, J. Anal. At. Spectrom., 13 (1998) 995–1000.

[536] K.A. Taylor, B.L. Sharp, D.J. Lewis, H.M. Crews, Design and characterisation of a microconcentric nebuliser interface for capillary electrophoresis-inductively coupled plasma mass spectrometry, J. Anal. At. Spectrom., 13 (1998) 1095–1100.

[537] I.I. Stewart, J.W. Olesik, The effect of nitric acid concentration and nebulizer gas flow rates on aerosol properties and transport rates in inductively coupled plasma sample introduction, J. Anal. At. Spectrom., 13 (1998) 1249–1256.

[538] I.B. Brenner, M. Zischka, B. Maichin, G. Knapp, Ca and Na interference effects in an axially viewed ICP using low and high aerosol loadings, J. Anal. At. Spectrom., 13 (1998) 1257–1264.

[539] C. Dubuisson, E. Poussel, J.M. Mermet, Comparison of ionic-line based internal standardization with axially and radially viewed inductively coupled plasma atomic emission spectrometry to compensate for sodium effets on accuracy, J. Anal. At. Spectrom., 13 (1998) 1265–1269.

[540] I.I. Stewart, J.W. Olesik, Steady state acid effects in ICP-MS, J. Anal. At. Spectrom., 13 (1998) 1313–1320.

[541] S.D. Lofthouse, G.M. Greenway, S.C. Stephen, Comparison of inductively coupled plasma mass spectrometry with a microconcentric nebuliser and total reflection X-ray spectrometry for the analysis of small liquid volume samples, J. Anal. At. Spectrom., 13 (1998) 1333–1335.

[542] A. Tangen, W. Lund, B. Josefsson, H. Borg, Interface for the coupling of capillary electrophoresis and inductively coupled plasma mass spectrometry, J. Chromatogr. A, 826 (1998) 87–94.

[543] S. Akita, M. Rovira, A.M. Sastre, H. Takeuchi, Cloud point extraction of gold(III) with nonionic surfactant—fundamental studies and application to gold recovery from printed substrates, Sep. Sci. Technol., 33 (1998) 2159–2177.

[544] J.W. Olesik, J.A. Kinser, E.J. Grunwald, K.K.T. Thaxton, S.V. Olesik, Gold(III) with nonionic surfactant—fundamental studies and application to gold recovery, Spectrochim. Acta, 53B (1998) 239–251.

[545] C. Dubuisson, E. Poussel, J.L. Todolí, J.M. Mermet, Effect of sodium during the aerosol transport and filtering in inductively coupled plasma atomic emission spectrometry, Spectrochim. Acta, 53B (1998) 593–600.

[546] J.A. Morales, E.H. van Veen, M.T.C. de Loos-Vollebregt, Practical implementation of survey analysis in inductively coupled plama optical emission spectrometry, Spectrochim. Acta, 53B (1998) 683–697.

[547] B.S. Duersch, Y. Chen, A. Ciocan, P.B. Farnsworth, Optical measurements of ion density in the second vacuum stage of an inductively coupled plasma mass spectrometer, Spectrochim. Acta, 53B (1998) 569–579.

[548] J.A. Horner, G.M. Hieftje, Computerized simulation of mixed-solute-particle vaporization in an inductively coupled plasma, Spectrochim. Acta, 53B (1998) 1235–1259.

[549] J. Liu, B. Huang, X. Zeng, Donut-shaped spray chamber for inductively coupled plasma spectrometry, Spectrochim. Acta, 53B (1998) 1469–1474.

[550] J.S. Becker, H.J. Dietze, Inorganic trace analysis by mass spectrometry, Spectrochim. Acta, 53B (1998) 1475–1506.

[551] V.L. Dressler, D. Pobezon, A.J. Curtius, Determination of heavy metals by inductively coupled plasma mass spectrometry after on-line separation and preconcentration, Spectrochim. Acta, 53B (1998) 1527–1539.

[552] A.S. Al-Ammar, R.M. Barnes, Correction for non-spectroscopic matrix effects in inductively coupled plasma-atomic emission spectroscopy by internal standardization using spectral lines of the same analyte, Spectrochim. Acta, 53B (1998) 1583–1593.

[553] A. Montaser, Ed., Inductively Coupled Plasma Mass Spectrometry, Wiley-VCH, New York (1998).

[554] M. Huang, T. Shirasaki, A. Hirabayashi, H. Koizuma, Microliter sample introduction technique for microwave-induced plasma mass spectrometry, Anal. Chem., 71 (1999) 427–432.

[555] J.S. Becker, H.J. Dietze, J.A. McLean, A. Montaser, Ultratrace and isotope analysis of long-lived radionuclides by inductively coupled plasma quadrupole mass spectrometry using direct injection high efficiency nebulizer, Anal. Chem., 71 (1999) 3077–3084.

[556] C. Barbante, G. Cozzi, G. Capodaglio, K. van de Velde, C. Ferrari, A. Veysseyre, C.F. Boutron, G. Scarponi, P. Cescon, Determination of Rh, Pd, and Pt in polar and alpine snow and ice by double-focusing ICPMS with microconcentric nebulizer, Anal. Chem., 71 (1999) 4125–4133.

[557] V.L. Dressler, D. Pozebon, A.J. Curtius, Introduction of alcohols in inductively coupled plasma mass spectrometry by a flow injection system, Anal. Chim. Acta, 379 (1999) 175–183.

[558] O.Y. Akinbo, J.W. Carnahan, The analysis of organic solutions and liquid chromatographic samples with low power helium microwave induced plasma, Anal. Chim. Acta, 390 (1999) 217–226.

[559] B. Wen, X. Shan, S. Xu, Preconcentration of ultratrace rare earth elements in seawater with 8-hydroxyquinoline immobilized polyacrylonitrile hollow fiber membrane for determination by inductively coupled plasma mass spectrometry, Analyst, 124 (1999) 621–626.

[560] A.C. Lazar, P.B. Farnsworth, Matrix effect studies in the inductively couped plasma with monodisperse droplets. Part I: The influence of matrix on the vertical analyte emission profile, Appl. Spectrosc., 53 (1999) 457–464.

[561] A.C. Lazar, P.B. Farnsworth, Matrix effect studies in the inductively couped plasma with monodisperse droplets. Part II: The influence of matrix on spatially integrated ion density, Appl. Spectrosc., 53 (1999) 465–470.

[562] S.A. Baker, N.J. Miller-Ihli, Comparison of a cross-flow and microconcentric nebulizer for chemical speciation measurements using CZE-ICP-MS, Appl. Spectrosc., 53 (1999) 471–478.

[563] Q. Xu, G. Mattu, G.R. Agnes, Influence of droplets with net charge in inductively coupled plasma atomic emission spectrometry and implications for the easily ionizable element chemical matrix effect, Appl. Spectrosc., 53 (1999) 965–973.

[564] A.G. Coedo, T. Dorado, I. Padilla, Evaluation of a desolvating microconcentric nebulizer in inductively coupled plasma mass spectrometry to improve the determination of arsenic in steels, Appl. Spectrosc., 53 (1999) 974–978.

[565] J.A. McLean, R.A. Huff, A. Montaser, Fundamental properties of aerosols produced in helium by a direct injection nebulizer, Appl. Spectrosc., 53 (1999) 1331–1340.

[566] J.S. Lee, H.B. Lim, Application of a membrane desolvator to the analysis of organic solvents in inductively coupled plasma atomic emission spectrometry, Bull. Korean Chem. Soc., 20 (1999) 1040–1044.

[567] D. Schaumlöffel, A. Prange, A new interface for combining capillary electrophoresis with inductively coupled plasma-mass spectrometry, Fresenius J. Anal. Chem., 364 (1999) 452–456.

[568] M. Hamester, D. W Iederin, J. Wills, W. Kerl, C.B. Douthitt, Strategies for isotope ratio measurements with a double focusing sector field ICP-MS, Fresenius J. Anal. Chem., 364 (1999) 495–498.

[569] O. Keil, J. Dahmen, D.A. Volmer, Automated matrix separation and preconcentration for the trace level determination of metal impurities in ultrapure inorganic salts by high-resolution ICP-MS, Fresenius J. Anal. Chem., 364 (1999) 694–699.

[570] S. Maestre, J. Mora, J.L. Todolí, A. Canals, Evaluation of several commercially available spray chambers for use in inductively coupled plasma atomic emission spectrometry, J. Anal. At. Spectrom., 14 (1999) 61–67.

[571] K. Krengel-Rothensee, U. Richter, P. Heitland, Low-level determination of non-metals (Cl, Br, I, S, P) in waste oils by inductively coupled plasma optical emission spectrometry using prominent spectral lines in the 130–190 nm, J. Anal. At. Spectrom., 14 (1999) 699–702.

[572] E.H. van Veen, M.T.C. de Loos-Vollebregt, On the use of line intensity ratios and power adjustments to control matrix effects in inductively coupled plasma optical emission spectrometry, J. Anal. At. Spectrom., 14 (1999) 831–838.

[573] S. Jai Kumar, S. Gangadharan, Determination of trace elements in naphta by inductively coupled plasma mass spectrometry using water-in-oil emulsions, J. Anal. At. Spectrom., 14 (1999) 967–971.

[574] E.H. Evans, S. Chenery, A. Fisher, J. Marshall, K. Sutton, Impact of membrane desolvation on the effects of sodium on response in inductively coupled plasma atomic emission spectrometry with ultrasonic sample introduction, J. Anal. At. Spectrom., 14 (1999) 977–1004.

[575] C. Shuqin, C. Hangtin, Z. Xianjin, Determination of mercury in biological samples using organic compounds as matrix modifiers by inductively coupled plasma mass spectrometry, J. Anal. At. Spectrom., 14 (1999) 1183–1186.

[576] J.L. Todolí, V. Hernandis, A. Canals, J.M. Mermet, Comparison of characteristics and limits of detection of pneumatic micronebulizers and a conventional nebulizer operating at low uptake rates in ICP-AES, J. Anal. At. Spectrom., 14 (1999) 1289–1295.

[577] A. Prange, D. Schaumlöffel, Determination of element species at trace levels using capillary electrophoresis-inductively coupled plasma sector field mass spectrometry, J. Anal. At. Spectrom., 14 (1999) 1329–1332.

[578] M.P. Field, J.T. Cullen, R.M. Sherrell, Direct determination of 10 trace metals in 50 μL samples of coastal seawater using desolvating micronebulization sector field ICP-MS, J. Anal. At. Spectrom., 14 (1999) 1425–1431.

[579] C. Barbante, G. Cozzi, G. Capodaglio, K. van de Velde, C. Ferrari, C. Boutron, P. Cescon, Trace element determination in alpine snow and ice double focusing inductively coupled plasma mass spectrometry with microconcentric nebulization, J. Anal. At. Spectrom., 14 (1999) 1433–1438.

[580] J.S. Becker, H.J. Dietze, Application of double-focusing sector field ICP mass spectrometry with shielded torch using different nebulizers for ultratrace and precise isotope analysis of long-lived radionuclides, J. Anal. At. Spectrom., 14 (1999) 1493–1500.

[581] S.D. Lofthouse, G.M. Greenway, S.C. Stephen, Miniaturisation of a matrix separation/ preconcentration procedure for inductively coupled plasma mass spectrometry using 8-hydroxyquinoline immobilised on a microporous silica frit, J. Anal. At. Spectrom., 14 (1999) 1839–1842.

[582] V. Majidi, J. Qvarnström, Q. Tu, W. Frech, Y. Thomassen, Improving sensitivity for CE-ICP-MS using multicapillary parallel separation, J. Anal. At. Spectrom., 14 (1999) 1933–1935.

[583] I.I. Stewart, J.W. Olesik, Time-resolved measurements with single droplet introduction to investigate space-charge effects in plasma mass spectrometry, J. Am. Soc. Mass Spectrom., 10 (1999) 159–174.

[584] I.B. Brenner, A. Le Marchand, C. Deraed, L. Chauvet, Compensation of Ca and Na interference effects in axially and radially viewed ICPs, Microchem. J., 63 (1999) 344–353.

[585] L. Gras, J. Mora, J.L. Todolí, A. Canals, V. Hernandis, Microwave desolvation for acid sample introduction in inductively coupled plasma atomic emission spectrometry, Spectrochim. Acta, 54B (1999) 469–480.

[586] P. Masson, Matrix effect during trace element analysis in plant samples by inductively coupled plasma atomic emission spectrometry with axial view configuration and pneumatic nebulizer, Spectrochim. Acta, 54B (1999) 603–612.

[587] J.L. Todolí, J.M. Mermet, Acid interferences in atomic spectrometry: analyte signal effects and subsequent reduction, Spectrochim. Acta, 54B (1999) 895–929.

[588] L. Gras, J. Mora, J.L. Todolí, V. Hernandis, A. Canals, Desolvation of acid solutions in inductively coupled plasma atomic emission spectrometry by infrared radiation. Comparison with a system based on microwave radiation, Spectrochim. Acta, 54B (1999) 1321–1333.

[589] A.S. Al-Ammar, E. Reiznerová, R.M. Barnes, Feasibility of using beryllium as internal reference to reduce non-spectroscopic carbon species matrix effect in the inductively coupled plasma-mass spectrometry (ICP-MS) determination of boron in biological samples, Spectrochim. Acta, 54B (1999) 1813–1820.

[590] J. Singh, D.E. Pritchard, D.L. Carlisle, J.A. Mclean, A. Montaser, J.M. Orenstein, S.R. Patierno, Internalization of carcinogenic lead chromate particles by cultured normal human lung epithelial cells: Formation of intracellular lead-inclusion bodies and induction of apoptosis, Toxicol. Appl. Pharmacol., 161 (1999) 240–248.

[591] E. Debrah, G. Légère, Design and performance of a novel high efficiency sample introduction system for plasma source spectrometry, Plasma Source Mass Spectrometry, J.G. Holland, S.D. Tanner, Eds., Royal Society of Chemistry, Cambrigde (1999) 20–26.

[592] H.S. Tan, Pneumatic concentric nebulizer with an adjustable capillary, US Patent, 5884846 (1999).

[593] B.W. Acon, J.A. McLean, A. Montaser, A large bore-direct injection high efficiency nebulizer for inductively coupled plasma spectrometry, Anal. Chem., 72 (2000) 1885–1893.

[594] N. Praphairaksit, R.S. Houk, Attenuation of matrix effects in inductively coupled plasma mass spectrometry with a supplemental electron source inside the skimmer, Anal. Chem., 72 (2000) 2351–2355.

[595] N. Praphairaksit, R.S. Houk, Reduction of space charge effects in inductively coupled plasma mass spectrometry using a supplemental electron source inside the skimmer: ion transmission and mass spectral characteristics, Anal. Chem., 72 (2000) 2356–2361.

[596] M. Huang, A. Hirabayashi, T. Shirazaki, H. Koizumi, A multimicrospray nebulizer for microwave-induced plasma mass spectrometry, Anal. Chem., 72 (2000) 2463–2467.

[597] N. Praphairaksit, R.S. Houk, Reduction of mass bias and matrix effects in inductively coupled plasma mass spectrometry with a supplemental electron source in a negative extraction lens, Anal. Chem., 72 (2000) 4435–4440.

[598] R. Ma, D. Bellis, C.W. McLeod, Isotopic analysis of uranium in tree bark by ICP mass spectrometry: a strategy for assessment of airborne contamination, Anal. Chem., 72 (2000) 4878–4881.

[599] A.R. Elwaer, C.W. McLeod, K.C. Thompson, On-line separation and determination of bromate in drinking waters using flow injection ICP mass spectrometry, Anal. Chem., 72 (2000) 5725–5730.

[600] M. Huang, H. Kojima, T. Shirasaki, A. Hirabayashi, H. Koisumi, Study on solvent loading effect on inductively coupled plasma and microwave-induced plasma sources with a microliter nebulizer, Anal. Chim. Acta, 413 (2000) 217–222.

[601] F.R. Lauritzen, M.A. Mendes, T. Aggerholm, Direct detection of large fat-soluble biomolecules in solution using membrane inlet mass spectrometry and desorption chemical ionization, Analyst, 125 (2000) 211–215.

[602] Q. Tu, J. Qvarnstrom, W. Frech, Determination of mercury species by capillary zone electrophoresis – inductively coupled plasma mass spectrometry: a comparison of two spray chamber- nebulizer combinations, Analyst, 125 (2000) 705–710.

[603] M. Haldimann, B. Zimmerli, A. Eastgate, Improved measurement of iodine in food samples using inductively coupled plasma isotope dilution mass spectrometry, Analyst, 125 (2000) 1977–1982.

[604] J. Mora, J.L. Todolí, F.J. Sempere, A. Canals, V. Hernandis, Determination of metals in lubricating oils by flame atomic absorption spectrometry using a single-bore high-pressure pneumatic nebulizer, Analyst, 125 (2000) 2344–2349.

[605] Q. Xu, G.R. Agnes, Use of laser light scatter signals to study the effect of a direct-current biased mesh screen in a spray chamber on aerosols generated for use in atomic spectroscopy, Appl. Spectrosc., 54 (2000) 94–98.

[606] I.I. Stewart, C.E. Hensman, J.W. Olesik, Influence of gas sampling on analyte transport within the ICP and ion sampling for ICP-MS studied using individual isolated sample droplets, Appl. Spectrosc., 54 (2000) 164–174.

[607] J.A. McLean, B.W. Acon, A. Montaser, J. Singh, D.E. Pritchard, S.R. Patierno, Determination of chromium in human lung fibroblast cells using a large bore-direct injection high-efficiency nebulizer with inductively coupled plasma mass spectrometry, Appl. Spectrosc., 54 (2000) 659–663.

[608] M.G. Minnich, A. Montaser, Direct injection high efficiency nebulization in inductively coupled plasma mass spectrometry under cool and normal plasma conditions, Appl. Spectrosc., 54 (2000) 1261–1269.

[609] J.A. Horner, P. Yang, G.M. Hieftje, Effect of operating frequency on spatial features of the inductively coupled plasma, Appl. Spectrosc., 54 (2000) 1824–1830.

[610] C. Prohaska, K. Pomazal, I. Steffan, Determination of Ca, Mg, Fe, Cu, and Zn in blood fractions and whole blood of humans by ICP-OES, Fresenius J. Anal. Chem., 367 (2000) 479–484.

[611] R. Nehm, J.A.C. Broekaert, Noise power spectra and recovery rates obtained with different nebulizer systems in ICP atomic emission spectrometric analyses in the case of different types of salts and salt contents, Fresenius J. Anal. Chem., 368 (2000) 156–161.

[612] B. Budic, Matrix interferences in the determination of trace elements in waste waters by inductively coupled plasma atomic emission spectrometry with ultrasonic nebulization, Fresenius J. Anal. Chem., 368 (2000) 371–377.

[613] J.L. Todolí, S. Maestre, J. Mora, A. Canals, V. Hernandis, Comparison of several spray chambers operating at very low flow rates in inductively coupled plasma atomic emission spectrometry, Fresenius J. Anal. Chem., 368 (2000) 773–779.

[614] M. Grotti, E. Magi, R. Frache, Multivariate investigation of matrix effects in inductively coupled plasma atomic emission spectrometry using pneumatic or ultrasonic nebulization, J. Anal. At. Spectrom., 15 (2000) 89–95.

[615] C.S. Kim, C.K. Kim, J.I. Lee, K.J. Lee, Rapid determination of Pu isotopes and atom ratios in small amounts of environmental samples by an on-line sample pre-treatment system and isotope dilution high resolution inductively coupled plasma mass spectrometry, J. Anal. At. Spectrom., 15 (2000) 247–255.

[616] L. Ding, F. Liang, Y. Huan, Y. Cao, H. Zhang, G. Jin, A low-powered microwave thermo-spray nebulizer for inductively coupled plasma atomic emission spectrometry, J. Anal. At. Spectrom., 15 (2000) 293–296.

[617] T. Prohaska, G. Köllensperger, M. Krachler, K. De Winne, G. Stingeder, L. Moens, Determination of trace elements in human milk by inductively coupled plasma sector field mass spectrometry (ICP-SFMS), J. Anal. At. Spectrom., 15 (2000) 335–340.

[618] P.K. Appelblad, I. Rodushkin, D.C. Baxter, The use of PT guard electrode in inductively coupled plasma sector field mass spectrometry: advantages and limitations, J. Anal. At. Spectrom., 15 (2000) 359–364.

[619] B. Gammelgaard, O. Jons, Comparison of an ultrasonic nebulizer with a cross-flow nebulizer for selenium speciation by ion-chromatography and inductively coupled plasma mass spectrometry, J. Anal. At. Spectrom., 15 (2000) 499–505.

[620] J.W. Tromp, R.T. Tremblay, J.M. Mermet, E.D. Salin, Matrix interference diagnostics for the automation of inductively coupled plasma mass spectrometry (ICP-MS), J. Anal. At. Spectrom., 15 (2000) 617–625.

[621] S. Hann, T. Prohaska, G. Köllensperger, C. Latkoczy, G. Stingeder, Separation of spectral and non-spectral interferences by on-line high performance ion chromatography inductively coupled plasma sector filed mass spectrometry (HPIC-ICP-SFMS) for accurate determination of ^{234}U, ^{235}U and ^{232}Th in industrial ores, J. Anal. At. Spectrom., 15 (2000) 721–725.

[622] R.S. Olofsson, I. Rodushkin, M.D. Axelsson, Performance characteristics of a tandem spray chamber arrangement in double focusing sector field ICP-MS, J. Anal. At. Spectrom., 15 (2000) 727–729.

[623] J.L. Todolí, J.M. Mermet, Effect of the spray chamber design on steady and transient acid interferences in inductively coupled plasma atomic emission spectrometry, J. Anal. At. Spectrom., 15 (2000) 863–867.

[624] M. Villaneuva, M. Catasús, E.D. Salin, M. Pomares, Study of mixed-matrix effects induced by Ca and Mg in ICP-AES, J. Anal. At. Spectrom., 15 (2000) 877–881.

[625] I. Rodushhkin, F. Ödman, R. Olofsson, M.D. Axelsson, Determination of 60 elements in whole blood by sector field inductively coupled plasma mass spectrometry, J. Anal. At. Spectrom., 15 (2000) 937–944.

[626] M.L. Griffiths, D. Svozil, P.J. Worsfold, S. Denham, E.H. Evans, Comparison of traditional and multivariate calibration techniques applied to complex matrices using inductively coupled plasma atomic emission spectrometry, J. Anal. At. Spectrom., 15 (2000) 967–972.

[627] J.M. Costa-Fernández, N.H. Bings, A.M. Leach, G.M. Hieftje, Rapid simultaneous multielemental speciation by capillary electrophoresis coupled to inductively coupled plasma time-of-flight mass spectrometry, J. Anal. At. Spectrom., 15 (2000) 1063–1068.

[628] K.L. Ackley, K.L. Sutton, J.A. Caruso, A comparison of nebulizers for microbore LC-ICP-MS with mobile phases containing methanol, J. Anal. At. Spectrom., 15 (2000) 1069–1073.

[629] M. Catasus, J.L. Todolí, L. Gras, V. Hernandis, Selection of the operating conditions for overcoming acids and sodium chloride non-spectroscopic interferences in inductively coupled plasma atomic emission spectrometry: effect of the liquid-to-vapor mass ratios, J. Anal. At. Spectrom., 15 (2000) 1203–1206.

[630] R. Wang, R.L. McLaughlin, I.D. Brindle, A concentric capillary nebulizer (CCN) for inductively coupled plasma atomic emission spectrometry, J. Anal. At. Spectrom., 15 (2000) 1303–1312.

[631] J.A. Day, J.A. Caruso, J.S. Becker, H.J. Dietze, Application of capillary electrophoresis interfaced to double focusing sector field ICP-MS for nuclide abundance determination of lanthanides produced via spallation reactions in an irradiated tantalum target, J. Anal. At. Spectrom., 15 (2000) 1343–1348.

[632] T. Hirata, Development of a flushing spray chamber for inductively coupled plasma-mass spectrometry, J. Anal. At. Spectrom., 15 (2000) 1447–1450.

[633] M. Rehkämper, K. Mezger, Investigation of matrix effects for Pb isotope ratio measurements by multiple collector ICP-MS: verification and application of optimized analytical protocols, J. Anal. At. Spectrom., 15 (2000) 1451–1460.

[634] M.E. Ketterer, D.D. Hudson, Preliminary characterization of a laboratory-constructed, easily disassembled concentric pneumatic nebulizer for inductively coupled plasma mass spectrometry, J. Anal. At. Spectrom., 15 (2000) 1574–1577.

[635] A. Tangen, W. Lund, Capillary electrophoresis–inductively coupled plasma mass spectrometry interface with minimised dead volume for high separation efficiency, J. Chromatog. A, 891 (2000) 129–138.

[636] Y. Sung, H.B. Lim, Double membrane desolvator for direct analysis of isopropyl alcohol in inductively coupled plasma atomic emission spectrometry (ICP-AES) and inductively coupled plasma mass spectrometry (ICP-MS), Microchem. J., 64 (2000) 51–57.

[637] M. Villanueva, M. Pomares, M. Catasus, J. Diaz, Application of factorial designs for the description and correction of combined matrix effects in ICP-AES, Quim. Anal., 19 (2000) 39–42.

[638] D.S. de Jesus, M. das Graças Korn, S.L.C. Ferreira, M.S. Carvalho, A separation method to overcome the interference of aluminium on zinc determination by inductively coupled plasma atomic emission spectroscopy, Spectrochim. Acta, 55B (2000) 389–394.

[639] N. Praphairaksit, D.R. Wiederin, R.S. Houk, An externally air-cooled low-flow torch for inductively coupled plasma mass spectrometry, Spectrochim. Acta, 55B (2000) 1279–1293.

[640] Y.C. Sun, C.H. Hsieh, T.S. Lin, J.C. Wen, Determination of trace impurities in high purity gold by inductively coupled plasma mass spectrometry with prior removal by electrodeposition, Spectrochim. Acta, 55B (2000) 1481–1489.

[641] M.M. Fraser, D. Beauchemin, Effect of concomitant elements on the distribution of ions in inductively coupled plasma-mass spectrometry. Part I. Elemental ions, Spectrochim. Acta, 55B (2000) 1705–1731.

[642] J. Li, T. Umemura, T. Odake, Kin-ichi Tsunoda, A high-efficiency cross-flow micronebulizer for inductively coupled plasma mass spectrometry, Anal. Chem., 73 (2001) 1416–1424.

[643] A.D. Anbar, K.A. Knab, J. Barling, Precise determination of mass-dependent variations in the isotopic composition of molybdenum using MC-ICPMS, Anal. Chem., 73 (2001) 1425–1431.

[644] A. Makishima, E. Nakamura, Determination of total sulfur at microgram per gram levels in geological materials by oxidation of sulfur into sulfate with in situ generation of bromine using isotope dilution high-resolution ICPMS, Anal. Chem., 73 (2001) 2547–2553.

[645] J. Li, T. Umemura, T. Odake, K. Tsunoda, A high-efficiency cross-flow micronebulizer interface for capillary electrophoresis and inductively coupled plasma mass spectrometry, Anal. Chem., 73 (2001) 5992–5999.

[646] X. Dai, C. Koeberl, H. Fröschl, Determination of platinum group elements in impact breccias using neutron activation analysis and ultrasonic nebulization inductively coupled plasma mass spectrometry after anion exchange preconcentration, Anal. Chim. Acta, 436 (2001) 79–85.

[647] S. Nakai, S. Fukuda, S. Nakada, Thorium isotopic measurements on silicate rock samples with a multi-collector inductively coupled plasma mass spectrometer, Analyst, 126 (2001) 1707–1710.

[648] J.A. McLean, J.S. Becker, S.F. Boulyga, H.J. Dietze, A. Montaser, Ultratrace and isotopic analysis of long-lived radionuclides by double-focusing sector field inductively coupled plasma mass spectrometry using direct liquid sample introduction, Int. J. Mass Spectrom., 208 (2001) 193–204.

[649] E. Björn, W. Frech, Non-spectral interference effects in inductively coupled plasma mass spectrometry using direct injection high efficiency and microconcentric nebulisation, J. Anal. At. Spectrom., 16 (2001) 4–11.

[650] L. Bendahl, B. Gammelgaard, O. Jøns, O. Farver, S.H. Hansen, Interfacing capillary electrophoresis with inductively coupled plasma mass spectrometry by direct injection nebulization for selenium speciation, J. Anal. At. Spectrom., 16 (2001) 38–42.

[651] R. Garciá, M. Gomez, M.A. Palacios, J. Bettmer, C. Camara, On-line removal of mass interferences in palladium determination by ICP-MS using modified capillaries coupled to micro-flow nebulizers, J. Anal. At. Spectrom., 16 (2001) 481–486.

[652] J.L. Todolí, J.M. Mermet, Evaluation of a direct injection high-efficiency nebulizer (DIHEN) by comparison with a high-efficiency nebulizer (HEN) coupled to a cyclonic spray chamber as a liquid sample introduction system for ICP-AES, J. Anal. At. Spectrom., 16 (2001) 514–520.

[653] K. Polec, J. Szpunar, O. Palacios, P. González-Duarte, S. Atrian, R. Lobinski, Investigation of metal binding by recombinant and native metallothioneins by capillary zone electrophoresis (CZE) coupled with inductively coupled plasma mass spectrometry (ICP-MS) via a self-aspirating total consumption burner, J. Anal. At. Spectrom., 16 (2001) 567–574.

[654] Q. Xu, D. Balik, G.R. Agnes, Aerosol static electrification and its effects in inductively coupled plasma spectroscopy, J. Anal. At. Spectrom., 16 (2001) 715–723.

[655] B.W. Acon, J.A. McLean, A. Montaser, A direct injection high efficiency nebulizer interface for microbore high-performance liquid chromatography-inductively coupled plasma mass spectrometry, J. Anal. At. Spectrom., 16 (2001) 852–857.

[656] R. Garcia-Sanchez, R. Feldhaus, J. Bettmer, L. Ebdon, Lead speciation in rainwater samples by modified fused silica capillaries coupled to a direct injection nebulizer (DIN) for sample introduction in ICP-MS, J. Anal. At. Spectrom., 16 (2001) 1028–1034.

[657] H.P. Longerich, W. Diegor, Evalutation of a water-jacketed spray chamber using time-of-flight inductively coupled plasma mass spectrometry (TOF-ICP-MS), J. Anal. At. Spectrom., 16 (2001) 1196–1201.

[658] J. Wang, E.H. Hansen, Interfacing sequential injection on-line preconcentration using a renewable micro-column incorporated in a lab-on-valve system with direct injection nebulization inductively coupled plasma mass spectrometry, J. Anal. At. Spectrom., 16 (2001) 1349–1355.

[659] A.P. Packer, M.F. Giné, Analysis of undigested honey samples by isotope dilution inductively coupled plasma mass spectrometry with direct injection nebulization (ID-ICP-MS), Spectrochim. Acta, 56B (2001) 69–75.

[660] J. Dennaud, A. Howes, E. Poussel, J.M. Mermet, Study of ionic-to-atomic line intensity ratios for two axial viewing-based inductively coupled plasma atomic emission spectrometers, Spectrochim. Acta, 56B (2001) 101–112.

[661] E. Vassileva, M. Hoenig, Determination of arsenic in plant samples by inductively coupled plasma atomic emission spectrometry with ultrasonic nebulization: a complex problem, Spectrochim. Acta, 56B (2001) 223–232.

[662] M. Stepan, P. Musil, E. Poussel, J.M. Mermet, Matrix-induced shift effects in axially viewed inductively coupled plasma atomic emission spectrometry, Spectrochim. Acta, 56B (2001) 443–453.

[663] M.A.M. da Silva, V.L.A. Frescura, A.J. Curtius, Determination of trace elements in water samples by ultrasonic nebulization inductively coupled plasma mass spectrometry after cloud point extraction, Spectrochim. Acta, 55B (2000) 803–813.

[664] C.M. Benson, S.F. Gimelshein, D.A. Levin, A. Montaser, Simulation of droplet heating and desolvation in an inductively coupled plasma—Part I, Spectrochim. Acta, 56B (2001) 1097–1112.

[665] M.G. Minnich, J.A. McLean, A. Montaser, Spatial aerosol characteristics of a direct injection high efficiency nebulizer via optical patternation, Spectrochim. Acta, 56B (2001) 1113–1126.

[666] J.W. Olesik, C. Hensman, S. Rabb, D. Rago, Sample introduction, plasma—sample interactions, ion transport and ion—molecule reactions: fundamental understanding and practical improvements in ICP-MS, Plasma Source Mass Spectrometry, G. Holland, S.D. Tanner, Eds, Royal Society of Chemistry, Cambridge (2001) 3–16.

[667] Y.C. Sun, C.H. Hsieh, T.S. Lin, J.C. Wen, Determination of trace impurities in high purity gold by inductively coupled plasma mass spectrometry with prior removal by electrodeposition, Spectrochim. Acta, 55B (2000) 1481–1489.

[668] J.R. Chirinos, K. Kahen, S.A.E. O'Brien, A. Montaser, Mixed-gas inductively coupled plasma atomic emission spectrometry using a direct injection nebulizer, Anal. Bioanal. Chem., 372 (2002) 128–135.

[669] P. Krystek, R. Ritsema, Determination of uranium in urine-measurement of isotope ratios and quantification by use of inductively coupled plasma mass spectrometry, Anal. Bioanal. Chem., 374 (2002) 226–229.

[670] D.R. Bandura, V.I. Baranov, S.D. Tanner, Detection of ultratrace phosphorous and sulphur by quadrupole ICPMS with dynamic reaction cell, Anal. Chem., 74 (2002) 1497–1502.

[671] C.S. Kim, C.K. Kim, K.J. Lee, Determination of Pu isotopes in seawater by an on-line sequential injection technique with sector field inductively coupled plasma mass spectrometry, Anal. Chem., 74 (2002) 3824–3832.

[672] D. Beauchemin, K. Kyser, D. Chipley, Inductively coupled plasma mass spectrometry with on-line leaching: a method to assess the mobility and fractionation of elements, Anal. Chem., 74 (2002) 3924–3928.

[673] E. Svantesson, J. Capala, K.E. Markides, J. Petterson, Determination of boron-containing compounds in urine and blood plasma from boron neutron capture therapy patients. The Importance of using coupled techniques, Anal. Chem., 74 (2002) 5358–5360.

[674] C.M. Petibon, H.P. Longerich, I. Horn, M.N. Tubrett, Neon inductively coupled plasma for laser ablation-inductively coupled plasma mass spectrometry, Appl. Spectrosc., 56 (2002) 658–664.

[675] S.A.E. O'Brien, J.A. McLean, B.W. Acon, B.J. Eshelman, W.F. Bauer, A. Montaser, Determination of memory-prone elements using direct injection high efficiency nebulizer inductively coupled plasma mass spectrometry, Appl. Spectrosc., 56 (2002) 1006–1012.

[676] S. Mounicou, S. McSheehy, J. Szpunar, M. Potin-Gautier, R. Lobinski, Analysis of selenized yeast for selenium speciation by size-exclusion chromatography and capillary zone electrophoresis with inductively coupled plasma mass spectrometric detection (SEC-ICP-MS), J. Anal. At. Spectrom., 17 (2002) 15–20.

[677] M. Wind, A. Eisenmenger, W.D. Lehmann, Modified direct injection high efficiency nebulizer with minimized dead volume for the analysis of biological samples by micro- and nano-LC-ICP-MS, J. Anal. At. Spectrom., 17 (2002) 21–26.

[678] J.M. Cano, J.L. Todolí, V. Hernandis, J. Mora, The role of the nebulizer on the sodium interferent effects in inductively coupled plasma atomic emission spectrometry, J. Anal. At. Spectrom., 17 (2002) 57–63.

[679] J.L. Todolí, L. Gras, V. Hernandis, J. Mora, Elemental matrix effects in ICP-AES, J. Anal. At. Spectrom., 17 (2002) 142–169.

[680] J.L. Todolí, J.M. Mermet, Influence of the spray chamber design for vapor-based liquid sample introduction at room temperature in ICP-AES, J. Anal. At. Spectrom., 17 (2002) 211–218.

[681] A. Canals, L. Gras, H. Contreras, Elimination of nitric acid interference in ICP-AES by using a cyclonic spray chamber/Nafion membrane-based desolvation system, J. Anal. At. Spectrom., 17 (2002) 219–226.

[682] M. Edlund, H. Visser, P. Heitland, Analysis of biodiesel by argon–oxygen mixed-gas inductively coupled plasma optical emission spectrometry, J. Anal. At. Spectrom., 17 (2002) 232–235.

[683] R. Kucharkowski, C. Vogt, Simultaneous ICP atomic emission spectrometry for accurate stoichiometric determination: application to a YNi_2B_2C superconducting material system, J. Anal. At. Spectrom., 17 (2002) 263–269.

[684] Y. Okamoto, F. Nakata, Y. Obata, T. Takahashi, T. Fujiwara, M. Yamamoto, Optimisation of on-line liquid–liquid extraction for beryllium determination by inductively coupled plasma atomic emission spectrometry (ICP-AES), J. Anal. At. Spectrom., 17 (2002) 277–279.

[685] G. Schaldach, L. Berger, I. Razilov, H. Berndt, Characterization of a cyclone spray chamber for ICP spectrometry by computer simulation, J. Anal. At. Spectrom., 17 (2002) 334–344.

[686] J.L. Todolí, J.M. Mermet, New torch design with an in-built chamber for liquid sample analysis by ICP-AES, J. Anal. At. Spectrom., 17 (2002) 345–351.

[687] L. Gras, M.L. Alvarez, A. Canals, Evaluation of new models for drop size distribution prediction of aerosols in atomic spectrometry: pneumatic nebulizers, J. Anal. At. Spectrom., 17 (2002) 524–529.

[688] B. Gammelgaard, L. Bendahl, U. Sidenius, O. Jøns, Selenium speciation in urine by ion-pairing chromatography with perfluorinated carboxylic acids and ICP-MS detection, J. Anal. At. Spectrom., 17 (2002) 570–575.

[689] C.S. Westphal, J.A. McLean, B.W. Acon, L.A. Allen, A. Montaser, Axial inductively coupled plasma time-of-flight mass spectrometry using direct liquid sample introduction, J. Anal. At. Spectrom., 17 (2002) 669–675.

[690] J.L. Todolí, J.M. Mermet, Study of matrix effects using an adjustable chamber volume in a torch-integrated sample introduction system (TISIS) in ICP-AES, J. Anal. At. Spectrom., 17 (2002) 913–921.

[691] S.F. Boulyga, J.L. Matusevich, V.P. Mironov, V.P. Kudrjashov, L. Halicz, I. Segal, J.A. McLean, A. Montaser, J. Sabine Becker, Determination of $^{236}U/^{238}U$ isotope ratio in contaminated environmental samples using different ICP-MS instruments, J. Anal. At. Spectrom., 17 (2002) 958–964.

[692] G. Koellensperger, J. Nurmi, S. Hann, G. Stingeder, W.J. Fitz, W.W. Wenzel, CE-ICP-SFMS and HPIC-ICP-SFMS for arsenic speciation in soil solution and soil water extracts, J. Anal. At. Spectrom., 17 (2002) 1042–1047.

[693] S.F. Boulyga, J.S. Becker, Isotopic analysis of uranium and plutonium using ICP-MS and estimation of burn-up of spent uranium in contaminated environmental samples, J. Anal. At. Spectrom., 17 (2002) 1143–1147.

[694] E. Björn, T. Jonsson, D. Goitom, Noise characteristics and analytical precision of a direct injection high efficiency and micro concentric nebuliser for sample introduction in inductively coupled plasma mass spectrometry, J. Anal. At. Spectrom., 17 (2002) 1257–1263.

[695] J. Wang, E.H. Hansen, Coupling sequential injection on-line preconcentration using a PTFE beads packed column to direct injection nebulization inductively coupled plasma mass spectrometry, J. Anal. At. Spectrom., 17 (2002) 1278–1283.

[696] E. Björn, T. Jonsson, D. Goitom, The origin of peristaltic pump interference noise harmonics in inductively coupled plasma mass spectrometry, J. Anal. At. Spectrom., 17 (2002) 1390–1393.

[697] T.T. Hoang, S.W. May, R.F. Browner, Developments with the oscillating capillary nebulizer-effects of spray chamber design, droplet size and turbulence on analytical signals and analyte transport efficiency of selected biochemically important organoselenium compounds, J. Anal. At. Spectrom., 17 (2002) 1575–1581.

[698] J.A. Horner, S.A. Lehn, G.M. Hieftje, Computerized simulation of aerosol-droplet desolvation in an inductively coupled plasma, Spectrochim. Acta, 57B (2002) 1025–1042.

[699] G. Schaldach, L. Berger, I. Razilov, H. Berndt, Characterization of a double-pass spray chamber for ICP spectrometry by computer simulation (CFD), Spectrochim. Acta, 57B (2002) 1505–1520.

[700] S. Maestre, J. Mora, J.-L. Todolí, Studies about the origin of the non-spectroscopic interferences caused by sodium and calcium in inductively coupled plasma atomic emission spectrometry. Influence of the spray chamber design, Spectrochim. Acta, 57B (2002) 1753–1770.

[701] C. Duyck, N. Miekeley, C.L. Porto da Silveira, P. Szatmari, Trace element determination in crude oil and its fractions by inductively coupled plasma mass spectrometry using ultrasonic nebulization of toluene solutions, Spectrochim. Acta, 57B (2002) 1979–1990.

[702] J. Nölte, ICP Emission Spectrometry. A practical guide, Wiley-VCH, Weinheim (2002).

[703] S.E. Maestre, Studies about the role of the spray chamber on the signal and matrix effects caused by inorganic species in ICP-AES, Ph. D. Thesis, University of Alicante (2002).

[704] H.A. Contreras, Evaluation of aerosol desolvation systems based on the use of NafionTM membranes for ICP-AES, Ph. D. Thesis, University of Alicante (2002).

[705] E. Björn, W. Frech, Introduction of high carbon content solvents into inductively coupled plasma mass spectrometry by a direct injection high efficiency nebulizer, Anal. Bioanal. Chem., 377 (2003) 274–278.

[706] D. Pröfock, P. Leonhard, A. Prange, Determination of sulfur and selected trace elements in metallothionein-like proteins using capillary electrophoresis hyphenated to inductively coupled plasma mass spectrometry with an octopole reaction cell, Anal. Bioanal. Chem., 377 (2003) 132–139.

[707] Y.-L. Feng, R.E. Sturgeon, J.W. Lam, Generation of atomic and molecular cadmium species from aqueous media, Anal. Chem., 75 (2003) 635–640.

[708] A.J. Baca, A.B. de la Ree, F. Zhou, A.Z. Mason, Anodic stripping voltammetry combined on-line with inductively coupled plasma-MS via a direct-injection high-efficiency nebulizer, Anal. Chem., 75 (2003) 2507–2511.

[709] B. Kuczewski, C.M. Marquardt, A. Seibert, H. Geckeis, J.Volker Kratz, N. Trautmann, Separation of plutonium and neptunium species by capillary electrophoresis-inductively coupled plasma-mass spectrometry and application to natural groundwater samples, Anal. Chem., 75 (2003) 6769–6774.

[710] D. Schaumlöffel, J.R. Encinar, R. Lobinski, Development of a sheathless interface between reverse-phase capillary HPLC and ICPMS via a microflow total consumption nebulizer for selenopeptide mapping, Anal. Chem., 75 (2003) 6837–6842.

[711] H. Matsuura, T. Hasegawa, H. Nagata, K. Takatani, M. Asano, A. Itoh, H. Haraguchi, Speciation of small molecules and inorganic ions in salmon egg cell cytoplasm by surfactant-mediated HPLC/ICP-MS, Anal. Sci., 19 (2003) 117–121.

[712] H. Isoyama, T. Uchida, T. Nagashima, O. Ohira, Modified Babington nebulizer for inductively coupled plasma atomic emission spectrometry, Anal. Sci., 19 (2003) 593–597.

[713] C. Wolf, D. Schaumloffel, A.N. Richarz, A. Prange, P. Batter, CZE-ICP-MS separation of metallothioneins in human brain cytosols: comparability of electropherograms obtained from different sample matrices, Analyst, 128 (2003) 576–580.

[714] P. Deschamps, R. Doucelance, B. Ghaleb, J.-L. Michelot, Further investigations on optimized tail correction and high-precision measurement of uranium isotopic ratios using multi-collector ICP-MS, Chem. Geol., 201 (2003) 141–160.

[715] V. Huxter, J. Hamier, E.D. Salin, Tandem calibration methodology: dual nebulizer sample introduction for ICP-MS, J. Anal. At. Spectrom., 18 (2003) 71–75.

[716] B. Langlois, J.-L. Dautheribes, J.-M. Mermet, Comparison of a direct injection nebulizer and a micronebulizer associated with a spray chamber for the determination of iodine in the form of volatile CH_3I by inductively coupled plasma sector field mass spectrometry, J. Anal. At. Spectrom., 18 (2003) 76–79.

[717] D. Cardinal, L.Y. Alleman, J. de Jong, K. Ziegler, L. André, Isotopic composition of silicon measured by multicollector plasma source mass spectrometry in dry plasma mode, J. Anal. At. Spectrom., 18 (2003) 213–218.

[718] S.E. O'Brien, B.W. Acon, S.F. Boulyga, J.S. Becker, H.-J. Dietze, A. Montaser, Reduction of molecular ion interferences with hexapole collision cell in direct injection nebulization–inductively coupled plasma mass spectrometry, J. Anal. At. Spectrom., 18 (2003) 230–238.

[719] M.P. Field, R.M. Sherrell, Direct determination of ultra-trace levels of metals in fresh water using desolvating micronebulization and HR-ICP-MS: application to Lake Superior waters, J. Anal. At. Spectrom., 18 (2003) 254–259.

[720] M. Grotti, E. Magi, R. Leardi, Selection of internal standards in inductively coupled plasma atomic emission spectrometry by principal component analysis, J. Anal. At. Spectrom., 18 (2003) 274–281.

[721] G. Álvarez-Llamas, M. del Rosario Fernández de la Campa, A. Sanz-Medel, Sample stacking capillary electrophoresis with ICP-(Q)MS detection for Cd, Cu and Zn speciation in fish liver metallothioneins, J. Anal. At. Spectrom., 18 (2003) 460–466.

[722] H.R. Hansen, A. Raab, J. Feldmann, New arsenosugar metabolite determined in urine by parallel use of HPLC-ICP-MS and HPLC-ESI-MS, J. Anal. At. Spectrom., 18 (2003) 474–479.

[723] D. Pröfrock, P. Leonhard, A. Prange, Determination of phosphorus in phosphorylated deoxyribonucleotides using capillary electrophoresis and high performance liquid chromatography hyphenated to inductively coupled plasma mass spectrometry with an octopole reaction cell, J. Anal. At. Spectrom., 18 (2003) 708–713.

[724] G. Schaldach, H. Berndt, B.L. Sharp, An application of computational fluid dynamics (CFD) to the characterisation and optimisation of a cyclonic spray chamber for ICP-AES, J. Anal. At. Spectrom., 18 (2003) 742–750.

[725] B.G. Ting, R.S. Pappas, D.C. Paschal, Rapid analysis for plutonium-239 in urine by magnetic sector inductively coupled plasma-mass spectrometry using Aridus desolvation introduction system, J. Anal. At. Spectrom., 18 (2003) 795–797.

[726] B. Packert Jensen, B. Gammelgaard, S.H. Hansen, J. Vanggaard Andersen, Comparison of direct injection nebulizer and desolvating microconcentric nebulizer for analysis of chlorine-, bromine- and iodine-containing compounds by reversed phase HPLC with ICP-MS detection, J. Anal. At. Spectrom., 18 (2003) 891–896.

[727] S.A.E O'Brien, J.R. Chirinos, K. Jorabchi, K. Kahen, M.E. Cree, A. Montaser, Investigation of the direct injection high efficiency nebulizer for axially and radially viewed inductively coupled plasma atomic emission spectrometry, J. Anal. At. Spectrom., 18 (2003) 910–916.

[728] D.B. Aeschliman, S.J. Bajic, D.P. Baldwin, R.S. Houk, High-speed digital photographic study of an inductively coupled plasma during laser ablation: comparison of dried solution aerosols from a microconcentric nebulizer and solid particles from laser ablation, J. Anal. At. Spectrom., 18 (2003) 1008–1014.

[729] J.L. Todolí, J.M. Mermet, Optimization of the evaporation cavity in a torch integrated sample introduction system based ICP-AES system. Applications to matrix and transient effects, analysis of microsamples and analysis of certified solid samples, J. Anal. At. Spectrom., 18 (2003) 1185–1191.

[730] Y.-L. Feng, R.E. Sturgeon, J.W. Lam, Chemical vapor generation characteristics of transition and noble metals reacting with tetrahydroborate(III), J. Anal. At. Spectrom., 18 (2003) 1435–1442.

[731] T. Wang, X. Jia, J. Wu, Direct determination of metals in organics by inductively coupled plasma atomic emission spectrometry in aqueous matrices, J. Pharm. Biomed. Anal., 33 (2003) 639–646.

[732] K. Kahen, A. Strubinger, J.R. Chirinos, A. Montaser, Direct injection high efficiency nebulizer-inductively coupled plasma mass spectrometry for analysis of petroleum samples, Spectrochim. Acta, 58B (2003) 397–413.

[733] E.G. Yanes, N.J. Miller-Ihli, Characterization of microconcentric nebulizer uptake rates for capillary electrophoresis inductively coupled plasma mass spectrometry, Spectrochim. Acta, 58B (2003) 949–955.

[734] C.M. Benson, J. Zhong, S.F. Gimelshein, D.A. Levin, A. Montaser, Simulation of droplet heating and desolvation in inductively coupled plasma-part II: coalescence in the plasma, Spectrochim. Acta, 58B (2003) 1453–1471.

[735] S.A. Lehn, K.A. Warner, M. Huang, G.M. Hieftje, Effect of sample matrix on the fundamental properties of the inductively coupled plasma, Spectrochim. Acta, 58B (2003) 1785–1806.

[736] G. Schaldach, I. Razilov, H. Berndt, Optimization of the geometry of a double-path spray chamber for inductively coupled plasma-atomic emission spectrometry by computer simulation and an evolutionary strategy, Spectrochim. Acta, 58B (2003) 1807–1819.

[737] Y. Shan, J. Mostaghimi, Numerical simulation of aerosol droplets desolvation in a radio frequency inductively coupled plasma, Spectrochim. Acta, 58B (2003) 1959–1977.

[738] J. Olesik, Micronebulization with high spectral or chemical resolution ICP-MS, Federation of Analytical Chemistry and Spectroscopy Societies, Fort Lauderdale, FL, paper 464 (2003).

[739] J. Burgener, Federation of Analytical Chemistry and Spectroscopy Societies, Fort Lauderdale, FL, paper 267 (2003).

[740] J. Bergmann, S. Lassen, A. Prange, Determination of the absolute configuration of seleno-methionine from antarctic krill by RP-HPLC/ICP-MS using chiral derivatization agents, Anal. Bioanal. Chem., 378 (2004) 1624–1629.

[741] L. Marjanovic, R.I. McCrindle, B.M. Botha, H.J. Potgieter, Use of a simplified generalized standard additions method for the analysis of cement, gypsum and basic slag by slurry nebulization ICP-OES, Anal. Bioanal. Chem., 379 (2004) 104–107.

[742] S.E. Maestre, J.L. Todolí, J.M. Mermet, Evaluation of several pneumatic micronebulizers with different designs for use in ICP–AES and ICP–MS. Future directions for further improvement, Anal. Bioanal. Chem., 379 (2004) 888–899.

[743] J. Heilmann, S.F. Boulyga, K.G. Heumann, Accurate determination of sulfur in gasoline and related fuel samples using isotope dilution ICP–MS with direct sample injection and microwave-assisted digestion, Anal. Bioanal. Chem., 380 (2004) 190–197.

[744] S.F. Boulyga, V. Loreti, J. Bettmer, K.G. Heumann, Application of SEC-ICP-MS for comparative analyses of metal-containing species in cancerous and healthy human thyroid samples, Anal. Bioanal. Chem., 380 (2004) 198–203.

[745] J. Koch, G. Schaldach, H. Berndt, K. Niemax, Numerical simulation of aerosol transport in atomic spectrometry, Anal. Chem., 76 (2004) 130A–136A.

[746] N.H. Bings, A. Bogaerts, J.A.C. Broekaert, Atomic spectroscopy, Anal. Chem., 76 (2004) 3313–3336.

[747] D. Beauchemin, Inductively coupled plasma mass spectrometry, Anal. Chem., 76 (2004) 3395–3416.

[748] M. Regelous, S.P. Turner, T.R. Elliott, K. Rostami, C.J. Hawkesworth, Measurement of femtogram quantities of protactinium in silicate rock samples by multicollector inductively coupled plasma mass spectrometry, Anal. Chem., 76 (2004) 3584–3589.

[749] N. Dauphas, P.E. Janney, R.A. Mendybaev, M. Wadhwa, F.M. Richter, A.M. Davis, M. van Zuilen, R. Hines, C.N. Foley, Chromatographic separation and multicollection-ICPMS analysis of iron. Investigating mass-dependent and –independent isotope effects, Anal. Chem., 76 (2004) 5855–5863.

[750] K. Kahen, K. Jorabchi, C. Gray, A. Montaser, Spatial mapping of droplet velocity and size for direct and indirect nebulization in plasma spectrometry, Anal. Chem., 76 (2004) 7194–7201.

[751] C.S. Westphal, J.A. McLean, S.J. Hakspiel, W.E. Jackson, D.E. McClain, A. Montaser, Determination of depleted uranium in urine via isotope ratio measurements using large-bore direct injection high efficiency nebulizer-inductively coupled plasma mass spectrometry, Appl. Spectrosc., 58 (2004) 1044–1050.

[752] A.G. Coedo, I. Padilla, M.T. Dorado, Comparison of laser ablation and dried solution aerosol as sampling systems in inductively coupled plasma mass spectrometry, Appl. Spectrosc., 58 (2004) 1481–1487.

[753] P. Bienvenu, E. Brochard, E. Excoffier, M. Piccione, Determination of iodine 129 by ICP-QMS in environmental samples, Can. J. Anal. Sci. Spectrosc., 49 (2004) 423–428.

[754] J. Baker, D. Peate, T. Waight, C. Meyzen, Pb isotopic analysis of standards and samples using a ^{207}Pb-^{204}Pb double spike and thallium to correct for mass bias with a double-focusing MC-ICP-MS, Chem. Geol., 211 (2004) 275–303.

[755] J.N. Christensen, P.E. Dresel, M.E. Conrad, K. Maher, D.J. Depaolo, Identifying the sources of subsurface contamination at the handford site in Washington using high-precision uranium isotopic measurements, Environ. Sci. Technol., 38 (2004) 3330–3337.

[756] M.V. Zoriy, C. Pickhardt, P. Ostapczuk, R. Hille, J.S. Becker, Determination of Pu in urine at ultratrace level by sector field inductively coupled plasma mass spectrometry, Int. J. Mass Spectrom., 232 (2004) 217–224.

[757] M. Kovacevic, R. Leber, S.D. Kohlwein, W. Goessler, Application of inductively coupled plasma mass spectrometry to phospholipid analysis, J. Anal. At. Spectrom., 19 (2004) 80–84.

[758] B. Gammelgaard, L. Bendahl, Selenium speciation in human urine samples by LC- and CE-ICP-MS-separation and identification of selenosugars, J. Anal. At. Spectrom., 19 (2004) 135–142.

[759] J. Zheng, H. Hintelmann, Hyphenation of high performance liquid chromatography with sector field inductively coupled plasma mass spectrometry for the determination of ultra-trace level anionic and cationic arsenic compounds in freshwater fish, J. Anal. At. Spectrom., 19 (2004) 191–195.

[760] W.R.L. Cairns, C. Barbante, G. Capodaglio, P. Cescon, A. Gambaro, A. Eastgate, Performance characteristics of a low volume spray chamber with a micro-flow nebulizer for ICP-MS, J. Anal. At. Spectrom., 19 (2004) 286–291.

[761] M.V. Zoriy, L. Halicz, M.E. Ketterer, C. Prickhardt, P. Ostapczuk, J.S. Becker, Reduction of UH$^+$ formation for ^{236}U/^{238}U isotope ratio measurements at ultratrace level in double focusing sector field ICP-MS using D$_2$O as solvent, J. Anal. At. Spectrom., 19 (2004) 362–367.

[762] S.-A.E. O'Brien Murdock, K. Kahen, J.R. Chirinos, M.E. Ketterer, D.D. Hudson, A. Montaser, Aerosol diagnostics and inductively coupled plasma mass spectrometry with demountable concentric nebulizers, J. Anal. At. Spectrom., 19 (2004) 666–674.

[763] J.L. Todolí, S.E. Maestre, J.M. Mermet, Compensation for matrix effects in ICP-AES by using air segmented liquid microsample introduction. The role of the spray chamber, J. Anal. At. Spectrom., 19 (2004) 728–737.

[764] P. Gabrielli, A. Varga, C. Barbante, C. Boutron, G. Cozzi, V. Gaspari, F. Planchon, W. Cairns, S. Hong, C. Ferrari, G. Capodaglio, Determination of Ir and Pt down to the sub-femtogram per gram level in polar ice by ICP-SFMS using preconcentration and a desolvation system, J. Anal. At. Spectrom., 19 (2004) 831–837.

[765] M.E. Wieser, D. Buhl, C. Bouman, J. Schwieters, High precision calcium isotope ratio measurements using a magnetic sector multiple collector inductively coupled plasma mass spectrometer, J. Anal. At. Spectrom., 19 (2004) 844–851.

[766] M. Krachler, J. Zheng, D. Fisher, W. Shotyk, Novel calibration procedure for improving trace element determinations in ice and water samples using ICP-SMS, J. Anal. At. Spectrom., 19 (2004) 1017–1019.

[767] M. Chu, D. Beauchemin, Simple method to assess the maximum bio-accessibility of elements from food using flow injection and inductively coupled plasma mass spectrometry, J. Anal. At. Spectrom., 19 (2004) 1213–1216.

[768] L. Halicz, D. Günther, Quantitative analysis of silicates using LA-ICP-MS with liquid calibration, J. Anal. At. Spectrom., 19 (2004) 1539–1545.

[769] A.C.S. Bellato, M.F. Giné, A.A. Menegário, Determination of B in body fluids by isotope dilution inductively coupled mass spectrometry with direct injection nebulization, Microchem. J., 77 (2004) 119–122.

[770] G.C.-Y. Chan, G.M. Hieftje, Using matrix effects as a probe for the study of the charge-transfer mechanism in inductively coupled plasma-atomic emission spectrometry, Spectrochim. Acta, 59B (2004) 163–183.

[771] C.S. Westphal, K. Kahen, W.F. Rutkowski, B.W. Acon, A. Montaser, Demountable direct injection high efficiency nebulizer for inductively coupled plasma mass spectrometry, Spectrochim. Acta, 59B (2004) 353–368.

[772] H. Matusiewicz, B. Golik, Simultaneous determination of macro and trace elements in biological reference materials by microwave induced plasma optical emission spectrometry with slurry sample introduction, Spectrochim. Acta, 59B (2004) 749–754.

[773] E.G. Yanes, N.J. Miller-Ihli, Use of a parallel path nebulizer for capillary-based microseparation techniques coupled with an inductively coupled plasma mass spectrometer for speciation measurements, Spectrochim. Acta, 59B (2004) 883–890.

[774] E.G. Yanes, N.J. Miller-Ihli, Cobalamin speciation using reversed-phase micro-high-performance liquid chromatography interfaced to inductively coupled plasma mass spectrometry, Spectrochim. Acta, 59B (2004) 891–899.

[775] M. Grotti, C. Lagomarsino, R. Frache, A new nebulization device with exchangeable aerosol generation mode as a useful tool to investigate sample introduction processes in inductively coupled plasma atomic emission spectrometry, Spectrochim. Acta, 59B (2004) 1001–1006.

[776] Z. Hu, S. Hu, S. Gao, Y. Liu, S. Lin, Volatile organic solvent-induced signal enhancements in inductively coupled plasma-mass spectrometry: a case study of methanol and acetone, Spectrochim. Acta, 59B (2004) 1463–1470.

[777] D. Pröfrock, P. Leonhard, W. Ruck, A. Prange, Development and characterisation of a new interface for coupling capillary LC with collision cell ICP-MS and its application for phosphorylation profiling of tryptic protein digests, Anal. Bioanal. Chem., 381 (2005) 194–204.

[778] V. Loreti, D. Toncelli, E. Morelli, G. Scarano, J. Bettmer, Biosynthesis of Cd-bound phytochelatins by Phaeodactylum tricornutum and their speciation by size-exclusion chromatography and ion-pair chromatography coupled to ICP–MS, Anal. Bioanal. Chem., 383 (2005) 398–403.

[779] K. Jorabchi, K. Kahen, C. Gray, A. Montaser, In Situ Visualization and Characterization of aerosol droplets in an inductively coupled plasma, Anal. Chem., 77 (2005) 1253–1260.

[780] S.D. Richardson, T.A. Temes, Water analysis: Emerging contaminants and current issues, Anal. Chem., 77 (2005) 3807–3838.

[781] D. Larivière, V.N. Epov, K.M. Reiber, R.J. Cornett, R.D. Evans, Micro-extraction procedures for the determination of Ra-226 in well waters by SF-ICP-MS, Anal. Chim. Acta, 528 (2005) 175–182.

[782] A.N. Anthemidis, V. Arvanitidis, J.A. Stratis, On-line emulsion formation and multi-element analysis of edible oils by inductively coupled plasma atomic emission spectrometry, Anal. Chim. Acta, 537 (2005) 271–278.

[783] Q.Z. Bian, P. Jacob, H. Berndt, K. Niemax, Online flow digestion of biological and environmental samples for inductively coupled plasma–optical emission spectroscopy (ICP–OES), Anal. Chim. Acta, 538 (2005) 323–329.

[784] S. Weyer, A.D. Anbar, G.P. Brey, C. Münker, K. Mezger, A.B. Woodland, Iron isotope fractionation during planetary differentiation, Earth Planet. Sci. Lett., 240 (2005) 251–264.

[785] A.D. Brandon, M. Humayun, I.S. Puchtel, M.E. Zolensky, Re-Os isotopic systematics and platinum group element composition of the Tagish Lake carbonaceous chondrite, Geochim. Cosmochim. Acta, 69 (2005) 1619–1631.

[786] R. Schoenberg, F. von Blanckenburg, An assessment of the accuracy of stable Fe isotope ratio measurements on samples with organic and inorganic matrices by high-resolution multicollector ICP-MS, Int. J. Mass Spectrom., 242 (2005) 257–272.

[787] D. Schaumlöffel, P. Giusti, M.V. Zoriy, C. Pickhardt, J. Szpunar, R. Lobinski, J. Sabine Becker, Ultratrace determination of uranium and plutonium by nano-volume flow injection double-focusing sector field inductively coupled plasma mass spectrometry (nFI-ICP-SFMS), J. Anal. At. Spectrom., 20 (2005) 17–21.

[788] S. Groom, G. Schaldach, M. Ulmer, P. Walzel, H. Berndt, Adaptation of a new pneumatic nebulizer for sample introduction in ICP spectrometry, J. Anal. At. Spectrom., 20 (2005) 169–175.

[789] Z. Wang, Z. Ni, D. Qiu, G. Tao, P. Yang, Characterization of stability of ceramic suspension for slurry introduction in inductively coupled plasma optical emission spectrometry and application to aluminium nitride analysis, J. Anal. At. Spectrom., 20 (2005) 315–319.

[790] L. Bendahl, B. Gammelgaard, Sample introduction systems for reversed phase LC-ICP-MS of selenium using large amounts of methanol-comparison of systems based on membrane desolvation, a spray chamber and direct injection, J. Anal. At. Spectrom., 20 (2005) 410–416.

[791] V.N. Epov, K. Benkhedda, R.J. Cornett, R.D. Evans, Rapid determination of plutonium in urine using flow injection on-line preconcentration and inductively coupled plasma mass spectrometry, J. Anal. At. Spectrom., 20 (2005) 424–430.

[792] G. Grindlay, S. Maestre, J. Mora, V. Hernandis, L. Gras, A microwave assisted desolvation system based on the use of a TM_{010} cavity for inductively coupled plasma based analytical techniques, J. Anal. At. Spectrom., 20 (2005) 455–461.

[793] K. Benkhedda, D. Larivière, S. Scott, D. Evans, Hyphenation of flow injection on-line preconcentration and ICP-MS for the rapid determination of ^{226}Ra in natural waters, J. Anal. At. Spectrom., 20 (2005) 523–528.

[794] W.C. Wetzel, J.A.C. Broekaert, G.M. Hieftje, A contribution to the study of cooling a vertical-rotary spray chamber in inductively coupled plasma time-of-flight mass spectrometry, J. Anal. At. Spectrom., 20 (2005) 621–625.

[795] C. Fragniere, M. Haldimann, A. Eastgate, U. Krahenbuhl, A direct ultratrace determination of platinum in environmental, food and biological samples by ICP-SFMS using a desolvation system, J. Anal. At. Spectrom., 20 (2005) 626–630.

[796] K. Kahen, B.W. Acon, A. Montaser, Modified Nukiyama–Tanasawa and Rizk–Lefebvre models to predict droplet size for microconcentric nebulizers with aqueous and organic solvents, J. Anal. At. Spectrom., 20 (2005) 631–637.

[797] D. Goitom, E. Björn, W. Frech, M.T.C. de Loos-Vollebregt, Radial ICP characteristics for ICP-AES using direct injection or microconcentric nebulisation, J. Anal. At. Spectrom., 20 (2005) 645–651.

[798] B. Gammelgaard, L. Bendahl, N.W. Jacobsen, S. Stürup, Quantitative determination of selenium metabolites in human urine by LC-DRC-ICP-MS, J. Anal. At. Spectrom., 20 (2005) 889–893.

[799] A.N. Anthemidis, V.G. Pliatsika, On-line slurry formation and nebulization for inductively coupled plasma atomic emission spectrometry. Multielement analysis of cocoa and coffee powder samples, J. Anal. At. Spectrom., 20 (2005) 1280–1286.

[800] C.P. Lienemann, Analyse de métaux traces dans les produits pétroliers, état de l'art, Oil Gas Sci. Technol., 60 (2005) 951–965.

[801] Z. Wang, Z. Ni, D. Qiu, G. Tao, P. Yang, Determination of impurities in titanium nitride by slurry introduction axial viewed inductively coupled plasma optical emission spectrometry, Spectrochim. Acta, 60B (2005) 361–367.

[802] E.G. Yanes, N.J. Miller-Ihli, Parallel path nebulizer: Critical parameters for use with microseparation techniques combined with inductively coupled plasma mass spectrometry, Spectrochim. Acta, 60B (2005) 555–561.

[803] R.T. Gettar, P. Smichowski, R.N. Garavaglia, S. Farías, D.A. Batistoni, Effect of nebulizer/spray chamber interfaces on simultaneous, axial view inductively coupled plasma optical emission spectrometry for the direct determination of As and Se species separated by ion exchange high-performance liquid chromatography, Spectrochim. Acta, 60B (2005) 567–573.

[804] L.C. Trevizan, E.C. Vieira, A.R.A. Nogueira, J.A. Nobrega, Use of factorial design for evaluation of plasma conditions and comparison of two liquid sample introduction systems for an axially viewed inductively coupled plasma optical emission spectrometer, Spectrochim. Acta, 60B (2005) 575–581.

[805] R.M. de Souza, B.M. Mathias, C.L.P. da Silveira, R.Q. Aucelio, Inductively coupled plasma optical emission spectrometry for trace multi-element determination in vegetable oils, margarine and butter after stabilization with propan-1-ol and water, Spectrochim. Acta, 60B (2005) 711–715.

[806] M. Krachler, N. Rausch, H. Feuerbacher, P. Klemens, A new HF-resistant tandem spray chamber for improved determination of trace elements and Pb isotopes using inductively coupled plasma-mass spectrometry, Spectrochim. Acta, 60B (2005) 865–869.

[807] P. Gaines, Sample introduction for ICPMS and ICP-OES, Spectroscopy, 20 (2005) 20–23.

[808] J.L. Todolí, J.M. Mermet, Elemental analysis of liquid microsamples through inductively coupled plasma spectrochemistry, Trends Anal. Chem., 24 (2005) 107–116.

[809] J.W. Olesik, Federation of Analytical Chemistry and Spectroscopy Societies, Quebec, Canada, paper 107 (2005).

[810] K. Pitcairn, P.E. Warwick, J.A. Milton, D.A.H. Teagle, Method for ultra-low-level analysis of gold in rocks, Anal. Chem., 78 (2006) 1290–1295.

[811] N.H. Bings, A. Bogaerts, J.A.C. Broekaert, Atomic spectroscopy, Anal. Chem., 78 (2006) 3917–3945.

[812] D. Beauchemin, Inductively coupled plasma mass spectrometry, Anal. Chem., 78 (2006) 4111–4135.

[813] S. D'Ilio, N. Violante, S. Caimi, M. Di Gregorio, F. Petrucci, O. Senofonte, Determination of trace elements in serum by dynamic reaction cell inductively coupled plasma mass spectrometry. Developing of a method with a desolvating system nebulizer, Anal. Chim. Acta, 573–574 (2006) 432–438.

[814] S. D'Ilio, N. Violante, M. Di Gregorio, O. Senofonte, F. Petrucci, Simultaneous quantification of 17 trace elements in blood by dynamic reaction cell inductively coupled plasma mass spectrometry (DRC-ICP-MS) equipped with a high-efficiency sample introduction system, Anal. Chim. Acta, 579 (2006) 202–208.

[815] O. Sato, M. Konno, Y. Wakui, H. Matsunaga, M. Shirai, Supercritical decomposition of polyethylene samples for the determination of their trace cadmium content, Anal. Sci., 22 (2006) 1461–1463.

[816] S.D. Maind, S.A. Kumar, N. Chattopadhyay, Ch. Gandhi, M. Sudersanan, Analysis of Indian blue ballpoint pen inks tagged with rare-earth thenoyltrifluoroacetonates by inductively coupled plasma–mass spectrometry and instrumental neutron activation analysis, Forensic Sci. Int., 159 (2006) 32–42.

[817] Z. Stefanka, K. Koellensperger, G. Stingederb, S. Hann, Down-scaling narrowbore LC-ICP-MS to capillary LC-ICP-MS: a comparative study of different introduction systems, J. Anal. At. Spectrom., 21 (2006) 86–89.

[818] D. Goitom, E. Björn, Noise characteristics and analytical precision of inductively coupled plasma mass spectrometry using a Vulkan direct injection nebuliser for sample introduction, J. Anal. At. Spectrom., 21 (2006) 168–176.

[819] D. Maldonado, J. Chirinos, Z. Benzo, C. Gómez, E. Marcano, Analytical evaluation of a dual micronebulizer sample introduction system for inductively coupled plasma spectrometry, J. Anal. At. Spectrom., 21 (2006) 743–749.

[820] B. Almagro, A.M. Gañán-Calvo, M. Hidalgo, A. Canals, Flow focusing pneumatic nebulizer in comparison with several micronebulizers in inductively coupled plasma atomic emission spectrometry, J. Anal. At. Spectrom., 21 (2006) 770–777.

[821] R.M. de Souza, A.L. Saraceno, C. Lúcia, P. da Silveira, R.Q. Aucélio, Determination of trace elements in crude oil by ICP-OES using ultrasound-assisted acid extraction, J. Anal. At. Spectrom., 21 (2006) 1345–1349.

[822] G. Grindlay, S. Maestre, L. Gras, J. Mora, Introduction of organic solvent solutions into inductively coupled plasma-atomic emission spectrometry using a microwave assisted sample introduction system, J. Anal. At. Spectrom., 21 (2006) 1403–1411.

[823] C. Huang, D. Beauchemin, Simultaneous determination of two conditional stability constants by IC-ICP-MS, J. Anal. At. Spectrom., 21 (2006) 1419–1422.

[824] S.F. Boulyga, K.G. Heumann, Determination of extremely low $^{236}U/^{238}U$ isotope ratios in environmental samples by sector-field inductively coupled plasma mass spectrometry using high-efficiency, sample introduction, J. Environ. Radioact., 88 (2006) 1–10.

[825] T.R. Brooks, T.E. Bodkin, G.E. Potts, S.A. Smullen, Elemental analysis of human cremains using ICP-OES to classify legitimate and contaminated cremains, J. Forensic Sci., 51 (2006) 967–973.

[826] R.M. de Souza, A.L.S. Meliande, C.L.P. da Silveira, R.Q. Aucélio, Determination of Mo, Zn, Cd, Ti, Ni, V, Fe, Mn, Cr and Co in crude oil using inductively coupled plasma optical emission spectrometry and sample introduction as detergentless microemulsions, Microchem. J., 82 (2006) 137–141.

[827] X. Song, T. Duan, P. Guo, H. Chen, Determination of Nb and Ta in Nb/Ta minerals by inductively coupled plasmas optical emission spectrometry using slurry sample introduction, Microchem. J., 84 (2006) 22–25.

[828] J.L. Todolí, J.M. Mermet, Sample introduction systems for the analysis of liquid microsamples by ICP-AES and ICP-MS, Spectrochim. Acta, 61B (2006) 239–283.

[829] E. Paredes, S.E. Maestre, J.L. Todolí, Use of stirred tanks for studying matrix effects caused by inorganic acids, easily ionized elements and organic solvents in inductively coupled plasma atomic emission spectrometry, Spectrochim. Acta, 61B (2006) 326–339.

[830] J. Borkowska-Burnecka, A. Lesniewicz, W. Zyrnicki, Comparaison of pneumatic and ultrasonic nebulization in inductively coupled plasma atomic emission spectrometry – matrix effects and plasma parameters, Spectrochim. Acta, 61B (2006) 579–587.

[831] J.W. Olesik, J.A. Kinzer, Measurement of monodisperse droplet desolvation in an inductively coupled plasma using droplet size dependent peaks in Mie scattering intensity, Spectrochim. Acta, 61B (2006) 696–704.

[832] B. Bouyssiere, Y. Nuevo Ordóñez, Ch.-P. Lienemann, D. Schaumlöffel, R. Łobiński, Determination of mercury in organic solvents and gas condensates by μflow-injection—inductively coupled plasma mass spectrometry using a modified total consumption micronebulizer fitted with single pass spray chamber, Spectrochim. Acta, 61B (2006) 1063–1068.

[833] N. Miskolczi, L. Bartha, J. Borszeki, P. Halmos, Determination of sulfur content of diesel fuels and diesel fuel-like fractions of waste polymer cracking, Talanta, 69 (2006) 776–780.

[834] G. Grindlay, Development of an integrated nebulization—desolvation system based on microwave radiation for atomic spectrometry, Ph. D. Thesis, University of Alicante (2006).

[835] A. Santos, T. Magalhaes, D.N. Vieira, A.A. Almeida, A.V. Sousa, Firing distance estimation through the analysis of the gunshot residue deposit pattern around the bullet entrance hole by inductively coupled plasma-mass spectrometry: an experimental study, Am. J. Forensic Med. Pathol., 28 (2007) 24–30.

[836] M.E. Lipschutz, S.F. Wolf, F.B. Culp, A.J.R. Kent, Geochem. and Cosmochem. Materials, Anal. Chem., 79 (2007) 4249–4274.

[837] T.A. Bretell, J.M. Buttler, J.R. Almirall, Forensic science, Anal. Chem., 79 (2007) 4365–4384.

[838] Z. Varga, Application of inductively coupled plasma sector field mass spectrometry for low-level environmental americium-241 analysis, Anal. Chim. Acta, 587 (2007) 165–169.

[839] Z. Varga, G. Suranyi, N. Vajda, Z. Stefanka, Rapid methods for the determination of long-lived radionuclides in environmental samples by ICP-SFMS and radioanalytical techniques, Czechoslovak J. Phys., 95 (2007) 81–87.

[840] J. Dulude, V. Dolic, R. Stux, Applications of a temperature controlled spray chamber, Glass Expansion, article 010407 (2007).

[841] E.K. Shibuya, J.E.S. Sarkis, O. Negrini-Neto, J. Ometto, Multivariate classification based on chemical and stable isotopic profiles in sourcing the origin of marijuana samples seized in Brazil, J. Braz. Chem. Soc., 18 (2007) 205–214.

[842] G.A. Zachariadis, C.E. Michos, Development of a slurry introduction method for multi-element analysis of antibiotics by inductively coupled plasma atomic emission spectrometry using various types of spray chamber and nebulizer configurations, J. Pharm. Biomed. Anal., 43 (2007) 951–958.

[843] B. Gammelgaard, B.P. Jensen, Application of inductively coupled plasma mass spectrometry in drug metabolism studies, J. Anal. At. Spectrom., 22 (2007) 235–249.

[844] D. Goitom, E. Björn, Comparison of aerosol properties and ICP-MS analytical performance of the Vulkan direct injection nebuliser and the Direct Injection High Efficiency Nebuliser, J. Anal. At. Spectrom., 22 (2007) 250–257.

[845] S. Dreyfus, C. Pécheyran, C.P. Lienemann, C. Magnier, A. Prinzhofer, O.F.X. Donard, Determination of lead isotope ratios in crude oils with Q-ICP/MS, J. Anal. At. Spectrom., 22 (2007) 351–360.

[846] C. Lagomarsino, M. Grotti, J.L. Todolí, J.M. Mermet, Study of the absence of recondensation with low liquid delivery rates by using a cavity sheathing gas in inductively coupled plasma-atomic emission spectrometry, J. Anal. At. Spectrom., 22 (2007) 523–531.

[847] G. Pearson, G. Greenway, A highly efficient sample introduction system for interfacing microfluidic chips with ICP-MS, J. Anal. At. Spectrom., 22 (2007) 657–662.

[848] S. Goyard, V. Lavoine, P. Gareil, M. Fromm, J.C. Hubinois, Evaluation of different nebulizers and spray chambers for the on-line coupling of ion chromatography with ICP-OES, J. Anal. At. Spectrom., 22 (2007) 1507–1511.

[849] Z. Varga, G. Suranyi, N. Vajda, Z. Stefanka, Determination of plutonium and americium in environmental samples by inductively coupled plasma sector field mass spectrometry and alpha spectrometry, Microchem. J., 85 (2007) 39–45.

[850] I. Rezic, I. Steffan, ICP-OES determination of metals present in textile materials, Microchem. J., 86 (2007) 46–51.

[851] M. Bauer, J.A.C. Broekaert, Investigations on the use of pneumatic cross-flow nebulizers with dual solution loading including the correction of matrix effects in elemental determinations by inductively coupled plasma optical emission spectrometry, Spectrochim. Acta, 62B (2007) 145–154.

[852] C. Duyck, N. Miekeley, C.L. Porto da Silveira, R.Q. Aucélio, R.C. Campos, P. Grinberg, G.P. Brandão, The determination of trace elements in crude oil and its heavy fractions by atomic spectrometry, Spectrochim. Acta, 62B (2007) 939–951.

[853] CETAC Technologies Inc., Application of the MCN-6000 for determination of trace elements in concentrated acids MCN-6000 with ICP-MS, http://www.cetac.com, (2008).

[854] J. Bosque, J.L. Todolí, J.M. Mermet, Unpublished results.

Index

www.ingramcontent.com/pod-product-compliance
Lightning Source LLC
Chambersburg PA
CBHW060339220326
41598CB00023B/2755